MW01493611

Algorithms for Worst-case Design and Applications to Risk Management

Algorithms for Worst-case Design and Applications to Risk Management

Berç Rustem

Department of Computing
Imperial College of Science, Technology & Medicine
180 Queen's Gate, London SW7 2BZ, UK

Melendres Howe

Imperial College and Asian Development Bank
6 ADB Avenue, Mandaluyong City 0401 MM
PO Box 789, 0980 Manila, Philippines

PRINCETON UNIVERSITY PRESS

PRINCETON AND OXFORD

Copyright © 2002 by Princeton University Press

Published by Princeton University Press,
41 William Street, Princeton, New Jersey 08540

In the United Kingdom: Princeton University Press,
3 Market Place, Woodstock, Oxfordshire OX20 1SY

All Rights Reserved

Library of Congress Cataloging-in-Publication Data applied for.
Rustem, Berç and Howe, Melendres
Algorithms for Worst-case Design and Applications to Risk Management / Berç
Rustem and Melendres Howe
p. cm.
Includes bibliographical references and index.
ISBN 0-691-09154-4 (alk. paper)

British Library Cataloguing-in-Publication Data

A catalogue record for this book is available from the British Library.

This book has been composed in Times and Abadi

Printed on acid-free paper. ∞

www.pup.princeton.edu

Printed in the United States of America

10 9 8 7 6 5 4 3 2 1

The gods to-day stand friendly, that we may,
Lovers in peace, lead on our days to age!
But, since the affairs of men rest still incertain
Let's reason with the worst that may befall.

William Shakespeare

Julius Caesar, Act 5 Scene 1.

Dedicated to those who have suffered the worst case.

Contents

Preface

The conventional approach to decisions under uncertainty is based on expected value optimization. The main problem with this concept is that it neglects the worst-case effect of the uncertainty in favor of expected values. While acceptable in numerous instances, decisions based on expected value optimization may often need to be justified in view of the worst-case scenario. This is especially important if the decision to be made can be influenced by such uncertainty that, in the worst case, might have drastic consequences on the system being optimized. On the other hand, given an uncertain effect, some worst-case realizations might be so improbable that dwelling on them might result in unnecessarily pessimistic decisions. Nevertheless, even when decisions based on expected value optimization are to be implemented, the worst-case scenario does provide an appropriate benchmark indicating the risks.

This book is intended for the dual role of proposing worst-case design for robust decisions and methods and algorithms for computing the solution to quantitative decision models. Actually, very little space is devoted to justify worst-case design. This is implicit in the optimality condition of minimax, discussed in Chapter 1. In Chapter 6, the robustness of worst-case optimal strategies are considered for discrete scenarios. Subsequent chapters illustrate the property. Basically, the performance of the minimax optimal strategy is noninferior for any scenario, and better for those other than the worst case. As such, worst-case design needs no further justification as a robust strategy than a deterministic optimal strategy requires in view of suboptimal alternatives.

In the book, we consider methods for optimal decisions which take account of the worst-case eventuality of uncertain events. The robust character of minimax, mentioned above, is central to the usefulness of the strategies discussed in this book. *The discrete minimax strategy ensures a guaranteed optimal performance in view of the worst case* and this is assured for all scenarios: if any scenario, other than the one corresponding to the worst case is realized, performance is assured to improve. The continuous minimax strategy provides a guaranteed optimal performance in view of a continuum of scenarios. If this continuum is taken as scenarios varying between upper and lower bounds, performance is assured over the worst case defined between upper and lower bounds. As such, *continuous minimax is a forecaster's dream as it provides the opportunity for specifying forecasts defined over a range, rather than point forecasts.* Despite all this, however, we stress that these are merely computational tools. If the forecaster tries to specify too many discrete forecasts, in an attempt to cover most possibilities, discrete minimax may yield too pessimistic strategies or even run into numerical, or computational, problems due to the resulting numerous scenarios. Similarly, as the upper and lower bounds on a range forecast get wider, to provide coverage to a wider set of possibilities, the minimax strategy may become pessimistic. Thus, scenarios have to be chosen with care, among genuinely likely values. The minimax strategy will then answer the legitimate question of what the best strategy should be, in view of the worst-case.

The stochastic characterization of uncertainty relies on the average or expected performance of the system in the presence of uncertain effects. The book provides the means for taking account of the effect of the worst case which would, in general, not be reflected in average or expected performance evaluation of the system. While expected performance optimization is often adequate, it is the realization of the worst case that mostly causes the failure of systems. Hence, all decisions need to take account of the worst case and the expected performance of the system. Through its inherent pessimism, the minimax strategy may lead to a serious deterioration of performance. Alternatively, the realization of the worst-case scenario may result in an unacceptable performance deterioration for the strategy based on expected value optimization. *Neither minimax nor expected value optimization provide a substitute to wisdom.* At best, they can be regarded as risk management tools for analyzing the effects of uncertain events.

The book is intended for graduate students, researchers in minimax and for practitioners of risk management in economics, engineering design, finance, management science, operations research. After an introductory Chapter 1 in which the basic concepts and fundamental results used later in the book are discussed, we have devoted Chapters 2–7 to algorithms and Chapters 8–11 to risk management applications in finance. Specifically, we survey continuous minimax algorithms in Chapter 2 and discuss the solution of a subclass, the saddle point problem, in Chapter 3. A quasi-Newton algorithm for continuous minimax is developed in Chapter 4 which has desirable convergence properties and has proved to be successful in practical applications. The latter is considered in Chapter 5 particularly from the point of view of justifying the practical use of a simplified version of the algorithm in Chapter 4. The discrete minimax problem is introduced in Chapter 6 and a quasi-Newton algorithm for the nonlinearly constrained problem is developed in Chapter 7. In Chapter 8, the application of continuous minimax to options and hedging is discussed. The application of mainly the discrete minimax problem to portfolio optimization is explored in Chapter 9. The worst-case analysis of the asset-liability management problem and exchange rate scenarios are considered in Chapters 10 and 11, respectively.

The focus of the discussion is static optimization. Specific problems of dynamical systems are omitted. For linear dynamical systems, the interested reader might wish to consult the H^{∞} control literature. Among other items omitted, we are acutely aware of the absence of engineering applications. Worst-case design in engineering is perhaps the most intuitively obvious area in which minimax can yield tangible benefits. This is not because of lack of trying to get several members of the UK manufacturing sector interested, rather the reluctance of some of those contacted to embark on new concepts and developments. Fortunately, EPSRC has recognized the potential of the area by funding a new research project on the subject. The results of this will be reported in future research papers.

Throughout the development of the risk management concepts in finance, the authors have benefited from extensive discussions with Michael Selby. We are indebted to Marc Hendriks for strong support from a practitioner's point of view and to Rudi Bogni for initially validating the usefulness of the approach. Last, but by no means least, Robin Becker has provided highly useful critical evaluations of worst-case design.

Berç Rustem (br@doc.ic.ac.uk)
Melendres Howe (mhowe@adb.org)

Algorithms for Worst-case Design and Applications to Risk Management

Chapter 1

Introduction to minimax

We consider the problem of minimizing a nondifferentiable function, defined by the maximum of an inner function. We refer to this objective function as the *max-function*. In practical applications of minimax, the max-function takes the form of a maximized error, or disutility, function. For example, portfolio selection models in finance can be formulated in a scenario-based framework where the max-function takes the form of a maximized risk measure across all given scenarios. To solve the minimax problem, algorithms requiring derivative information cannot be used directly and the usual methods that do not require gradients are inadequate for this purpose. Instead of gradients, we need to consider generalized gradients or subgradients to formulate smooth methods for nonsmooth problems.

The minimax notation is introduced with relevant concepts in convex analysis and nonsmooth optimization. We consider the basic theory of continuous minimax, characterized by continuous values of maximizing and minimizing variables, and associated optimality conditions. These need to be satisfied at the solution generated by all algorithms. The problem of discrete minimax, with continuous minimization but discrete maximization variables, and related conditions are considered in Chapters 6 and 7.

1 BACKGROUND AND NOTATION

Equation and section numbering follow the following rule: (1.2.3) refers to Equation 3 in Chapter 1, Section 2. In Chapter 1 only, this is referred to as (2.3), elsewhere as (1.2.3). Chapter 1, Section 2 is referred to in Chapter 1 only as Section 2, elsewhere as Section 1.2.

In this book, we consider strategies, algorithms, properties and applications of worst-case design problems. When taking decisions under uncertainty, it is desirable to evaluate the best policy in view of the worst-case uncertain effect. Essentially, this entails minimax formulations in which the best decision and the worst case is determined simultaneously. In this sense, optimality is defined over all possible values of the uncertain effects as opposed to certain likely realizations. Worst-case design is useful in all disciplines with rival representations of the same system. For example, in economics Chow

(1979) and Becker et al. (1986) consider rival macroeconomic models of the same economy. A similar approach to resource allocation is discussed in Pang and Yu (1989). The robustness property is explored in Hansen et al. (1998). In finance, Howe et al. (1994, 1996), Dert and Oldenkamp (1997), Howe and Rustem (1997), Ibanez (1998) and Rustem et al. (2000) consider worst-case decisions in options pricing and portfolio optimization. In engineering, Ben-Tal and Nemirovski (1993, 1994) discuss truss topology design under rival load scenarios. In control systems, H^∞-control theory is essentially an equivalent minimax formulation for the uncertainties in the system and this aspect is explored in Başar and Bernhard (1991). Rival representations can be characterized either in terms of a discrete choice, such as two or more models of the same system, or as values from a continuous range, such as all the values an uncertain variable may take within an upper and lower bound.

The worst-case design or minimax problem can thus be formulated as

$$\min_{x \in \Re^n} \max_{y \in \mathcal{Y}} f(x, y). \tag{1.1}$$

where $x \in \Re^n$ is a column vector of real numbers, denoting the decision variables in the n-dimensional Euclidian space. The vector y represents the uncertain variables and is defined over the feasible set $\mathcal{Y}$. An equivalent problem to the above formulation is given by

$$\min_{x \in \Re^n} \Phi(x)$$

where

$$\Phi(x) = \max_{y \in \mathcal{Y}} f(x, y)$$

is the *max-function*. If $\mathcal{Y}$ is a set of continuous variables, then the problem is known as *continuous minimax*. An example for such a set is

$$\mathcal{Y} = \left\{ y \in \Re^m \mid y_{\ell_i} \le y_i \le y_{u_i}, \ i = 1, ..., m \right\}$$

where y_{ℓ_i} and y_{u_i} are the lower and upper bounds on the ith element of y. Algorithms for this problem are discussed in detail in Chapters 2–5.

If $\mathcal{Y}$ consists of a discrete set of values, the corresponding problem is known as *discrete minimax*. We consider equality and inequality constraints on x in particular for the case of discrete min-max. The problem is expressed as

$$\min_{x} \max_{i \in \{1,2,...,m\}} \left\{ f^i(x) \mid g(x) = 0, \ h(x) \le 0 \right\}$$

where $f^i(x)$ is the value corresponding to the ith member of the discrete set $\{1, 2, ..., m\}$ over which the maximum is evaluated and g, h are vectors of equality and inequality constraints. Properties of, and algorithms for, this formulation are considered in Chapters 6 and 7.

In this section, we review some of the basic concepts used elsewhere in the book. Some of the more preliminary material is covered in Comments and Notes (CN 1–CN 11) at the end of the chapter. We start by considering the closed line segment joining x and z, denoted by $[x, z]$,

$$[x, z] \equiv \{w \in \Re^n \mid w = \lambda x + (1 - \lambda)z, \ 0 \le \lambda \le 1\} \qquad (1.2)$$

and by (x, z) the corresponding open line segment.

Convexity is invoked extensively in minimax. We define this property for sets and functions. A set $C \subset \Re^n$ is called *convex* if $[x, z] \subset C$ for all $x, z \in C$. The linear combination

$$\sum_{j=1}^{J} \lambda^j x_j$$

is called a *convex combination* of vectors $x_1, \ldots, x_J \in \Re^n$ if

$$\lambda^j \ge 0, \ j = 1, \ldots, J \quad \text{and} \quad \sum_{j=1}^{J} \lambda^j = 1.$$

The *convex hull* of a set $C \subset \Re^n$, denoted by conv C, is the set of all convex combinations of points in C.

Let C be a convex set. A function $f : C \subset \Re^n \to \Re^1$ is said to be *convex* if

$$f(\lambda x + (1 - \lambda)z) \le \lambda f(x) + (1 - \lambda)f(z) \qquad (1.3)$$

for $x, z \in C$ and $\lambda \in (0, 1)$. If strict inequality holds for $x \ne z$ and $\lambda \in (0, 1)$, then f is *strictly convex*. A function defined on C is said to be *(strictly) concave* if the function $f = -g$ is (strictly) convex.

Lemma 1.1   *Let $f(x) \in C^1$ (see CN 3). Then, f is convex over a convex set C if and only if*

$$f(z) \ge f(x) + \nabla f(x)(z - x), \qquad \forall z, x \in C.$$

The function is strictly convex if this inequality is strict.

Proof. Let f be convex. Then, we have

$$f(\lambda x + (1 - \lambda)z) \le \lambda f(x) + (1 - \lambda)f(z), \qquad \forall \lambda \in [0, 1]$$

and hence

$$\frac{f(\lambda x + (1 - \lambda)z) - f(z)}{\lambda} \le f(x) - f(z).$$

This expression yields the required inequality for $\lambda \to 0$. Let

$$f(z) \ge f(x) + \nabla f(x)(z - x), \qquad \forall z, x \in C$$

and

$$x = \lambda x_1 + (1 - \lambda)y$$

for some $x_1 \in C$ and $\lambda \in [0, 1]$. We have the inequalities

$$f(x_1) \geq f(x) + \nabla f(x)(x_1 - x)$$

$$f(y) \geq f(x) + \nabla f(x)(y - x).$$

Multiplying the first of these inequalities by λ and the second by $(1 - \lambda)$ and adding yields

$$\lambda f(x_1) + (1 - \lambda)f(y) \geq f(x) + \nabla f(x)(\lambda x_1 + (1 - \lambda)y - x).$$

Substituting $x = \lambda x_1 + (1 - \lambda)y$ yields

$$\lambda f(x_1) + (1 - \lambda)f(y) \geq f(\lambda x_1 + (1 - \lambda)y). \quad \square$$

Lemma 1.2 *Let $f(x) \in C^2$. Then, f is convex (strictly convex) over a convex set C containing an interior point if and only if the Hessian matrix $\nabla^2 f(x)$ of f is positive semi-definite (positive definite) (see CN 4) throughout C.*

Proof. Using the second order Taylor expansion of f (see CN 5), we have

$$f(y) = f(x) + \langle \nabla f(x),\ y - x \rangle + \frac{1}{2}\langle y - x,\ \nabla^2 f(x + \lambda(y - x))(y - x) \rangle$$

for some $\lambda \in [0, 1]$ ($\langle \cdot, \cdot \rangle$ is defined in CN 2). If the Hessian is positive semi-definite everywhere, we have

$$f(y) \geq f(x) + \langle \nabla f(x),\ y - x \rangle \quad (1.4)$$

and, in view of Lemma 1.1, this establishes the convexity of f.

We show that if $\nabla^2 f(x)$ is not positive semi-definite for some $x \in C$, then f is not convex. Let $\nabla^2 f(x)$ not be positive semi-definite for some $x \in C$. By continuity of $\nabla^2 f(x)$ it can be assumed, without loss of generality that x is an interior point of C. There is a $y \in C$ such that

$$\langle x - y, \nabla^2 f(x)(x - y) \rangle < 0.$$

Again, by continuity of the Hessian, y may be selected such that

$$\langle x - y, \nabla^2 f(x + \lambda(y - x))(x - y) \rangle < 0, \quad \forall \lambda \in [0, 1].$$

In view of the Taylor expansion above, (1.4) does not hold and by Lemma 1.1, f is not convex. $\square$

An important property of convexity of f is the uniqueness of its minimum, as stated in the following result.

Lemma 1.3 *Let f be a convex function defined over the convex set C. Then, any local minimum of f is also a global minimum (see CN 11).*

Proof. Suppose that x_1 is a local minimum. Suppose also that there is another point $x_2 \in C$, $x_1 \neq x_2$, with $f(x_2) < f(x_1)$. Then, for $\alpha x_1 + (1 - \alpha)x_2$, $\alpha \in (0, 1)$, we have

$$f(\alpha x_1 + (1 - \alpha)x_2) \leq \alpha f(x_1) + (1 - \alpha)f(x_2) < f(x_1)$$

which contradicts that x_1 is a local minimum point. $\square$

If $f(\lambda x) = \lambda f(x)$ for all $\lambda \geq 0$, then f is said to be positively homogeneous. If

$$f(x + w) \leq f(x) + f(w), \quad \forall x, w \in \Re^n$$

then f is *subadditive*. A positively homogeneous subadditive function is always convex. A convex function $f : \Re^n \to \Re^1$ is locally Lipschitz continuous (see CN 8) at x, for any $x \in \Re^n$ (Makela and Neittaanmaki, 1992, p. 8, Theorem 2.1.1).

The concept of linear (in)dependence of a set of vectors is used extensively, in particular for the analysis and characterization of the gradients of the *max-function*, in conjunction with Caratheodory's Theorem which we also cover in this section. We commence with a definition of linear independence.

1.1 Linear Independence

A set of vectors $\{x_1, ..., x_r\}$ in $\Re^n$ is linearly independent if and only if the equality

$$\sum_{j=1}^{r} \alpha^j x_j = 0 \tag{1.5}$$

is achieved for $\alpha^j = 0, j = 1, ..., r$. If $r \geq n + 1$, any vectors $x_1, ..., x_r$ in $\Re^n$ are linearly dependent. Hence there exist numbers $\beta^1, ..., \beta^r$ such that

$$\sum_{j=1}^{r} |\beta^j| > 0 \quad \text{and} \quad \sum_{j=1}^{r} \beta^j x_j = 0. \tag{1.6}$$

We note that β^j can be positive or nonpositive. If $r \geq n + 2$, then any vectors $x_1, ..., x_r$ in $\Re^n$ are linearly dependent with (1.6) including the additional condition that

$$\sum_{j=1}^{r} \beta^j = 0. \tag{1.7}$$

To verify that (1.7) holds for $r = n + 2$, we represent

$$\sum_{j=1}^{n+2} \beta^j x_j = 0$$

as

$$\sum_{j=1}^{n+1} \beta^j x_j + \beta^{n+2} x_{n+2} = 0. \tag{1.8}$$

Since $r = n + 2 > n + 1$, x_{n+2} can be represented as a linear combination of $\{x_1, \ldots, x_{n+1}\}$ as

$$\sum_{j=1}^{n+1} \alpha^j x_j = x_{n+2}.$$

We let

$$\alpha^j = \frac{\beta^j}{\sum\limits_{j=1}^{n+1} \beta^j}$$

which yields

$$\sum_{i=1}^{n+1} \frac{\beta^i x_i}{\sum\limits_{j=1}^{n+1} \beta^j} = x_{n+2}.$$

Returning to (1.8),

$$\sum_{i=1}^{n+1} \beta^i x_i + \beta^{n+2} \left\{ \sum_{i=1}^{n+1} \frac{\beta^i x_i}{\sum\limits_{j=1}^{n+1} \beta^j} \right\} = 0$$

which reduces to

$$\sum_{j=1}^{n+1} \beta^j + \beta^{n+2} = 0 = \sum_{j=1}^{n+2} \beta^j.$$

Thus, (1.7) holds for $r = n + 2$. The argument can be straightforwardly extended to $r > n + 2$.

An equivalent, and shorter, demonstration of (1.7) is done by considering the $n + 1$ dimensional vectors

$$\bar{x}_k = \left[x_1^k, \ldots, x_n^k, 1 \right]^{\mathrm{T}} \in \Re^{n+1}, \quad k = [1, \ldots, r], \ r \geq n + 2$$

where x_i^k is the ith element, $i = 1, ..., n$, of $\bar{x}_k$. There are numbers β^k, $\sum_{k=1}^{r} |\beta^k| > 0$, satisfying

$$\sum_{k=1}^{r} \beta^k \, \bar{x}_k = 0$$

which is equivalent to (1.7).

1.2 Tangent Cone, Normal Cone and Epigraph

The *tangent cone* of a convex set C, at $x \in C$, is given by

$$Z_C(x) \equiv \left\{ y \in \Re^n \mid \exists t^i \downarrow 0 \text{ and } y_i \to y \text{ with } (x + t^i y_i) \in C \right\}. \quad (1.9)$$

The *normal cone* of the convex set C, at $x \in C$, is given by

$$\hat{Z}_C(x) \equiv \{ z \in \Re^n \mid \langle y, z \rangle \le 0, \forall y \in Z_C(x) \}. \quad (1.10)$$

The *epigraph* of a function $f : \Re^n \to \Re^1$ is

$$\text{epi} f \equiv \left\{ (x, \rho) \in \Re^n \times \Re^1 \mid f(x) \le \rho \right\}. \quad (1.11)$$

The epigraph of a convex function $f : \Re^n \to \Re^1$ is a closed convex set (Makela and Neittaanmaki, 1992, Theorem 2.3.7).

1.3 Subgradients and Subdifferentials of Convex Functions

The *subdifferential* of a convex function $f : \Re^n \to \Re^1$, at $x \in \Re^n$, is the set

$$\partial f(x) \equiv \left\{ g \in \Re^n \mid f(x') \ge f(x) + g^T(x' - x), \forall x' \in \Re^n \right\}. \quad (1.12)$$

Each element $g \in \partial f(x)$ is called a *subgradient* of f at x. The *directional derivative* in the direction $v \in \Re^n$ satisfies

$$f'(x; v) = \lim_{t \downarrow 0} \frac{f(x + tv) - f(x)}{t}. \quad (1.13)$$

More generally, the Clarke (1983) generalized derivative of a locally Lipschitz function f at x in the direction $v \in \Re^n$ is defined by

$$f^0(x; v) \equiv \limsup_{\substack{y \to x \\ t \downarrow 0}} \frac{f(y + tv) - f(y)}{t}. \quad (1.14)$$

The following important result is used to define a subgradient of the *max-function* where the maximizer is nonunique, in terms of the subdifferential.

Theorem 1.1 (Caratheodory's Theorem) *Let G be a set in the finite dimensional Euclidian space, $G \subseteq \Re^n$. Then, any vector*

$$g \in \text{conv } G$$

may be expressed as a convex combination of at most $n + 1$ vectors in G.

Proof. This result is widely known (e.g., Demyanov and Malozemov, 1974, Appendix 2, Lemma 1.1). Its proof does provide some insight for the subsequent discussion. Consider the definition of the convex hull of G

$$\text{conv } G = \left\{ x = \sum_{k=1}^{r} \alpha^k x_k \in G, \ x_k \in G, \ \alpha^k \geq 0, \ \sum_{k=1}^{r} \alpha^k = 1, \ r = 1, 2, \dots \right\}.$$

(1.15)

Suppose that some $x \in G$ cannot be represented by (1.15) with less than $n + 2$ terms (after those with $\alpha^k = 0$ have been discarded), that is, $r \geq n + 2$ in any representation

$$x = \sum_{k=1}^{r} \alpha^k x_k, \quad x_k \in G, \quad \alpha^k > 0, \quad \sum_{k=1}^{r} \alpha^k = 1.$$

(1.16)

From the discussion on linear independence, if $r \geq n + 2$, there exist numbers β^k, $\sum_{k=1}^{r} |\beta^k| > 0$, such that

$$\sum_{k=1}^{r} \beta^k x_k = 0, \quad \sum_{k=1}^{r} \beta^k = 0.$$

(1.17)

Let

$$\epsilon = \min_{\{k | \beta^k > 0\}} \frac{\alpha^k}{\beta^k} > 0, \quad \overline{\alpha}^k = \alpha^k - \epsilon \beta^k, \quad k \in \{1, \dots, r\}.$$

Vector x can be expressed as

$$x = \sum_{k=1}^{r} \overline{\alpha}^k x_k, \quad \sum_{k=1}^{r} \overline{\alpha}^k = 1.$$

(1.18)

To show the equivalence between (1.16) and (1.18), we expand (1.18) to yield

$$x = \sum_{k=1}^{r} (\alpha^k - \epsilon \beta^k) x_k, \quad \sum_{k=1}^{r} (\alpha^k - \epsilon \beta^k) = 1$$

$$x = \sum_{k=1}^{r} \alpha^k x_k - \epsilon \sum_{k=1}^{r} \beta^k x_k, \quad \sum_{k=1}^{r} \alpha^k - \epsilon \sum_{k=1}^{r} \beta^k = 1.$$

In view of (1.17), we have

$$x = \sum_{k=1}^{r} \alpha^k x_k, \quad \sum_{k=1}^{r} \alpha^k = 1.$$

Considering positive and nonpositive β^k separately, we see that $\bar{\alpha}^k \geq 0$, $k \in \{1, \ldots, r\}$. At least one $\bar{\alpha}^{\bar{k}}$ must vanish and this corresponds to subscript

$$\bar{k} \in \left\{ k \left| \frac{\alpha^k}{\beta^k} = \epsilon \right. \right\}.$$

Thus, x is a convex combination of at most $r - 1$ vectors of G. Iterating this procedure sufficiently many times, we reach a representation $r \leq n + 1$, (1.17) is no longer satisfied, and we have the desired result. □

We invoke Caratheodory's Theorem in the context of the subdifferential

$$\partial \Phi(x) = \text{conv } G$$

where

$$G = \left\{ \nabla_x f(x, y) \mid y = \arg \max_{y \in Y} f(x, y) \right\}$$

and

$$\text{conv } G \equiv \left\{ g = \sum_y \alpha^y \nabla_x f(x, y) \mid \nabla_x f(x, y) \in G, \ \alpha^y \geq 0, \ \sum_y \alpha^y = 1 \right\}.$$

$$(1.19)$$

Using Caratheodory's Theorem, g can be characterized by the convex combination of at most $n + 1$ vectors $\nabla_x f(x, y)$. To verify this, suppose that a subgradient $g \in \partial \Phi(x) = \text{conv } G$ has no representation of type (1.19), for less than $n + 2$ vectors (after vectors $[\nabla_x f(x, y)]$ with associated $\alpha^y = 0$ are discarded). Hence, $r \geq n + 2$ in the representation

$$g = \sum_y \alpha^y \nabla_x f(x, y), \quad \nabla_x f(x, y) \in G, \quad \alpha^y > 0, \quad \sum_y \alpha^y = 1.$$

From the discussion on linear dependence, because $r \geq n + 2$, these exist numbers β^y such that

$$\sum_y |\beta^y| > 0 \quad \text{and} \quad \sum_y \beta^y \nabla_x f(x, y) = 0 \quad \text{and} \quad \sum_y \beta^y = 0.$$

Setting ϵ and $\bar{\alpha}^y$ as in the proof of Caratheodory's Theorem, we have

$$g = \sum_y \bar{\alpha}^y \nabla_x f(x, y), \ \sum_y \bar{\alpha}^y = 1.$$

Considering positive and nonpositive β^y separately, we see that $\bar{\alpha}^y \geq 0$, $y = 1, \ldots, r$, and at least one $\bar{\alpha}^y$ must vanish. Thus, g is a convex combination of at most $r - 1$ vectors of G. Iterating this procedure sufficiently many times, we reach a representation with $r \leq n + 1$.

2 CONTINUOUS MINIMAX

Consider the minimization of the real function $\Phi(x)$ given by

$$\min_{x \in \mathfrak{R}^n} \Phi(x) \tag{2.1}$$

where $\Phi(x)$ is the max-function introduced in Section 1.

$$\Phi(x) = \max_{y \in \mathcal{Y}} f(x, y) \tag{2.2}$$

and $\mathcal{Y}$ is a bounded closed subset of $\mathfrak{R}^m$ (see CN 7). We consider the case when $\mathcal{Y}$ is a set of continuous values. Hence (2.2) is a *continuous minimax* problem. In this chapter, and Chapters 2–5, we discuss various concepts, characteristics and algorithms for continuous minimax. The discrete minimax case is discussed in Chapters 6–7.

Let the *set of maximizers* at x be given by

$$\mathcal{Y}(x) \equiv \left\{ y(x) \in \mathcal{Y} \mid y(x) = \arg\max_{y \in \mathcal{Y}} f(x, y) \right\}. \tag{2.3}$$

In the continuous case, considered henceforth, the function $f(x, y)$ is continuous in x, y, and at least once continuously differentiable with respect to x. Thus, $\Phi(x)$ is piecewise $\mathbb{C}^1$, that is, $\mathfrak{R}^n$ consists of regions inside which the gradient $\nabla\Phi(x)$ exists and is continuous. At the boundary of the regions $\nabla\Phi(x)$ jumps, although $\Phi(x)$ itself is continuous. This boundary corresponds to a kink: a point where $\nabla\Phi(x)$ is nonunique.

Lemma 2.1 *For a given direction $d \in \mathfrak{R}^n$, we have the equality*

$$\max_{z \in \partial\Phi(x)} \langle d, z \rangle = \max_{\{\nabla_x f(x,y) \mid y \in \mathcal{Y}(x)\}} \langle d, \nabla_x f(x, y) \rangle. \tag{2.4}$$

Proof (Demyanov and Malozemov, 1974; Theorem 3.1, p. 195, Lemma 3.1, p. 60). $\square$

The significance of Lemma 2.1 is that, when the maximal directional derivative is evaluated, given d, it is not necessary to consider the entire convex hull. This directional derivative can be determined just by using the information on the subgradients that characterize the convex hull.

$\Phi(x)$ has a gradient whenever $f(x, y)$ is maximized by a single y. Differentiability fails when y is not unique. The subdifferential of $\Phi(x)$ at x, denoted by $\partial\Phi(x)$, is given by

$$\partial\Phi(x) = \text{conv}\{g \mid g = \nabla\Phi(x) = \nabla_x f(x, y), \ y \in \mathcal{Y}(x)\}. \tag{2.5}$$

(Rockafellar, 1981, Proposition 3H). This is also referred to as the Clarke generalized gradient in Polak (1989).

The subdifferential is a nonempty convex compact set (see CN 7) which reduces to the gradient in case $\Phi(x)$ has a unique derivative at x. The elements of $\partial\Phi(x)$ are called subgradients. As in (1.13) and (1.14), the directional derivative $\Phi'(x; d)$, is given by

$$\Phi'(x; d) = \lim_{t\downarrow 0} \frac{\Phi(x + td) - \Phi(x)}{t}. \tag{2.6}$$

(Demyanov and Malozemov, 1974, Theorem 2.1, p. 188). The directional derivative is the support function of $\partial\Phi(x)$, that is,

$$\Phi'(x; d) = \max_{g\in\partial\Phi(x)} \langle g, d \rangle. \tag{2.7}$$

(Polak, 1987, Definition 2.2.4). For a fixed x, $\Phi'(x; d)$ is convex in d. As such, it has a subdifferential which is $\partial\Phi(x)$.

If x_* is a local minimizer (see CN 11) of Φ, then the following statements hold:

(i) $\Phi'(x; d) \geq 0, \ \forall d \in \Re^n$

(ii) $0 \in \partial\Phi(x)$,

(Polak, 1989, Theorem 5.1). Hence, the set X_* of optimal points x_* is characterized by

$$X_* = \{x_* \in \Re^n \mid 0 \in \partial\Phi(x_*)\}. \tag{2.8}$$

3 OPTIMALITY CONDITIONS AND ROBUSTNESS OF MINIMAX

A fundamental condition for minimax is that, at the solution x_*, we have

$$\Phi(x_*) = \max_{y\in\mathcal{Y}} f(x_*, y) \geq f(x_*, y), \quad \forall y \in \mathcal{Y}.$$

The inequality simply indicates that the performance of the solution x_* is noninferior for any y. This is essentially the *robust* nature of minimax which yields a *guaranteed best lower-bound performance* in view of the worst-case. It is assured that this performance will improve if the worst-case is not realized.

We discuss the necessary condition for problem (2.1) and the related Haar condition.

Theorem 3.1 *A necessary condition for $\Phi(x)$ to achieve its minimum on $\Re^n$ at x_* is given by the inequality*

$$\min_{z \in \Re^n} \max_{y \in \mathcal{Y}(x_*)} \left\langle \frac{\partial f(x_*, y)}{\partial x}, z - x_* \right\rangle \geq 0 \tag{3.1a}$$

If $\Phi(x)$ is convex on $\Re^n$, this condition is also sufficient.

Remark An equivalent condition to (3.1a) is

$$\max_{y \in \mathcal{Y}(x_*)} \left\langle \frac{\partial f(x_*, y)}{\partial x}, d \right\rangle \geq 0, \quad \forall d \in \Re^n \tag{3.1b}$$

which simply states that at the solution x_*, there is no descent, or improvement, direction in view of the maximizers at x_*.

Proof (Demyanov and Malozemov, 1974, p. 191). To show the necessity of this condition, let $x_* \in \Re^n$ be a minimum point of $\Phi(x)$ and suppose that (3.1) fails to hold. There is thus a point $z_1 \in \Re^n$ such that

$$\max_{y \in \mathcal{Y}(x_*)} \left\langle \frac{\partial f(x_*, y)}{\partial x}, z_1 - x_* \right\rangle = -\rho < 0. \tag{3.2}$$

The expression on the left of (3.1) is always negative, having failed (3.2). Clearly from (3.2), $z_1 \neq x_*$. Let $g_1 \in \Re^n$ be given by

$$g_1 = \frac{z_1 - x_*}{\|z_1 - x_*\|}, \quad \|g_1\| = 1.$$

Consider the linear approximation of $\Phi(x)$

$$\Phi(x_* + \alpha g_1) = \Phi(x_*) + \alpha \Phi'(x, g_1) + o(\alpha g_1) \tag{3.3}$$

(see CN 10) and the directional derivative (2.6) and (2.7), for some vector $g \in \Re^n$

$$\Phi'(x; g) \equiv \lim_{\alpha \downarrow 0} \frac{\Phi(x + \alpha g) - \Phi(x)}{\alpha} = \max_{y \in \mathcal{Y}(x)} \left\langle \frac{\partial f(x, y)}{\partial x}, g \right\rangle. \tag{3.4}$$

It follows from (3.2), (3.3) and (3.4) that

$$\Phi'(x; g_1) = \frac{-\rho}{\|z_1 - x_*\|}. \tag{3.5}$$

Thus, for sufficiently small α, we have from (3.3) and (3.5)

$$\Phi(x_* + \alpha g_1) \leq \Phi(x_*) - \frac{\alpha \rho}{2\|z_1 - x_*\|} < \Phi(x_*). \tag{3.6}$$

Inequality (3.6) contradicts the assumption that x_* is a minimum point of $\Phi(x)$, since for all

$$\alpha \in [0, \|z_1 - x_*\|]$$

the points $x_* + \alpha g_1$ are in $\Re^n$.

To establish sufficiency, let $\Phi(x)$ be convex on $\mathfrak{R}^n$ and suppose that (3.1) is satisfied at $x_* \in \mathfrak{R}^n$. Suppose that x_* is not a minimum point of $\Phi(x)$ on $\mathfrak{R}^n$. Then there exists a point $z_1 \in \mathfrak{R}^n$ for which the inequality

$$\Phi(z_1) < \Phi(x_*) \tag{3.7}$$

holds. Clearly, we have $z_1 \neq x_*$. Setting again

$$g_1 = \frac{z_1 - x_*}{\|z_1 - x_*\|}, \quad \|g_1\| = 1$$

we evaluate

$$\Phi'(x; g_1) = \lim_{\alpha \downarrow 0} \frac{\Phi(x_* + \alpha g_1) - \Phi(x_*)}{\alpha}.$$

Since $\Phi(x)$ is convex, it follows that for $\beta \in [0, 1]$,

$$\Phi\big(x_* + \beta(z_1 - x_*)\big) = \Phi(\beta z_1 + (1 - \beta)x_*)$$

$$\leq \beta \Phi(z_1) + (1 - \beta)\Phi(x_*)$$

$$= \Phi(x_*) + \beta[\Phi(z_1) - \Phi(x_*)]. \tag{3.8}$$

Hence, for $\alpha \in [0, \|z_1 - x_*\|]$, we have

$$\frac{\Phi(x_* + \alpha g_1) - \Phi(x_*)}{\alpha} = \frac{\Phi\left(x_* + \dfrac{\alpha}{\|z_1 - x_*\|}(z_1 - x_*)\right) - \Phi(x_*)}{\alpha}$$

$$\leq \frac{1}{\|z_1 - x_*\|}[\Phi(z_1) - \Phi(x_*)]. \tag{3.9}$$

Thus, by (3.7), (3.8) and (3.9), we have

$$\Phi'(x; g_1) \leq \frac{\Phi(z_1) - \Phi(x_*)}{\|z_1 - x_*\|} < 0.$$

Thus, we observe that

$$\max_{y \in \mathcal{Y}(x_*)} \left\langle \frac{\partial f(x_*, y)}{\partial x}, z_1 - x_* \right\rangle = \|z_1 - x_*\|\Phi'(x, g_1) \leq \Phi(z_1) - \Phi(x_*) < 0.$$

This contradicts (3.1) and hence establishes the sufficiency for convex $\Phi(x)$. $\square$

3.1 The Haar Condition

We comment on the Haar condition which is usually discussed in the context of minimax. The necessary condition (3.1) can be expressed as

$$0 \in \partial \Phi(x_*) \tag{3.10}$$

which is also a sufficient condition when either $f(x, y)$ is strictly convex in $x \in \Re^n$, or the Haar condition is satisfied (Wierzbicki, 1982).

Let the vector e_0 be given by

$$e_0 = \begin{bmatrix} -1 \\ 0 \end{bmatrix} \in \Re^{n+1} \tag{3.11}$$

where $0 \in \Re^n$ denotes the zero vector. The *Haar sufficient condition* for a point x_* satisfying (3.10) to be a unique, or global (see CN 11) (local in the nonconvex case) solution of the original problem is expressed as

$$e_0 \in \hat{Z} \equiv \{h \in Z \mid \langle h, z \rangle < 0, \ \forall z \in Z, \ z \neq 0\} \tag{3.12}$$

where Z is the tangent cone and $\hat{Z}$ is the normal cone to the epigraph of $\Phi(x)$ at x_*.

The implication of the *Haar condition* is that there are at least $n + 1$ elements of $\mathcal{Y}(x_*)$, where the corresponding $\nabla_x f(x_*, y_i)$, $y_i \in \mathcal{Y}(x_*)$ can be linearly combined using nonzero λ^i such that

$$\sum_i \lambda^i \nabla_x f(x_*, y_i) = 0, \quad y_i \in \mathcal{Y}(x_*).$$

Hence the vectors $\nabla_x f(x_*, y_i)$, $y_i \in \mathcal{Y}(x_*)$, are linearly dependent. The Haar condition requires that all subsets of n of these $n + 1$ vectors, $\nabla_x f(x_*, y_i)$, $y_i \in \mathcal{Y}(x_*)$, are linearly independent. Suppose that there are less than this, no more than n [the cardinality of $\mathcal{Y}(x_*)$ is at most n (i.e., $|\mathcal{Y}(x_*)| \leq n$)]. According to (3.10), there exist nonzero λ^i such that the above equality is satisfied and all the gradients $\nabla_x f(x_*, y_i)$, $y_i \in \mathcal{Y}(x_*)$, are linearly dependent. If no more than n vectors of dimension n are linearly dependent, it would not be possible to form full-rank matrices from all collections of these vectors. Thus, there must be at least $n + 1$ vectors $\nabla_x f(x_*, y)$, $y \in \mathcal{Y}(x_*)$. and each collection of, say n, of them must be linearly independent. This implies in turn that at least $n + 1$ elements λ^i must be positive, since, if $|\mathcal{Y}(x_*)| = n + 1$, for each j we have

$$\sum_{y_i \in \mathcal{Y}(x_*), i \neq j} \lambda^i \nabla_x f(x_*, y_i) = -\lambda^j \nabla_x f(x_*, y_j) \neq 0$$

and $\lambda^j \geq 0$, which implies $\lambda^j > 0$. If at least $n + 1$ elements λ^i are positive, then e_0 is in the interior of the normal cone $\hat{Z} = \text{cone}\{-1, \nabla_x f(x_*, y); \ y \in \mathcal{Y}(x_*)\}$. Conversely, if e_0 is in the interior of $\hat{Z}$, then at least $n + 1$ vectors $\nabla_x f(x_*, y)$, $y \in \mathcal{Y}(x_*)$, sum up to zero with coefficients greater than zero and each collection of n, or less, of these vectors is linearly independent.

The Haar condition is rather restrictive: at least $n + 1$ vectors $\nabla_x f(x_*, y)$, $y \in \mathcal{Y}(x_*)$, must be binding at x_* (i.e., there must be at least $n + 1$ maximizers at x_*). However, there are cases where $|\mathcal{Y}(x_*)| \le n$. Then the normal cone $\hat{Z}$ does not have an interior and the Haar condition cannot be satisfied, and x_* must be determined on the basis of additional information using second order conditions.

4 SADDLE POINTS AND SADDLE POINT CONDITIONS

The special case for the minimax problem is the saddle point solution. This is of particular interest as saddle point solutions are more easily characterized and computed. Let $\mathcal{R} \subset \mathfrak{R}^n$ and $\mathcal{Y} \subset \mathfrak{R}^m$ be bounded closed sets (see CN 7) and $f(x, y)$ be a continuous function defined on $\mathcal{R} \times \mathcal{Y}$. A point $(x_*, y_*) \in \mathcal{R} \times \mathcal{Y}$ is said to be a *saddle point* of $f(x, y)$ on $\mathcal{R} \times \mathcal{Y}$ if

$$f(x_*, y) \le f(x_*, y_*) \le f(x, y_*) \tag{4.1}$$

for all $x \in \mathcal{R}, y \in \mathcal{Y}$. In this section we review important properties of saddle points. Algorithms for computing them are discussed in Chapter 3.

Lemma 4.1 *Let the function $f(x, y)$ be continuous on $\mathcal{R} \times \mathcal{Y}$ and $\mathcal{R}, \mathcal{Y}$ be closed and bounded sets as above. Then, $f(x, y)$ has a saddle point on $\mathcal{R} \times \mathcal{Y}$ if and only if*

$$\min_{x \in \mathcal{R}} \max_{y \in \mathcal{Y}} f(x, y) = \max_{y \in \mathcal{Y}} \min_{x \in \mathcal{R}} f(x, y). \tag{4.2}$$

Proof (Demyanov and Malozemov, 1974; Lemma 6.1, p. 222). □

The maxima and minima in (4.2) are achieved since $f(x, y)$ is continuous and $\mathcal{R}, \mathcal{Y}$ are closed and bounded. We show that x_*, y_* satisfying (4.1) are also solutions of both sides of equality (4.2). By (4.1), we have

$$\max_{y \in \mathcal{Y}} f(x_*, y) \le \min_{x \in \mathcal{R}} f(x, y_*).$$

Hence, we have

$$\min_{x \in \mathcal{R}} \max_{y \in \mathcal{Y}} f(x, y) \le \max_{y \in \mathcal{Y}} \min_{x \in \mathcal{R}} f(x, y).$$

The inverse is trivial: for any $x \in \mathcal{R}, y \in \mathcal{Y}$, we have

$$f(x, y) \le \max_{y' \in \mathcal{Y}} f(x, y') \tag{4.3}$$

hence

$$\min_{x \in \mathcal{R}} f(x, y) \le \min_{x \in \mathcal{R}} \max_{y' \in \mathcal{Y}} f(x, y')$$

and as this inequality is satisfied for all $y \in \mathcal{Y}$, we have

$$\max_{y \in \mathcal{Y}} \min_{x \in \mathcal{R}} f(x,y) \leq \min_{x \in \mathcal{R}} \max_{y \in \mathcal{Y}} f(x,y). \tag{4.4}$$

In view of (4.3) and (4.4), we have (4.2).

Let (4.2) be satisfied and

$$\Phi(x) = \max_{y \in \mathcal{Y}} f(x,y), \quad \phi(y) = \min_{x \in \mathcal{R}} f(x,y).$$

It follows from (4.2) and the continuity of $\Phi(x)$, $\phi(y)$ that there exist points $x_* \in \mathcal{R}$, $y_* \in \mathcal{Y}$ for which

$$\min_{x \in \mathcal{R}} \Phi(x) = \Phi(x_*) = \max_{y \in \mathcal{Y}} \phi(y) = \phi(y_*). \tag{4.5}$$

It follows from (4.5) that for any $(x,y) \in \mathcal{R} \times \mathcal{Y}$, we have

$$f(x_*,y) \leq \max_{y' \in \mathcal{Y}} f(x_*,y') = \Phi(x_*) = \phi(y_*) = \min_{x \in \mathcal{R}} f(x,y_*) \leq f(x_*,y_*) \tag{4.6}$$

and

$$f(x,y_*) \geq \min_{x' \in \mathcal{R}} f(x',y_*) = \phi(y_*) = \Phi(x_*) = \max_{y \in \mathcal{Y}} f(x_*,y) \geq f(x_*,y_*). \tag{4.7}$$

Combining (4.6) and (4.7) yields (4.1). $\square$

Hence, condition (4.2) is necessary and sufficient for the existence of a saddle point. In the next result, we discuss a special class of problems satisfying this condition.

Theorem 4.1 *Let $f(x,y)$ be continuous together with $\nabla_x f(x,y)$ on $\mathcal{R}' \times \mathcal{Y}$, where $\mathcal{R}' \subset \mathfrak{R}^n$, $\mathcal{Y} \subset \mathfrak{R}^m$. Assume that $\mathcal{R}'$ is open and let $\mathcal{R} \subset \mathcal{R}'$ and $\mathcal{Y}$ be bounded closed convex sets. Furthermore, let the function $f(x_0,y)$ be concave for every fixed $x_0 \in \mathcal{R}'$ and $f(x,y_0)$ be convex for every fixed $y_0 \in \mathcal{Y}$. Then, equality (4.2) is satisfied for $f(x,y)$.*

Proof (Demyanov and Malozemov, 1974; Theorem 5.2, p. 218). $\square$

A related result, establishing the saddle point of $f(x,y)$ convex in x, concave in y, is given in Corollary 3.1.1. This uses explicitly the condition on the Hessian in Lemma 1.2. As convexity in x and concavity in y of f assure (4.1) and (4.2), the main property used in subsequent chapters for the existence of a saddle point is that $f(x,y)$ is convex in x, concave in y. This subject is discussed further in Section 3.1, concerning the motivation, or justification, of algorithms for computing saddle points.

References

Apostol, T. (1981). *Mathematical Analysis*, 2nd Edition, Addison Wesley, Massachusetts.

Başar, T. and P. Bernhard (1991). *H^{∞}-Optimal Control and Related Minimax Design Problems*, Birkhäuser, Boston, MA.

Becker, R., B. Dwolatzky, E. Karakitsos and B. Rustem (1986). "The Simultaneous Use of Rival Models in Policy Optimization", *The Economics Journal*, 96, 89–101.

Ben-Tal, A. and A. Nemirovskii (1993). "A New Method for Optimal Truss Topology Design", *SIAM Journal on Optimization*, 3, 322–358.

Ben-Tal, A. and A. Nemirovskii (1994). "Potential Reduction Polynomial Time Method for Truss Topology Design", *SIAM Journal on Optimization*, 4, 596–612.

Dert, C. and B. Oldenkamp (1997). "Optimal Guaranteed Return Portfolios and the Casino Effect", Research Report, Department BFS, Free University of Amsterdam.

Chow, G.C. (1979). "Effective Use of Econometric Models in Macroeconomic Policy Formulation", in: S. Holly, B. Rustem and M. Zarrop (editors), *Optimal Control for Econometric Models*, Macmillan, London.

Clarke, F.H. (1983). *Optimization and Non-smooth Analysis*, Wiley, Canada.

Demyanov, V.F. and V.N. Malozemov (1974). *Introduction to Minimax*, Wiley, New York.

Golub, G.H and C. F. van Loan (1983). *Matrix Computations*, North Oxford Academic, Oxford.

Hansen, L.P., T.J. Sargent and D. Tallarini Jr. (1998). "Robust Permanent Income and Pricing", http://www.stanford.edu/sargent/research.html.

Horn, R.A. and C.R. Johnson (1993). *Matrix Analysis*, Cambridge University Press, Cambridge and New York.

Howe, M. and B. Rustem (1997). "A Robust Hedging Algorithm", *Journal of Economic Dynamics and Control*, 21, 1065–1092.

Howe, M.A., B. Rustem and M. Selby (1994). "Minimax Hedging Strategy", *Computational Economics*, 7, 245–275.

Howe, M., B. Rustem, M. Selby (1996). "Multi-period Minimax Hedging Strategies", *European Journal of Operational Research*, 93, 185–204.

Ibanez, A. (1998). "Hedging Portfolios of Derivatives Securities with Maxmin Strategies", Research Report, ITAM, Mexico.

Makela, M.M. and P. Neittaanmaki (1992). *Non-smooth Optimization: Analysis and Algorithms with Applications to Optimal Control*, World Scientific, London.

Pang, J.S. and C.S. Yu (1989). "A Min-max Resource Allocation Problem with Substitutions", *European Journal of Operational Research*, 41, 218–223.

Polak, E. (1987). "On the Mathematical Foundations of Non-differentiable Optimization in Engineering Design", *SIAM Review*, 29, 21–89.

Polak, E. (1989). "Basics of Minimax Algorithms", in: F.H. Clarke, V.F. Demyanov and F. Gianessi (editors), *Non-smooth Optimization and Related Topics*, Plenum, New York, pp. 343–369.

Rockafellar, R.T. (1981). *The Theory of Subgradients and Its Applications to Problems of Optimization*, Heldermann, Berlin.

Rustem, B., R.G. Becker and W. Marty (2000). "Robust Min-max Portfolio Strategies for Rival Forecast and Risk Scenarios", *Journal of Economic Dynamics and Control*, in press.

Wierzbicki, A.P. (1982). "Lagrangian Functions and Non-differentiable Optimization", in: E.A. Nurminski (editor), *Progress in Non-differentiable Optimization*, Publication CP-82-58, IIASA, 2361 Laxenburg, Austria.

COMMENTS AND NOTES

CN 1: Vectors and Matrices

Let $\Re^n$ be the set of n-dimensional vectors. Any $x \in \Re^n$, has n components, or coordinates. Generally, we refer to the ith component of x by x^i. A null vector refers to $x^i = 0$, $\forall i$, and is denoted by $x = 0$. Vectors in $\Re^n$ are viewed as column vectors, unless the contrary is stated explicitly. The transpose of x is a row vector and is denoted by x^{T}.

A matrix a is a rectangular array of numbers. For any A, we use a_{ij}, or $[A]_{ij}$ as the ijth element of A. For $i = 1, \ldots, n$ and $j = 1, \ldots, m$, A is referred to as an $(n \times m)$ matrix, with n rows and m columns. When $n = m$, matrix A is referred to as a *square* matrix and if $a_{ij} = 0$, $\forall i,j$, then $A = 0$. The transpose of the $(n \times m)$ matrix A is denoted by A^{T} and refers to an $(m \times n)$ matrix whose elements are given by a_{ji}. A matrix is *symmetric* if $A = A^{\mathrm{T}}$. The sum of two $(m \times n)$ matrices A and B is written as $A + B$ and is the $(m \times n)$ matrix whose elements are the sum of the corresponding elements in A and B. The product of a matrix A and scalar τ, $\tau A = A\tau$, is obtained by multiplying each element of A by τ. The product of an $(m \times n)$ matrix A and $(n \times q)$ matrix B is written as AB and is the $(m \times q)$ matrix F whose elements are given by $[F]_{ij} = \sum_{\ell=1}^{n} [A]_{i\ell}[B]_{\ell j}$. Finally, as in the vector case, an $(m \times n)$ matrix A is considered to be a member of the space $\Re^{n \times m}$ such that $A \in \Re^{n \times m}$.

CN 2: Inner Product, Vector and Matrix Norms

For any two vectors, $x, y \in \Re^n$, their *inner product* is defined by

$$\langle x, y \rangle = x^{\mathrm{T}} y = \sum_{i=1}^{n} x^i y^i.$$

Any two vectors $x, y \in \Re^n$ satisfying $\langle x, y \rangle = 0$ are called *orthogonal*.

A vector *norm* $\|\cdot\|$ on $\Re^n$ is a mapping that assigns a real scalar to every $x \in \Re^n$ and has the properties:

  (i) $\|x\| \geq 0$, $\forall x \in \Re^n$;

  (ii) $\|\tau x\| = |\tau| \|x\|$ for every $\tau \in \Re^1$ and every $x \in \Re^n$;

  (iii) $\|x\| = 0$, if and only if $x = 0$;

  (iv) $\|x + z\| \leq \|x\| + \|y\|$, $\forall x, y \in \Re^n$ (i.e., the triangle inequality).

The *Euclidian* norm is defined by

$$\|x\|_2 = \langle x, x \rangle^{1/2} = \left(\sum_{i=1}^{n} (x^i)^2 \right)^{1/2}$$

and satisfies the Cauchy–Schwartz inequality

$$|\langle x, y \rangle| \le \|x\|_2 \|y\|_2.$$

The *matrix norms* we consider belong to the class of norms that are induced by the corresponding vector norm. Given any norm $\|\cdot\|$ for vector x, the induced norm for the $(n \times m)$ matrix A is given by

$$\|A\| = \max_{\{x \in \Re^n \,|\, \|x\|=1\}} \|Ax\|.$$

We note that

$$\|A\| = \max_{\|x\| \ne 0} \frac{\|Ax\|}{\|x\|}$$

which follows because for any $x \ne 0$, x can be scaled so that its norm is equal to unity without changing the value of $\|Ax\|/\|x\|$. Furthermore, we have

$$\|Ax\| \le \|A\| \|x\| \quad \text{and} \quad \|AB\| \le \|A\| \|B\|$$

which follow from the above for B of appropriate dimension. A detailed discussion on matrix norms can be found in Golub and van Loan (1983) and Horn and Johnson (1993).

CN 3: Differentiable Function Definitions

If the function f has continuous first partial derivatives with respect to x, this is denoted by $f \in \mathbb{C}^1$. If f has continuous second partial derivatives with respect to x, this is denoted by $f \in \mathbb{C}^2$.

CN 4: Positive Semi-definite and Definite Matrices

The symmetric matrix H is *positive semi-definite* if for all $0 \ne v \in \Re^n$, $\langle v, Hv \rangle \ge 0$. The symmetric matrix H is positive-definite if for all $0 \ne v \in \Re^n$, $\langle v, Hv \rangle > 0$.

CN 5: First and Second Order Taylor Expansions

First order Taylor series expansion of a function $f : \Re^n \to \Re^1$, and $f \in \mathbb{C}^1$, for any $x, d \in \Re^n$, $\theta \in \Re^1$, is given by

$$f(x + \theta d) = f(x) + \theta \int_0^1 \langle \nabla f(x + t\theta d), d \rangle \, dt.$$

Second order Taylor series expansion of a function $f : \Re^n \to \Re^1$ and $f \in \mathbb{C}^2$, for any $x, d \in \Re^n$, $\theta \in \Re^1$, is given by

$$f(x + \theta d) = f(x) + \theta \langle \nabla f(x), d \rangle + \theta^2 \left(\int_0^1 (1 - t)\langle d, \nabla^2 f(x + t\theta d)d \rangle \, dt \right).$$

CN 6: Linear Independence of Vectors

A set of vectors $v_1, v_2, \ldots, v_s$ is said to be linearly dependent if there are scalars $\omega_1, \omega_2, \ldots, \omega_s$, not all zero, such that $\sum_{i=1}^s v_i \omega_i = 0$. If no such $\omega_1, \omega_2, \ldots, \omega_s$ exist, the vectors $v_1, v_2, \ldots, v_s$ are said to be *linearly independent*.

CN 7: Open Sets, Closed Sets, Bounded Sets, Compact Sets, Convergence of Sequences

The *open ball* with center x and radius $\epsilon > 0$ is denoted by

$$B(x, \epsilon) \equiv \{ z \in \Re^n \mid \|z - x\| < \epsilon \}$$

for some $\epsilon > 0$. A subset $S \subseteq \Re^n$ is open if around every point $x \in S$, there is an open ball contained in S (i.e., for $x \in S$, there is an $\epsilon > 0$ such that $\|y - x\| < \epsilon$, with $y \in S$). For example, the ball $\{y \in \Re^n \mid \|y\| < \epsilon\}$ is open. The interior of any set $S \subseteq \Re^n$ is the set of points $x \in S$ which are at the center of some ball contained in S. A set is said to be open if all its points are all interior points. Thus, the interior of a set is always open. The interior of the ball

$$\{y \in \Re^n \mid \|y\| \le \epsilon\}$$

is the open ball

$$\{y \in \Re^n \mid \|y\| < \epsilon\}.$$

A set S is closed if its complement $\Re^n - S$ is open. Alternatively, a set S is closed if every point arbitrarily close to S is a member of S. Equivalently, S is closed if $\{x_k\} \in S$ and $\{x_k\} \to x$ imply that $x \in S$. For example, the ball $\{y \in \Re^n \mid \|y\| \le \epsilon\}$ is closed. The closure of any set $S \subseteq \Re^n$ is the smallest closed set containing S. The boundary of a set is that part of the set that is not its interior.

A set is *bounded* if it lies entirely within a ball of some radius $\epsilon > 0$,

$$\{ y \in \Re^n \mid \|y - x\| \le \epsilon \}.$$

A set is *compact* if and only if it is closed and bounded. That is, if it is closed and contained within a sphere of finite radius.

For a compact set S, the Bolzano–Weierstrass theorem (Apostol, 1981) can be used to show that every infinite subset of S has an accumulation point in S. Thus, if $\{x_k\} \subset S$ (i.e., each member of the sequence is in S,) then $\{x_k\}$ has a limit point in S. This establishes that there is a subsequence of $\{x_k\}$ converging to a point in S. For example, the sequence defined by $x_{k+1} = -x_k, x_0 = 1$, has a subsequence $\{x_{2k}\}$ with limit point 1 and a subsequence $\{x_{2k+1}\}$ with limit point -1. Both sequences and their limits are clearly in the compact set

$$\{x \mid \|x\| \le 1\}.$$

The convergence of the algorithms in Chapters 2–7 depends on the Bolzano–Weierstrass theorem. We are concerned with algorithms that aim to solve for a fixed optimal point while generating a sequence $\{x_k\}$ within a compact set S such that a certain descent, or improvement, property is satisfied for a merit function (e.g., $f(x_{k+1}) < f(x_k)$ and for x_* solving the problem, $f(x_*) \le f(x_k)$, $\forall k$). There need not be a unique accumulation point for each subsequence. However, each accumulation point would need to satisfy the optimality condition of the problem.

CN 8: Lipschitz Continuity of a Function

A function f is said to be locally *Lipschitz continuous* with constant $c \ge 0$ at $x \in \Re^n$ if there exists some $\epsilon > 0$ such that

$$|f(w) - f(z)| \le c\|w - z\|, \quad \forall w, z \in B(x, \epsilon).$$

CN 9: Hyperplanes

A *hyperplane* is essentially the n-dimensional generalization of a three-dimensional plane. Let the column vector $a \in \Re^n$ and c be a scalar. Then, the set $\{x \in \Re^n \mid \langle a, x \rangle = c\}$ is a hyperplane in $\Re^n$.

CN 10: Order o($\cdot$), O($\cdot$)

Let f be a real-valued function of a real variable x. The notation $f(x) = O(x)$ signifies that $f(x)$ approaches zero at least as fast as x does. This implies that there exists a $\kappa \ge 0$ such that

$$\left| \frac{f(x)}{x} \right| \le \kappa, \text{ as } x \to 0.$$

The notation $f(x) = o(x)$ indicates that $f(x)$ approaches zero faster than x does, or equivalently, that $\kappa = 0$.

CN 11: Local and Global Minima

Let $g : \mathbb{G} \subset \mathfrak{R}^n \to \mathfrak{R}^e$, $h : \mathbb{H} \subset \mathfrak{R}^n \to \mathfrak{R}^i$ and the set of feasible points be defined by

$$\mathcal{R} \equiv \{ \, x \in \mathfrak{R}^n \mid g(x) = 0, \ h(x) \leq 0 \}.$$

A point x_* is a local minimum point of f over $\mathfrak{R}$ if there is no better solution within an $\epsilon > 0$ neighborhood of x_* such that,

$$f(x) \geq f(x_*), \quad \forall x \in \mathcal{R}, \ \|x - x_*\| < \epsilon$$

If $f(x) > f(x_*)$, for all $x_* \neq x \in \mathcal{R}$, $\|x - x_*\| < \epsilon$, then x_* is a *strict local minimum point* of f over $\mathcal{R}$.

A point x_* is a *global minimum point* of f over $\mathcal{R}$ if $f(x) \geq f(x_*)$ for all $x \in \mathcal{R}$. If $f(x) > f(x_*)$, for all $x_* \neq x \in \mathcal{R}$, then x_* is a *strict global minimum point* of f over $\mathcal{R}$.

Chapter 2

A survey of continuous minimax algorithms

We consider several continuous minimax algorithm models. All of these base their progress on gradient information. While some are implementable, others require substantial further development to be of practical use. In Chapter 4, we introduce and analyze in detail a quasi-Newton algorithm that builds upon some of the models introduced in the present chapter. In Chapter 5, we consider numerical experiments with a number of algorithms to justify empirically a simplified quasi-Newton algorithm.

Under the special assumption that $f(x, y)$ is convex in x and concave in y, continuous minimax can be formulated as a saddle point problem. This is an interesting special case for minimax problems and we discuss algorithms for computing saddle point solutions in Chapter 3. Another special case is discrete minimax and a superlinearly convergent quasi-Newton algorithm to solve this problem is discussed in Chapter 7.

1 INTRODUCTION

In this chapter, we survey algorithms for solving the continuous minimax problem introduced in Section 1.2,

$$\min_{x \in \Re^n} \max_{y \in \mathcal{Y}} f(x, y). \qquad (1.1)$$

A new quasi-Newton algorithm for this problem is developed and analyzed in detail in Chapter 4. Numerical experiments with a number of algorithms are discussed in Chapter 5.

Continuous minimax belongs to the general class of nonsmooth problems. The main reason for this nonsmooth character is the possible multiplicity of maximizers at any given point. The objective function has a different gradient, with respect to x, corresponding to each maximizer. As such, continuous minimax problems may be solved using nonsmooth optimization methods such as subgradient and bundle methods. Subgradient methods require at least one subgradient to be evaluated at each iteration to find a direction of descent (see CN 1). Bundle methods use subgradient information from successive iterations, within a ball of radius $r > 0$. These methods have been developed to solve either the general class or particular types of nonsmooth problems. Those addressing the general class lack explicit steps to deal with the maximization subproblem, while those addressing particular types, such as discrete minimax, are too specialized to be applicable to continuous minimax. For example, the nonsmooth algorithm due to Shor (1980) can be applied to discrete minimax problems but not to continuous minimax.

A number of algorithms have been proposed for (1.1). The method of centers by Chaney (1982) requires that the maximum in (1.1) is attained at a unique point $y(x)$, for all x. The algorithm also requires that, for all maximization involving x_k, there is a globally convergent procedure generating a sequence $\{y_j\}$ convergent to a unique $y(x_k)$. Hence, Chaney's method does not handle kinks in the *max-function* $\Phi(x)$. In this type of problem, $\Phi(x)$ is differentiable and smooth optimization techniques can be applied to solve it. Klessig and Polak (1973) consider a first order, feasible directions method that, like Chaney (1982), requires the maximizer to be unique.

Demyanov and Malozemov (1974) treat problem (1.1) in an indirect way by solving an infinite sequence of discrete minimax problems of the form

$$\min_{x \in \mathfrak{R}^n} \max_{y \in \mathcal{Y}^t} f(x, y) \tag{1.2}$$

for $t = 1, 2, ...$, where $\mathcal{Y}^t$ are finite subsets of $\mathcal{Y}$. The method assumes that for any $\epsilon > 0$, there exists t_0 such that for $t > t_0$, the distance between any point $y \in \mathcal{Y}$ and the point of $\mathcal{Y}^t$, nearest to y, is less than ϵ. Thus, as $t \to \infty$, these discrete minimax problems approximate continuous minimax with increasing accuracy.

Panin (1981) suggests the use of an approximation to $\Phi(x)$ at x_k, given by

$$\Phi_k^\ell(d) = \max_{y \in \mathcal{Y}} f_k^\ell(d, y) \tag{1.3}$$

$$f_k^\ell(d, y) = f(x_k, y) + \langle \nabla_x f(x_k, y), d \rangle. \tag{1.4}$$

The method is based on the assumption that for any $x \in \mathfrak{R}^n$, one can determine

$$\hat{d} = \arg\min_{d \in \mathfrak{R}^n} \Phi_k^\ell(d) + \frac{1}{2} \|d\|^2 \tag{1.5}$$

where $\|.\| = \|.\|_2 \equiv \langle \cdot, \cdot \rangle^{1/2}$. No procedure for solving (1.5) is given by Panin.

The method of Kiwiel (1987) is based on Panin (1981). At the kth iteration of Kiwiel's algorithm, the change in the objective

$$\{\Phi(x_k + d) - \Phi(x_k)\}$$

is approximated by

$$\{\Phi_k^\ell(d) - \Phi(x_k)\}.$$

At x_k, a descent direction for $\Phi(x_k)$ can be found by solving (1.5) using an auxiliary algorithm. Since the objective in (1.5) is strongly convex in d, $\hat{d}$ exists and is uniquely determined by

$$0 \in \hat{d} + \text{conv}\left\{ \nabla_x f(x_k, y) \mid y \in \mathcal{Y}_{k+1}^\ell \right\} \qquad (1.6)$$

where

$$\mathcal{Y}_{k+1}^\ell \equiv \mid \left\{ y \in \mathcal{Y} \mid y = \arg\max f_k^\ell(d_k, y) \right\}. \qquad (1.7)$$

Kiwiel's method finds, at each x, a linear combination of the vectors $\nabla_x f(x, y_i)$, $y_i \in \mathcal{Y}_{k+1}^\ell$.

In Sections 2, 3 and 4, we describe the algorithms of Chaney, Panin and Kiwiel, respectively. The latter two algorithms are of particular interest as they are related to the quasi-Newton algorithm in Chapter 5.

2 THE ALGORITHM OF CHANEY

In Chaney (1982), an algorithm is developed for solving the constrained continuous minimax problem of the form

$$\min_{x \in X} \max_{y \in \mathcal{Y}} f(x, y). \qquad (2.1)$$

The sets X and $\mathcal{Y}$ are determined by constraints of the form

$$X \equiv \left\{ x \in \mathfrak{R}^n \mid g_j(x) \le 0, \ j = 1, \dots, e \right\}$$

$$\mathcal{Y} \equiv \{ y \in \mathfrak{R}^m \mid h_i(y) \le 0, \ i = 1, \dots, i \}.$$

At each fixed x,

$$\Phi(x) = \max_{y \in \mathcal{Y}} f(x, y) \qquad (2.2)$$

and (2.1) is reformulated as

$$\min_{x \in X} \Phi(x). \qquad (2.3)$$

The critical assumption underlying Chaney's algorithm is that the problem of maximizing $f(x, y)$ over $\mathcal{Y}$ has a unique solution so that the set of maximizers at x has a single element, $\mathcal{Y}(x)$. Therefore, the max-function is given by

$$\Phi(x) = f(x, \mathcal{Y}(x))$$

for each x in X. Chaney proposes a modified version of the Pironneau and Polak (1972) method of centers for solving (2.3). The algorithm for solving (2.2) is left unspecified at the start. The combined algorithm seeks to intertwine the Pironneau–Polak minimizing algorithm and the maximizing algorithm, using an adaptive procedure that forces the maximizing problem to be solved at a progressively more rapid rate than the minimizing problem.

The Pironneau–Polak method of centers for solving (2.3) is not strictly implementable as it requires knowledge of $\mathcal{Y}(x_k)$ as soon as x_k is known. This algorithm is stated below.

Pironneau–Polak Method of Centers

Step 0. Given scalars c, $\sigma \in (0, 1)$, select $x_0 \in X$ and set $k = 0$.

Step 1. Obtain a solution $D_k \in \mathfrak{R}^1$ and $d_k \in \mathfrak{R}^n$ to the quadratic programming problem

$$\min_{d,D}\left\{D + \frac{1}{2}\|d\|^2 \,\middle|\, D \in \mathfrak{R}^1, \ d \in \mathfrak{R}^n, \ \langle\nabla\Phi(x_k), d\rangle \le D, \right.$$

$$\left. g_i(x_k) + \langle\nabla g_i(x_k), d\rangle \le D, \ i = 1, \ ..., \ t\right\} \tag{2.4}$$

Step 2. If $D_k = 0$, then Stop. Else, compute

$$\alpha_k = \max\left\{\alpha \,\middle|\, \Phi(x_k + \alpha d_k) - \Phi(x_k) \le c\alpha D_k, \right.$$

$$\left. g_i(x_k + \alpha d_k) \le c\alpha D_k, \ i = 1, ..., t, \ \alpha = (\sigma)^\nu, \ \nu = 0, 1, 2, ...\right\}$$

Step 3. Set $x_{k+1} = x_k + \alpha_k d_k$, $k = k + 1$ and go to Step 1.

Remarks (i) In Step 1, d_k is the direction of descent and D_k is an upper bound on the directional derivative. (ii) In Step 2, an Armijo-type line search strategy is adopted to determine the stepsize α_k (see Armijo, 1966). (iii) It is assumed that the solution of (2.4), satisfying the fist order Karush–Kuhn–Tucker optimality conditions exist (see CN 2).

In the development of the main algorithm, it is recognized that $\mathcal{Y}(x)$ cannot be obtained at once, when x is known. The Pironneau–Polak method of centers is modified so that $\mathcal{Y}(x_k)$ can be replaced by successively closer approximations in such a way that the convergence of the algorithm is maintained. An auxiliary algorithm, $A(x, y)$, designed to give a suitably close approximation to $\mathcal{Y}(x_k)$ is appended to the Pironneau–Polak method of centers.

Let $x \in X$, $y \in \mathfrak{R}^m$ and $QP(x, y)$ denote the following quadratic programming problem:

$$\min\left\{D + \frac{1}{2}\|d\|^2 \,\middle|\, D \in \mathfrak{R}^1, \ d \in \mathfrak{R}^n, \ \langle\nabla_x f(x, y), d\rangle \le D, \right.$$

$$\left. g_i(x) + \langle\nabla g_i(x), d\rangle \le D, \ i = 1, ..., t\right\}. \tag{2.5}$$

Let $D(x, y)$, $d(x, y)$ denote a solution to (2.5).

It is assumed that, given $x \in X$, $y \in \Re^m$, there is a globally convergent auxiliary algorithm $A(x, y)$ that generates a sequence $\{z_j\}$, convergent to $\mathcal{Y}(x)$, with $z_j = y$. Here, we initialize A with the current value of x and an arbitrary y. $A(x, y)$ then generates $\{z_j\} \rightarrow \mathcal{Y}(x)$. It is further assumed that, given $x \in X$, a test function $t(x, \cdot, c)$, on $(\cdot) \in \Re^m$ is known, for some scalar $c > 0$, so that positive numbers p_1, p_2 and $\rho \geq 1$ exist for which

$$p_1|y - \mathcal{Y}(x)|^\rho \leq |t\,(x, y,\ c)| \leq p_2|y - \mathcal{Y}(x)|^\rho, \quad x \in X,\ y \in W^* \quad (2.6)$$

where W^* is a bounded neighborhood of $\mathcal{Y}$ such that

$$\|\nabla^2_{xx}f(x, y)\| \leq M \quad \text{and} \quad \|\nabla^2_{xy}f(x, y)\| \leq M.$$

It is also assumed that the test function t is continuous on $X \times W^*$. Hence, for $x \in X$, and a sequence

$$\{z_j\}^\infty_{j=1} \subset W^*$$

it follows that $\{z_j\}^\infty_{j=1}$ converges to $\mathcal{Y}(x)$ if and only if

$$\{t(x,\ z_j,\ c)\}^\infty_{j=1} \rightarrow 0.$$

The test function is discussed further later in this section.

Chaney's Minimax Algorithm

Step 0. Initialization: choose the following parameters: $\beta \in (0, 1)$, $\delta_0 \in (0, 1)$,

   $\{\epsilon_j\}^\infty_{j=0}$ a strictly decreasing sequence convergent to zero,

   $\{\gamma_j\}^\infty_{j=0}$ a strictly decreasing sequence convergent to zero, with $\epsilon_0 = o(\gamma_j)$, $z_{-1} \in W^*$, $x_0 \in X$.

   Set $k = 0, j = 0, \bar{z} = z_{-1}, \mu_{-1} = 1$.

Step 1. Apply algorithm $A(x_k, \bar{z})$ to obtain z_j so that

$$|\ t(x_k, z_j, c)\ | \leq \epsilon_j.$$

Step 2. Solve problem $QP(x_k, z_j)$ to obtain $D(x_k, z_j)$ and $d(x_k, z_j)$. If

$$D(x_k, z_j) + \frac{1}{2}\|d(x_k, z_j)\|^2 > -\gamma_j$$

then set $\bar{z} = z_j$, $\mu_{k+1} = \mu_j$, $j = j + 1$, go to Step 1.
Else, set $\bar{z} = z_j$, $y_k = z_j$, go to Step 3.

Step 3. Perform the following steps:

(a) Set $\alpha = 1$,

(b) Apply algorithm $A(x_k + \alpha d(x_k, y_k), \bar{z})$ to obtain z^* such that

$$\left| t((x_k + \alpha d(x_k, y_k)), z^*, c) \right| \leq \epsilon_j.$$

If

$$g_i(x_k + \alpha d(x_k, y_k)) \leq \frac{1}{2} \alpha D(x_k, y_k), \quad i = 1, \dots, t$$

and

$$f(x_k + \alpha d(x_k, y_k), z^*) - f(x_k, y_k) \leq \frac{1}{2} \alpha D(x_k, y_k)$$

then set $\alpha_k = \alpha$, $\mu_{j+1} = \mu_j$,

$$x_{k+1} = x_k + \alpha_k d(x_k, y_k)$$

$\bar{z} = z^*$, $k = k + 1$, $j = j + 1$, go to Step 1. Else, go to Step 3c.

(c) If $\alpha \leq \mu_j$, then go to Step 3d. Else, set $\bar{z} = z^*$, $\alpha = \beta \alpha$, go to Step 3b.

(d) Set $\bar{z} = z_j$, $\mu_{j+1} = \delta_0 \mu_j$, $\alpha_k = 0$, $x_{k+1} = x_k$, $k = k + 1$, $j = j + 1$, go to Step 1.

Remarks (i) The purpose of $\epsilon_j = o(\gamma_j)$ is to assist the solution of the maximization problem (2.2) to be computed faster than that of the minimization problem. For a definition of orders, see CN 1.10. (ii) Step 1 is intended to give, at any stage, an improved solution to the current maximization problem. (iii) In Step 2, a decision is made as to whether the current point x_k can be improved. The algorithm proceeds to Step 3 only when the test in Step 2 suggests that we are now much closer to a solution $\mathcal{Y}(x_k)$ of the current maximization problem than we are to the solution of the minimization problem. (iv) In Step 3, an Armijo-type line search is performed from the current iterate x_k, along the direction $d(x_k, y_k)$. For each prospective value $x_k + \alpha_k d(x_k, y_k)$, it is necessary to search for a point z^* which is suitably close to $\mathcal{Y}(x_k + \alpha_k d(x_k, y_k))$. In case the algorithm arrives at Step 3d, the line search is abandoned as a failure. It is shown in Chaney (1982) that this failure can occur a finite number of times.

The Test Function

The test function $t(x, \cdot, c)$ serves as a monitoring device for the process of solving the maximization problem for finding $\mathcal{Y}(x)$. Given $x \in X$, consider the problem

$$\min\{ -f(x,y) \mid y \in \mathcal{Y}\}. \tag{2.7}$$

The function $t(x, \cdot, c)$ is defined in terms of the augmented Lagrangian associated with the maximization problem (2.2). Let $x \in X$, $y \in \Re^m$ and

$$\lambda(x,y) \equiv \begin{bmatrix} \lambda_1 \\ \vdots \\ \lambda_i \end{bmatrix} \in \Re^i$$

be the minimizer of the expression

$$\left\| -\nabla_y f(x,y) + \sum_{j=1}^{i} \lambda_j \nabla h_j(y) \right\|^2 + \sum_{j=1}^{i} \lambda_j^2 h_j (y)^2.$$

Let $c > 0$ and let $a(x,y,c) \in \Re^i$ be given by

$$a_j(x,y,c) = h_j(y) \quad \text{if } ch_j(y) + \lambda_j \geq 0$$

$$a_j(x,y,c) = \frac{-\lambda_j}{c} \quad \text{if } ch_j(y) + \lambda_j < 0.$$

Consider the augmented Lagrangian

$$L^a(x,y,c) = -f(x,y) + \frac{1}{2c} \sum_{j=1}^{i} \left\{ \max\left(0, ch_j(y) + \lambda_j\right)^2 - \lambda_j^2 \right\}.$$

Finally, the test function

$$t(x,y,c) = -\|\nabla_y L^a(x,y,c)\|^2 + \frac{\|a(x,y,c)\|^2}{c}$$

is used in Step 1 and Step 3b of the algorithm.

Given $x \in X$, y satisfies the first order Karush–Kuhn–Tucker necessary and sufficient conditions of optimality (see CN 2) for (2.7) if and only if, for any $c > 0$, the equalities

$$\nabla_y L^a(x,y,c) = 0$$

$$a(x,y,c) = 0$$

are satisfied. Sufficiency follows in Chaney's algorithm in view of the assumed uniqueness of the maximizer $\mathcal{Y}(x)$, given x. The function $t(\cdot, \cdot, c)$ is continuous on $X \times \Re^m$. It is demonstrated in Chaney (1982) that, for large c, the function $t(\cdot, \cdot, c)$ satisfies condition (2.6).

3 THE ALGORITHM OF PANIN

The algorithm discussed in Panin (1981) is intended for solving constrained continuous minimax problems similar to those addressed by Chaney (1982). Consider the problem

$$\min_{x \in X} \max_{y \in Y} f(x, y) \tag{3.1}$$

where $X \subseteq \Re^n$ and $Y \subseteq \Re^m$ are convex compact sets and $f(x, y)$, $\nabla_x f(x, y)$ are continuous with respect to X and Y (see CN 3). It is assumed that, for $y \in Y$ and $x_1, x_2 \in X$, $\nabla_x f$ is Lipschitz continuous in x,

$$\|\nabla_x f(x_1, y) - \nabla_x f(x_2, y)\| \le K \|x_1 - x_2\|$$

where $K > 0$ is a constant. Problem (3.1) is reformulated as in (2.2) and (2.3)

$$\min_{x \in X} \Phi(x) \tag{3.2}$$

where

$$\Phi(x) = \max_{y \in Y} f(x, y). \tag{3.3}$$

The algorithm for solving (3.2) is based on the iteration

$$x_{k+1} = x_k + \alpha_k d_k. \tag{3.4}$$

The stepsize, α_k, is determined below, $d_k = \bar{x} - x_k$, $\bar{x}$ is the solution to the problem

$$\min_{x \in X} \Phi_k(x) \tag{3.5}$$

where

$$\Phi_k(x) = \max_{y \in Y} f_k(x, y)$$

$$f_k(x, y) = f(x_k, y) + \langle \nabla_x f(x_k, y), x - x_k \rangle + \frac{a}{2} \|x - x_k\|^2$$

and $a > 0$ is a constant. Let the function Ψ_k be defined by

$$\Psi_k = \min_{x \in X} \Phi_k(x) - \Phi(x_k). \tag{3.6}$$

The algorithm is defined as follows.

Panin's Minimax Algorithm

Step 0. Select $x_0 \in X$, $a > 0$, termination accuracy $1 \gg \xi > 0$.

Step 1. Solve subproblems (3.5) and (3.6). If $\Psi_k \ge -\xi$, then stop.

Step 2. (a) Set $\alpha = 1$, $c = 1$. If the inequality

$$\Phi(x_k + \alpha d_k) - \Phi(x_k) \le c\alpha \Psi_k$$

is satisfied, then set $\alpha_k = \alpha$, $c_k = c$, x_{k+1} with (3.4).

Go to Step 3.

(b) Else, set $\alpha = \alpha/2$, $c = c/2$ and return to Step 2a.

Step 3. If $\alpha_k < 1$, set $a = 2a$.

Set $k = k + 1$ and go to Step 1.

Panin's algorithm is strictly not implementable for two reasons. First, the sets X and $\mathcal{Y}$ are not specified. Second, in Step 1, the method for solving (3.5) is not specified. These reasons confine the algorithm within a conceptual framework only. Kiwiel (1987) has developed this method and the resulting implementable algorithm is discussed in Section 4 below.

4 THE ALGORITHM OF KIWIEL

Kiwiel's (1987) development is based on the conceptual algorithm in Section 3. It uses an auxiliary algorithm to solve the subproblems in Step 1 of Panin's algorithm.

We consider the continuous minimax problem, constrained in y but unconstrained in x

$$\min_{x \in \mathfrak{R}^n} \max_{y \in \mathcal{Y}} f(x, y) \tag{4.1}$$

and reformulate it as

$$\min_{x \in \mathfrak{R}^n} \Phi(x)$$

$$\Phi(x) = \max_{y \in \mathcal{Y}} f(x, y). \tag{4.2}$$

Based on Panin's method, Kiwiel has proposed the linear approximation to the max-function

$$f_k^\ell(d, y) = f(x_k, y) + \langle \nabla_x f(x_k, y), d \rangle$$

$$\Phi_k^\ell(d) = \max_{y \in \mathcal{Y}} f_k^\ell(d, y). \tag{4.3}$$

A descent direction is computed at x_k. The auxiliary algorithm evaluates, in finite number of iterations, the descent direction d_k which solves

$$\min_{d \in \mathfrak{R}^n} \Phi_k^\ell(d) + \frac{1}{2}\|d\|^2. \tag{4.4}$$

In addition to the set of maximizers at x_k, given by $\mathcal{Y}(x_k)$ in (1.2.3), the algorithm utilizes a set of maximizers of $f_k^\ell(d, y)$, defined as

$$\mathcal{Y}_{k+1} \equiv \left\{ y_{k+1} \in \mathcal{Y} \,\middle|\, y_{k+1} = \arg\max_{y \in \mathcal{Y}} f_k^\ell(d, y) \right\}$$

with the termination criterion for the algorithm determined in terms of the function

$$\Psi_k^\ell = -\left\{ \|d_k\|^2 + \Phi(x_k) - f(x_k, y_{k+1}) \right\}.$$

The algorithm is discussed below. The default parameter values are those used in the numerical experiments in Chapter 5.

Kiwiel's Minimax Algorithm (Kiw)

Step 0. Initialization: select x_0, y_0; set $k = 0$ and

termination accuracy	$1 \gg \xi \geq 0, \quad (\xi = 10^{-6})$;
line search parameter	$c \in (0, 1), \quad (c = 10^{-4})$;
stepsize factor	$\sigma \in (0, 1), \quad (\sigma = 0.5)$;
linear approximation parameter	$m \in (0, 1), \quad (m = 2 \times 10^{-4})$.

Step 1. Solve the maximization at current point x_k:

$$\Phi(x_k) = \max_{y \in \mathcal{Y}} f(x_k, y).$$

Step 2. Direction-finding subproblem: Set $x = x_k$ and use auxiliary algorithm (AA) with parameters $\xi \geq 0$ and m until it terminates, returning d_k and Ψ_k^ℓ. If

$$\Psi_k^\ell \geq -\xi$$

the solution has been reached: stop.

Step 3. Line search: compute the stepsize α_k using

$$\alpha_k = \max\left\{ \alpha \mid \Phi(x_k + \alpha d_k) - \Phi(x_k) \leq c\alpha\Psi_k, \; \alpha = (\sigma)^i, \; i = 0, 1, 2, \dots \right\}$$

Set $x_{k+1} = x_k + \alpha_k d_k$, $k = k + 1$, go to Step 1.

Auxiliary Algorithm (AA) (requires input values: $x_k \in \mathfrak{R}^n$, $\Phi(x_k)$, $\xi \geq 0$, $m \in (0, 1)$)

Step 0. Initialization: set $x = x_k$, $\Phi(x) = \Phi(x_k)$, select any $y_1 \in \mathcal{Y}$, set $p_0 = \nabla_x f(x, y_1)$, $\Theta_0 = f(x, y_1)$, $i = 1$.

Step 1. Find the number μ_i that solves

$$\min_{\mu\in\Re^1} \left\{ \frac{1}{2}\|(1-\mu)p_{i-1} + \mu\nabla_x f(x,y_i)\|^2 - (1-\mu)\Theta_{i-1} - \mu f(x,y_i) \right\}.$$

Set

$$p_i = (1-\mu_i)p_{i-1} + \mu_i \nabla_x f(x,y_i), \quad \Theta_i = (1-\mu_i)\Theta_{i-1} + \mu_i f(x,y_i);$$

$$\Psi_i = -\left\{\|p_i\|^2 + \Phi(x) - \Theta_i\right\}.$$

If $\Psi_i \geq -\xi$ then go to Step 3.

Step 2. Primal optimality testing: set $d_i = -p_i$. Compute

$$y_{i+1} = \arg\max_{y\in Y} \{f(x,y) + \langle \nabla_x f(x,y), d_i\rangle\}.$$

If

$$f(x,y_{i+1}) + \langle \nabla_x f(x,y_{i+1}), d_i\rangle - \Phi(x) \leq m\Psi_i$$

then, go to Step 3. Else, set $i = i + 1$, and go to Step 1.

Step 3. Stop returning $d_k = -p_i$ and $\Psi_k^\ell = \Psi_i$.

Remarks   (i) At x_k, AA evaluates, in a finite number of iterations, a descent direction d that solves (4.4). At each iteration of AA, y_i yields a new estimate of the maximizer of $f(x_k, y_i)$ and $\nabla_x f(x_k, y_i)$ and these are combined linearly with old estimates to find a new direction d_i. (ii) AA attempts to find the minimum-norm subgradient that is an element of the subdifferential of the approximating function (4.3) and uses this subgradient in finding the descent direction. (iii) The algorithm is refined by using inexact evaluations (Kiwiel, 1987). This involves the assumption that for $d \in \Re^n$ and $\xi > 0$, it is possible to find a point $y \in Y$ such that

$$f(x,y) + \langle \nabla_x f(x,y), d\rangle \geq \Phi_k(d) - \xi.$$

The revised algorithm assumes that a finite process can find ξ-accurate solutions to the maximization subproblem. Inexact line searches and a mechanism for decreasing ξ ensure global convergence of the method to stationary points of $\Phi(x)$ (i.e., x satisfying the optimality condition of the original problem).

References

Armijo, L. (1966). "Minimization of Functions having Lipschitz-continuous First Partial Derivatives", *Pacific Journal of Mathematics*, 16, 1–3.

Chaney, R.W. (1982). "A Method of Centers Algorithm for Certain Minimax Problems" *Mathematical Programming*, 22, 206–226.

Demyanov, V.F. and V.N. Malozemov (1974). *Introduction to Minimax*, Wiley, New York.

Karush, W. (1939). "Minima of Functions of Several Variables with Inequalities as Side Constraints", M.S. Thesis, Department of Mathematics, University of Chicago.

Kiwiel, K.C. (1987). "A Direct Method of Linearization for Continuous Minimax Problems", *Journal of Optimization Theory and Applications*, 55, 271–287.

Klessig, R. and E. Polak (1973). "An Adaptive Precision Gradient Method for Optimal Control" *SIAM Journal on Control*, 11, 80–93.

Kuhn, H.W. and A.W. Tucker (1951). "Nonlinear Programming", in: J. Neyman (editor), *Proceedings of the Second Berkeley Symposium on Mathematical Statistics and Probability*, University of California Press, Berkeley and Los Angeles, CA.

Panin, V.M. (1981). "Linearization Method for Continuous Min-max Problems", *Kibernetika*, 2, 75–78.

Pironneau, O. and E. Polak (1972). "On the Rate of Convergence of Certain Methods of Centers", *Mathematical Programming*, 2, 230–257.

Rustem, B. (1998). *Algorithms for Nonlinear Programming and Multiple Objective Decisions*, Wiley, Chichester.

Shor, N.Z. (1980). *Minimization Methods for Nondifferentiable Functions*, Springer Verlag, Berlin.

COMMENTS AND NOTES

CN 1: Motivation for Descent

In general, at a point x_k, d_k is a descent direction such that the objective function decreases for small steps taken along d_k. Consider a linear approximation of $f(x)$,

$$f(x_k + \alpha_k d_k) \cong f(x_k) + \alpha_k \langle \nabla f(x_k), d_k \rangle.$$

For small α_k, this approximation is fairly accurate and

$$f(x_k + \alpha_k d_k) \leq f(x_k)$$

provided

$$\langle \nabla f(x_k), d_k \rangle \leq 0.$$

Consider the problem of determining the direction d_k that satisfies the last inequality. Let

$$d_k = \arg\min\{f(x_k) + \langle \nabla f(x_k), d \rangle \mid \| d \|_2 = 1\}$$

where d is required to be of unit, or fixed, length to ensure a well-defined solution. The first order optimality conditions for this problem are

$$\nabla f(x_k) + 2\lambda d_k = 0, \quad \|d_k\|_2^2 = 1.$$

Premultiplying the first by d_k and applying the constraint, we have

$$\lambda = -\frac{\langle \nabla f(x_k), d_k \rangle}{2}.$$

Since $2\lambda \langle \nabla f(x_k), d_k \rangle = -\langle \nabla f(x_k), \nabla f(x_k) \rangle$,

$$\lambda^2 = \frac{\langle \nabla f(x_k), \nabla f(x_k) \rangle}{4}$$

and, as the minimization of $f(x_k) + \langle \nabla f(x_k), d \rangle$ is being considered, this yields

$$\lambda = \frac{1}{2} \|\nabla f(x_k)\|_2 \quad \text{and} \quad d_k = -\frac{\nabla f(x_k)}{\|\nabla f(x_k)\|_2}$$

where d_k is known as the steepest descent direction.

If, instead of the linear approximation, we assume a quadratic approximation to $f(x)$, we have

$$f(x_k) + \langle \nabla f(x_k), d \rangle + \frac{1}{2} \langle d, \nabla^2 f(x_k), d \rangle.$$

For $\nabla^2 f(x_k)$ positive definite, the quadratic is minimized by d_k satisfying the first order optimality condition

$$\nabla f(x_k) + \nabla^2 f(x_k) d_k = 0$$

or $d_k = -(\nabla^2 f(x_k))^{-1} \nabla f(x_k)$. This is the *Newton direction* whose descent property,

$$\langle d_k, \nabla f(x_k) \rangle \leq 0$$

can easily be verified.

CN 2: Karush–Kuhn–Tucker Conditions of Optimality

Consider the nonlinear programming problem

$$\min\{f(x) \mid h(x) \leq 0, g(x) = 0\}$$

where f is the scalar objective function, g, h are fixed dimensional vectors of equality and inequality constraints. Let $f, g, h \in \mathbb{C}^2$ (see CN 1.3) and let the Lagrangian associated with this problem be given by

$$L(x, \mu^e, \mu^i) = f(x) + \langle g(x), \mu^e \rangle + \langle h(x), \mu^i \rangle$$

where μ^e, μ^i are the multipliers associated with the equality and inequality constraints. Let x_* denote the solution of the nonlinear programming problem and let $\nabla g(x_*)$ and $\nabla \overline{h}(x_*)$ be linearly independent, where $\overline{h}(x_*)$ is the set of inequality constraints satisfied as equalities at the solution.

Then there are vectors (μ_*^e, μ_*^i) which satisfy the first order necessary conditions of optimality

$$\nabla_x L(x_*, \mu_*^e, \mu_*^i) = 0$$

$$g(x_*) = 0, \quad h(x_*) \leq 0$$

$$\langle h(x_*), \mu_*^i \rangle = 0, \quad \mu_*^i \geq 0$$

(Karush, 1939; Kuhn and Tucker, 1951).

Let the superscripts ι, j denote the ιth and jth equality and inequality constraints, respectively. The second order necessary condition is satisfied if, in addition, the Hessian

$$\nabla^2 L(x_*, \mu_*^e, \mu_*^i) = \nabla^2 f(x_*) + \sum_\iota \nabla^2 g^\iota(x_*)(\mu_*^e)^\iota + \sum_j \nabla^2 h^j(x_*)(\mu_*^i)^j$$

is positive semi-definite on the subspace (see CN 1.4),

$$\mathcal{V} = \left\{ v \mid v \neq 0, \ \nabla g^T(x_*)v = 0, \ \nabla \bar{h}^T(x_*)v = 0 \right\}.$$

The second order sufficiency condition is ensured if, in addition to the first order conditions, the above Hessian is positive definite on the subspace

$$\mathcal{V}_0 = \left\{ v \mid v \neq 0, \ \nabla g^T(x_*)v = 0, \ \langle \nabla \bar{h}^j(x_*), v \rangle = 0, \ j \in I \right\}$$

where

$$I = \left\{ j \mid \nabla \bar{h}^j(x_*) = 0, \ (\mu_*^i)^j > 0 \right\}.$$

Furthermore, it can be shown that when f is a convex function and

$$\{ x \in \Re^n \mid h(x) \leq 0, \ g(x) = 0 \}$$

is a convex set, the first order necessary conditions are also sufficient. By Lemma 1.1.2, the Hessian of a convex (strictly convex) function f is positive semi-definite (positive definite). The local minimum of a convex function is also its global minimum (Rustem, 1998, Chapter 1).

CN 3: Continuous and Differentiable Functions

Let $\mathcal{D}$ be a domain in $\Re^n$ and $\Gamma \in \Re^m$ be a function such that $\Gamma(x) : D \subset \Re^n \to \Re^m$. Each element of $\Gamma(x)$ is continuous at x if $x_k \to x$ implies $\Gamma(x_k) \to \Gamma(x)$, elementwise. If, in addition, the elements of Γ have first (or second) partial derivatives which are continuous on D, this is denoted by $\Gamma(x) \in \mathbb{C}^1$ (or $\mathbb{C}^2$) (see CN 1.3).

Chapter 3

Algorithms for computing saddle points

Consider the basic continuous minimax problem when the underlying function $f(x, y)$ is convex in x and concave in y. Then, there is a unique solution to minimax which can be computed using specialized algorithms. Saddle point solutions like this are also used by the decision maker to assess the worst-case strategy of an opponent and compute the optimal response to the worst case. In the present context, the opponent can be interpreted as nature choosing the worst-case value of the uncertainty. The solution is an equilibrium strategy which ensures an optimal response to the worst case. Neither the decision maker nor nature would benefit by deviating from this equilibrium.

We consider the computation of saddle points with gradient-based algorithms such as steepest descent and Newton-type algorithms. These aim to satisfy the optimality conditions for the decision maker and for nature.

Gradient-based algorithms approximate, and periodically evaluate numerically, the Jacobian of the simultaneous system which characterizes the equilibrium condition. The approximation uses the Broyden update to improve the Jacobian. Global convergence is maintained by an Armijo-type stepsize strategy.

The global and local convergence of the quasi-Newton algorithm is discussed. Convergence under relaxed descent directions is established. The achievement of unit stepsizes is related to the accuracy of the Jacobian approximation. Furthermore, a simple derivation of the Dennis–Moré characterization of the Q-superlinear convergence rate is given.

1 COMPUTATION OF SADDLE POINTS

1.1 Saddle Point Equilibria

We extend the discussion in Chapter 2 to a special case of minimax characterized by the saddle point condition introduced in Section 1.4. We are specifically concerned with the equilibrium condition for the decision maker and nature such that the decision maker and nature seek to optimize their respective parts of the objective. We thus have the problem of simultaneously computing the solutions of

$$\min_{x \in \mathcal{R}} \{f(x,y)\} \tag{1.1a}$$

$$\max_{y \in \mathcal{Y}} \{f(x,y)\} \tag{1.1b}$$

and consequently,

$$\min_{x \in \mathcal{R}} \max_{y \in \mathcal{Y}} \{f(x,y)\} \tag{1.1c}$$

where x is the vector of decision variables, y is the vector of scenarios and $\mathcal{R} \subset \mathfrak{R}^n$, $\mathcal{Y} \subset \mathfrak{R}^m$ are defined in Section 1.4.

In this chapter, we consider the constrained problem (1.1a–c) mainly in terms of transformations into the unconstrained problem in x and y. The *unconstrained* problem is given by

$$\min_{x \in \mathfrak{R}^n} \max_{y \in \mathfrak{R}^m} \{f(x,y)\}. \tag{1.1d}$$

Assumption 1.1 *The function f is convex in x and concave in y.*

Hence, f has an unconstrained minimum in x and an unconstrained maximum y. Thus, for given x, $f(x,y)$ has an unconstrained maximum with respect to y, and given y, it has an unconstrained minimum with respect to x. We define the vector ζ by

$$\zeta \equiv \begin{bmatrix} x \\ y \end{bmatrix}.$$

The necessary and sufficient conditions for a joint optimum, or min-max solution, are satisfied by $\zeta_*^T = [x_*^T, y_*^T]$ which solve the simultaneous system of equations

$$\mathcal{F}(\zeta) \equiv \begin{bmatrix} \nabla_x f(x,y) \\ -\nabla_y f(x,y) \end{bmatrix} = 0. \tag{1.4}$$

At the saddle point, the solution, x_*, y_* should also satisfy the sufficiency conditions for a minimum with respect to x and maximum with respect to y, given by

$$\langle v, \nabla_x^2 f(x_*, y_*)v \rangle > 0, \quad \langle u, \nabla_y^2 f(x_*, y_*)u \rangle < 0, \quad \forall 0 \neq v \in \mathfrak{R}^n, \, 0 \neq u \in \mathfrak{R}^m \tag{1.5}$$

(e.g., Rustem, 1998, Chapter 1). Clearly, an algorithm solving (1.4) should yield solutions that also satisfy (1.5). The latter is a difficult condition to ensure. For this reason, we assume that $f(x,y)$ is convex with respect to x and concave with respect to y satisfying the conditions

$$m\|u\|^2 \le \langle u, \nabla_x^2 f(x, y)u \rangle \le M\|u\|^2 \tag{1.6a}$$

$$m\|v\|^2 \le -\langle v, \nabla_y^2 f(x, y)v \rangle \le M\|v\|^2 \tag{1.6b}$$

$$\forall 0 \ne u \in \Re^n, \quad 0 \ne v \in \Re^m, \quad M \ge m > 0.$$

We denote the Jacobian matrix by

$$\nabla_\zeta \mathcal{F}(\zeta) \equiv \begin{bmatrix} \nabla_x^2 f(x, y) & \nabla_{x,y}^2 f(x, y) \\ -\nabla_{y,x}^2 f(x, y) & -\nabla_y^2 f(x, y) \end{bmatrix}$$

and note that (1.6) ensures that for $\hat{\zeta} \in \Re^n \times \Re^m$

$$m\|\hat{\zeta}\|^2 \le \langle \hat{\zeta}, \nabla_\zeta \mathcal{F}(\zeta)\hat{\zeta} \rangle \le M\|\hat{\zeta}\|^2. \tag{1.7}$$

We now discuss the uniqueness condition for saddle points and show that this is satisfied for convex $f(x, y)$. We let

$$w \equiv \begin{bmatrix} u \\ v \end{bmatrix}$$

and define $\mathcal{D}$ by

$$\mathcal{D}(\zeta, w) \equiv f(u, y) - f(x, v)$$

such that

$$\mathcal{F}(\zeta) = \nabla_w \mathcal{D}(\zeta, \zeta).$$

It can be verified that if $\zeta_* \in \mathcal{R} \times \mathcal{Y}$ and

$$\min_{w \in \mathcal{R} \times \mathcal{Y}} \mathcal{D}(\zeta_*, w) = \mathcal{D}(\zeta_*, \zeta_*) = 0 \tag{1.8}$$

then ζ_* is a saddle point of f on $\mathcal{R} \times \mathcal{Y}$ since the inequalities

$$f(x_*, y) \le f(x_*, y_*) \le f(x, y_*)$$

are satisfied for all $(x_*, y) \in \mathcal{R} \times \mathcal{Y}, (x, y_*) \in \mathcal{R} \times \mathcal{Y}$.

Theorem 1.1 *Let $\mathcal{R}, \mathcal{Y}$ be convex and compact (see CN 2), $f \in \mathbb{C}^1$ and the condition*

$$\langle \mathcal{F}(\zeta + w) - \mathcal{F}(\zeta), w \rangle > 0, \quad w \ne 0, \quad \zeta, \zeta + w \in \mathcal{R} \times \mathcal{Y} \tag{1.9}$$

be satisfied. Then the saddle point of f on $\mathcal{R} \times \mathcal{Y}$ is unique.

Proof (Demyanov and Pevnyi, 1972). We note that the inequality

$$\min_{w \in \mathcal{R} \times \mathcal{Y}} \langle \mathcal{F}(\zeta_*), w - \zeta_* \rangle > 0$$

holds for saddle point $\zeta_* \in \mathcal{R} \times \mathcal{Y}$, as the contrary would imply the existence of a feasible descent direction for $\mathcal{D}(\zeta_*, w)$, contradicting (1.8). Assume that two saddle points ζ_1 and ζ_2 exist. Then, we have

$$\langle \mathcal{F}(\zeta_1), \zeta_2 - \zeta_1 \rangle \geq 0, \quad \langle \mathcal{F}(\zeta_2), \zeta_1 - \zeta_2 \rangle \geq 0$$

$$\langle \mathcal{F}(\zeta_1) - \mathcal{F}(\zeta_2), \zeta_1 - \zeta_2 \rangle \leq 0.$$

Hence, by (1.8), $\zeta_1 = \zeta_2$ which yields the required result. $\square$

The following corollary establishes saddle points for $f(x, y)$ convex in x, concave in y and is associated with the discussion in Lemma 1.4.1 and Theorem 1.4.1.

Corollary 1.1 *Let $f(x, y) \in \mathbb{C}^2$, $(x, y) \in \mathcal{R} \times \mathcal{Y}$, and f be strictly convex in x, concave in y, such that at least the left side inequalities of (1.6) hold. Then, condition (1.9) is satisfied.*

Proof. We have, for $\hat{\zeta} = \zeta + w$, $\zeta \in \mathcal{R} \times \mathcal{Y}$, $u \in \mathcal{R}$, $v \in \mathcal{Y}$,

$$\langle \hat{\zeta}, \nabla_\zeta \mathcal{F}(\zeta)\hat{\zeta} \rangle = \langle u, \nabla_x^2 f(x, y)u \rangle - \langle v, \nabla_y^2 f(x, y)v \rangle \geq m\Big[\|u\|^2 + \|v\|^2\Big] = m\|\hat{\zeta}\|^2.$$

In order to characterize the difference $\mathcal{F}(\zeta + w) - \mathcal{F}(\zeta)$, we need to invoke a special case of the mean value theorem (see CN 1) that holds for $\mathcal{F}$. Provided each element of the vector $\mathcal{F}$, with the jth element given by $\mathcal{F}^j : \mathbb{H} \subset \mathfrak{R}^n \to \mathfrak{R}^1$, is differentiable on an open convex set $\mathbb{H}_0 \subset \mathbb{H}$, then for any two points $\hat{\zeta}, w \in \mathbb{H}_0$, there exist $t^1, t^2, ..., t^j, ... \in (0, 1)$ such that

$$\nabla_\zeta \mathcal{F}(\hat{\zeta}, w) \equiv \left[\nabla \mathcal{F}_\zeta^1(w + t^1(\hat{\zeta} - w)) \vdots \nabla \mathcal{F}_\zeta^2(w + t^2(\hat{\zeta} - w)) \vdots \cdots \right] \quad (1.10)$$

where $\nabla \mathcal{F}^j$ denotes the gradient of the jth element of $\mathcal{F}$, and

$$\mathcal{F}(\hat{\zeta}) - \mathcal{F}(w) = \nabla_\zeta \mathcal{F}(\hat{\zeta}, w)(\hat{\zeta} - w). \quad (1.11)$$

Hence, we have

$$\langle w, \mathcal{F}(\zeta + w) - \mathcal{F}(\zeta) \rangle = \langle w, \nabla_\zeta \mathcal{F}(\zeta + w, \zeta)w \rangle \geq m\|w\|^2. \quad \square$$

1.2 Solution of Systems of Equations

From the above discussion, it is clear that for $f(x, y)$ convex in x, concave in y, the condition $\mathcal{F}(\zeta_*) = 0$ is necessary and sufficient for ζ_* to be a saddle point. Thus, the search for a saddle point is equivalent to solving the system of equations $\mathcal{F}(\zeta) = 0$. The remaining discussion in this chapter is concerned with the solution of a system of nonlinear equations arising from (1.4). Sometimes, it is convenient to characterize the latter problem as

$$\min_{\zeta} \left\{ \frac{1}{2} \|\mathcal{F}(\zeta)\|_2^2 \right\}. \tag{1.12}$$

We study the properties of $\frac{1}{2}\|\mathcal{F}(\zeta)\|_2^2$. By definition of the function, it follows that

$$\nabla_\zeta \left[\frac{1}{2} \|\mathcal{F}(\zeta)\|_2^2 \right] = \nabla_\zeta \mathcal{F}^{\mathrm{T}}(\zeta) \mathcal{F}(\zeta)$$

$$\nabla_\zeta^2 \left[\frac{1}{2} \|\mathcal{F}(\zeta)\|_2^2 \right] = \nabla_\zeta \mathcal{F}^{\mathrm{T}}(\zeta) \nabla_\zeta \mathcal{F}(\zeta) + \sum_{j=1}^{n+m} \nabla_\zeta^2 \mathcal{F}^j(\zeta) \mathcal{F}^j(\zeta) \tag{1.13}$$

When (1.6a,b) hold, the matrix $\nabla_\zeta \mathcal{F}(\zeta)$ is strictly positive definite and thus (1.12) has a single solution ζ_*. Assume that there exists a scalar $\chi \in (0, \infty)$, such that

$$\|\nabla_\zeta^2 \mathcal{F}^j(\zeta)\| \le \chi, \quad 1 \le j \le n + m. \tag{1.14}$$

From (1.13) and (1.14) and the continuity of $\mathcal{F}(\zeta)$, it follows that for some sequence $\{\zeta_k\} \to \zeta_*$, we have

$$\left\{ \nabla_\zeta^2 \left[\frac{1}{2} \|\mathcal{F}(\zeta_k)\|_2^2 \right] \right\} \to \nabla_\zeta^2 \left[\frac{1}{2} \|\mathcal{F}(\zeta_*)\|_2^2 \right] = \nabla_\zeta \mathcal{F}^{\mathrm{T}}(\zeta_*) \nabla_\zeta \mathcal{F}(\zeta_*).$$

In view of (1.7), there exist constants $m_1, M_1, M_1 \ge m_1 > 0$, so that in the neighborhood of ζ_*, we have for $\hat{\zeta} \in \mathfrak{R}^n \times \mathfrak{R}^m$

$$m_1 \|\hat{\zeta}\|^2 \le \left\langle \hat{\zeta}, \left[\nabla_\zeta^2 \left[\frac{1}{2} \|\mathcal{F}\zeta_k)\|_2^2 \right] \right] \hat{\zeta} \right\rangle \le M_1 \|\hat{\zeta}\|^2 \tag{1.15}$$

and the function $\frac{1}{2}\|\mathcal{F}(\zeta)\|_2^2$ is smooth and strictly convex in some neighborhood of ζ_*.

Definition The system $\mathcal{F}$, evaluated at point ζ_k, is denoted by $\mathcal{F}_k = \mathcal{F}(\zeta_k)$ and similarly, the Jacobian is denoted by $\nabla \mathcal{F}_k = \nabla_\zeta \mathcal{F}_k = \nabla_\zeta \mathcal{F}(\zeta_k)$.

Algorithms for solving (1.12) follow the iterative process

$$\zeta_{k+1} = \zeta_k + \alpha_k d_k \tag{1.16}$$

where d_k is the direction of progress and α_k is the stepsize and the choice of d_k, α_k determines the minimization method. For example, the steepest descent algorithm for the unconstrained minimax problem is based on

$$d_k = -\nabla_\zeta \left[\frac{1}{2} \|\mathcal{F}(\zeta_k)\|_2^2 \right] = -\nabla_\zeta \mathcal{F}_k^{\mathrm{T}} \mathcal{F}_k \qquad (1.17)$$

$$\alpha_k = \max\left\{ \alpha \left| \frac{1}{2} \|\mathcal{F}(\zeta_k + \alpha d_k)\|_2^2 - \frac{1}{2} \|\mathcal{F}_k\|_2^2 \le -\rho\alpha \|\nabla_\zeta \mathcal{F}_k^{\mathrm{T}} \mathcal{F}_k\|_2^2, \right.$$

$$\left. \alpha = (\overline{\alpha})^i, i = 0, 1, 2, ... \right\} \qquad (1.18)$$

where $\rho \in (0, \frac{1}{2})$ and (1.18) is a stepsize strategy that ensures sufficient progress at every iteration. While the convergence of (1.17)–(1.18) can be demonstrated using the discussion in Sections 2–3, an important drawback of the algorithm is that it requires the second derivatives of $f(x, y)$, which can cause difficulties. In subsequent sections, we consider modifications which do not require second derivatives of $f(x, y)$ but use only $\mathcal{F}(x_k)$, or the first derivatives of $f(x, y)$ as long as possible.

In Section 2, we discuss algorithms for computing saddle points. These are based on gradient information, $\nabla_\zeta \mathcal{F}_k^{\mathrm{T}} \mathcal{F}_k$ or an approximation to this. In Section 3, we establish the global convergence properties of a quasi-Newton algorithm which utilizes approximations to $\nabla_\zeta \mathcal{F}_k$. In Section 4 we discuss the local convergence properties of this algorithm in a neighborhood of the solution.

2 THE ALGORITHMS

In this section, we consider four algorithms. Two are intended for unconstrained problems and two for constrained problems.

2.1 A Gradient-based Algorithm for Unconstrained Saddle Points

We first consider an algorithm for the unconstrained saddle point problem, based on the direction

$$d_k = -\mathcal{F}(\zeta_k) \qquad (2.1)$$

(see Demyanov and Pevnyi, 1971; Danilin and Panin, 1974). The algorithm consists of the iteration step in (1.16), with d_k given by (2.1), and a stepsize strategy to determine α_k such that the algorithm is ensured sufficient progress at each iteration.

The stepsize α_k of the algorithm is determined by the stepsize strategy

$$\alpha_k = \max\left\{ \alpha \left| \frac{1}{2} \|\mathcal{F}(\zeta_k + \alpha d_k)\|_2^2 - \frac{1}{2} \|\mathcal{F}_k\|_2^2 \le -\rho\alpha \|\mathcal{F}_k\|_2^2, \right.$$

$$\left. \alpha = (\overline{\alpha})^i, i = 0, 1, 2, ... \right\} \qquad (2.2)$$

where $\overline{\alpha}, \rho \in (0, 1)$. The ultimate convergence of the algorithm is ensured by the monotonic decrease of the sequence $\{\|\mathcal{F}_k\|_2^2\}$. We provide an informal discussion to this end, leaving the more formal treatment to the subsequent full quasi-Newton case.

To demonstrate that choosing α_k based on (2.2) for this algorithm is justified, we show that there exists an α_k that satisfies (2.2) and consequently $\{\|\mathcal{F}_k\|_2^2\}$ is monotonically decreasing. Expanding the function $\|\mathcal{F}_k\|_2^2$ in a second order Taylor series (see CN 1.5), we obtain

$$\frac{1}{2}\|\mathcal{F}(\zeta_k + \alpha\, d_k)\|_2^2 - \frac{1}{2}\|\mathcal{F}_k\|_2^2 = \alpha\langle \nabla_\zeta \mathcal{F}_k^{\mathrm{T}} \mathcal{F}_k, d_k\rangle$$

$$+ \alpha^2\left(\int_0^1 (1-t)\left\langle d_k, \nabla_\zeta^2\left[\frac{1}{2}\|\mathcal{F}(\zeta_k(\alpha))\|_2^2\right] d_k\right\rangle dt\right) \tag{2.3}$$

where $\zeta_k(\alpha) = \zeta_k + \alpha t d_k$. With $d_k = -\mathcal{F}_k$, the right side can be expressed as

$$-\alpha\|\mathcal{F}_k\|_2^2\left[\frac{\langle\nabla_\zeta \mathcal{F}_k^{\mathrm{T}} \mathcal{F}_k, \mathcal{F}_k\rangle}{\|\mathcal{F}_k\|_2^2} - \frac{\alpha\left(\int_0^1 (1-t)\left\langle \mathcal{F}_k, \nabla_\zeta^2\left[\frac{1}{2}\|\mathcal{F}(\zeta_k(\alpha))\|_2^2\right]\mathcal{F}_k\right\rangle dt\right)}{\|\mathcal{F}_k\|_2^2}\right]$$

Using (1.15) yields the inequality

$$\alpha^2\left(\int_0^1 (1-t)\left\langle d_k, \nabla_\zeta^2\left[\frac{1}{2}\|\mathcal{F}(\zeta_k(\alpha))\|_2^2\right] d_k\right\rangle dt\right) \leq \frac{\alpha^2}{2} M_1\|d_k\|^2.$$

In view of (1.7), we have

$$\langle\nabla_\zeta \mathcal{F}_k^{\mathrm{T}} \mathcal{F}_k, d_k\rangle = -\langle\nabla_\zeta \mathcal{F}_k^{\mathrm{T}} \mathcal{F}_k, \mathcal{F}_k\rangle \leq -m\|\mathcal{F}_k\|_2^2$$

which is interesting in that it establishes the descent property of $d_k = -\mathcal{F}_k$ (see CN 2.1). Hence, it follows that for $\alpha \leq m/M_1$,

$$\frac{\langle\nabla_\zeta \mathcal{F}_k^{\mathrm{T}} \mathcal{F}_k, \mathcal{F}_k\rangle}{\|\mathcal{F}_k\|_2^2} - \frac{\alpha\left(\int_0^1 (1-t)\left\langle \mathcal{F}_k, \nabla_\zeta^2\left[\frac{1}{2}\|\mathcal{F}(\zeta_k(\alpha))\|_2^2\right]\mathcal{F}_k\right\rangle dt\right)}{\|\mathcal{F}_k\|_2^2}$$

$$\geq m - \frac{\alpha}{2} M_1 \geq \frac{m}{2}$$

and (2.3) yields

$$\frac{1}{2}\|\mathcal{F}(\zeta_k + \alpha d_k)\|_2^2 - \frac{1}{2}\|\mathcal{F}_k\|_2^2 \leq -\alpha\frac{m}{2}\|\mathcal{F}_k\|_2^2.$$

Since in (2.2), $\alpha = (\bar{\alpha})^i$, $i = 0, 1, 2, \ldots$, it is possible to obtain $\rho \leq m/2$, the choice of α_k is justified. Furthermore, there exist $\hat{\alpha} > 0$, $\rho \in (0, 1)$ such that $\alpha_k \geq \hat{\alpha} > 0$, $\rho \geq \hat{\rho} > 0$.

2.2 Quadratic Approximation Algorithm for Constrained Minimax Saddle Points

If there are constraints on x, y given by $h(x) \leq 0$ and $\mathfrak{H}(y) \leq 0$, such that the saddle point becomes

$$\min_{x} \max_{y} \{ f(x, y) \mid h(x) \leq 0, \mathfrak{H}(y) \leq 0 \} \tag{2.4}$$

where $h \in \mathfrak{R}^{n_h}$, $\mathfrak{H} \in \mathfrak{R}^{n_{\mathfrak{H}}}$ are vector functions of x, y, respectively.

Assumption 2.1 *The functions h, $\mathfrak{H}$ are convex in x and f, h, $\mathfrak{H} \in C^2$.*

Consider the Lagrangian associated with (2.4)

$$L(x, \zeta, \mu^x, \mu^y) = f(x, y) + \langle h(x), \mu^x \rangle - \langle \mathfrak{H}(y), \mu^y \rangle \tag{2.5}$$

where $\mu^x \in \mathfrak{R}^{n_h}$, $\mu^y \in \mathfrak{R}^{n_{\mathfrak{H}}}$ are the shadow prices of h, $\mathfrak{H}$ (see CN 2.2). Given that $f(x, y)$ is convex in x, concave in y, $h(x)$ is convex and $\mathfrak{H}(y)$ is concave, x_*, y_* are said to satisfy the necessary and sufficient conditions of optimality for (2.4) if there are vectors μ^x, μ^y such that

$$\nabla_x L(x_*, y_*, \mu^x, \mu^y) = 0 \tag{2.6}$$

$$\nabla_y L(x_*, y_*, \mu^x, \mu^y) = 0 \tag{2.7}$$

$$h(x_*) \leq 0; \quad \mathfrak{H}(y_*) \leq 0 \tag{2.8}$$

$$\langle h(x_*), \mu^x \rangle = 0, \quad \langle \mathfrak{H}(y_*), \mu^y \rangle = 0, \quad \mu^x \geq 0, \quad \mu^y \geq 0. \tag{2.9}$$

An approach to the constrained saddle point problem is suggested by Qi and Sun (1995). This is based on the generation of a search direction which solves a quadratic minimax saddle point problem, subject to linear approximations to the inequality constraints. The search direction at point x_k, y_k, is defined as

$$d_k \equiv \begin{bmatrix} d_k^h \\ d_k^{\mathfrak{H}} \end{bmatrix} = \begin{bmatrix} x - x_k \\ y - y_k \end{bmatrix} \tag{2.10}$$

and the quadratic function

$$L_k(d) \equiv L_k(d^h, d^{\mathfrak{H}}) = \left\langle \begin{bmatrix} \nabla_x f(x_k, y_k) \\ \nabla_y f(x_k, y_k) \end{bmatrix}, \begin{bmatrix} d^h \\ d^{\mathfrak{H}} \end{bmatrix} \right\rangle$$

$$+ \frac{1}{2} \left\langle \begin{bmatrix} d^h \\ d^{\mathfrak{H}} \end{bmatrix}, \begin{bmatrix} \nabla_{xx}^2 L & -\nabla_{xy}^2 L \\ -\nabla_{yx}^2 L & -\nabla_{yy}^2 L \end{bmatrix} \begin{bmatrix} d^h \\ d^{\mathfrak{H}} \end{bmatrix} \right\rangle$$

where $\nabla_{xx}^2 L, -\nabla_{xy}^2 L, -\nabla_{yy}^2 L$ are the corresponding approximations of the

Hessian of the Lagrangian at x_k, y_k, μ_k^x, μ_k^y and $\nabla_{xx}^2 L$, $\nabla_{yy}^2 L$ are assumed to be positive definite. We also define the linear approximations

$$h(x_k) + [\nabla h(x_k)]^T d^h \leq 0, \quad \mathfrak{H}(y_k) + [\nabla \mathfrak{H}(y_k)]^T d^{\mathfrak{H}} \leq 0$$

where $[\nabla h(x_k)]$, $[\nabla \mathfrak{H}(y_k)]$ are the Jacobian matrices associated with $h(x)$, $\mathfrak{H}(y)$, evaluated at x_k, y_k, respectively. The vector d_k is then given by

$$d_k = \arg\min_{d^h} \max_{d^{\mathfrak{H}}} \left\{ L_k(d^h, d^{\mathfrak{H}}) \mid h(x_k) + [\nabla h(x_k)]^T d^h \leq 0, \right.$$

$$\left. \mathfrak{H}(y_k) + [\nabla \mathfrak{H}(y_k)]^T d^{\mathfrak{H}} \leq 0 \right\}. \tag{2.11}$$

It can be verified from CN 2.2 that, if $\{\|d_k\|\} \to 0$, the first order necessary conditions of this quadratic minimax, expressed as a mathematical programming problem, coincide with the corresponding first order conditions for the original saddle point in (1.1).

The algorithm thus consists of the iteration

$$x_{k+1} = x_k + d_k$$

where d_k is given by (2.11), and suffers from two potential problems. The first is that, although the quadratic minimax (2.11) can be formulated and solved as a linear complementarity problem (see Wright, 1977), this solution can be greatly improved by applying an interior-point algorithm to solve the optimality conditions of the original problem (2.4). Hence the solution of subproblem (2.11) can be bypassed. The second is a stepsize strategy such as the one given in (2.2) which ensures sufficient progress during each iteration, stabilizing the algorithm and thereby ensuring its global convergence. Both of these points are addressed in the algorithm discussed in Section 2.3.

2.3 Interior Point Saddle Point Algorithm for Constrained Problems

Problem (2.4) can be transformed into an unconstrained formulation such as (1.1d) which penalizes the transgression of the constraints simultaneously with the evaluation of the saddle point. To this end, slack variables, $0 \leq s^y \in \mathfrak{R}^{n_{\mathfrak{H}}}, 0 \leq s^x \in \mathfrak{R}^{n_h}$, are introduced to reduce the nonlinear inequalities to equalities such that

$$\min_x \max_y \{ f(x, y) \mid h(x) + s^x = 0, \mathfrak{H}(y) + s^y = 0, s^x \geq 0, s^x \geq 0 \}.$$

Given that $f(x, y)$ is convex in x and concave in y, the first-order optimality conditions for this problem are

$$\nabla_x f(x, y) + \nabla_x h(x) \mu^x = 0$$

$$\nabla_y f(x, y) - \nabla_y \mathfrak{H}(y) \mu^y = 0$$

$$\mu^x - w^x = 0, \quad -\mu^y + w^y = 0$$

$$h(x) + s^x = 0, \quad \mathfrak{H}(y) + s^y = 0, \quad s^x \geq 0, \quad s^y \geq 0$$

$$\langle w^x, s^x \rangle = 0, \quad \langle w^y, s^y \rangle = 0, \quad w^x, s^x, w^y, s^y \geq 0$$

where $\mu^x \in \mathfrak{R}^{n_h}$, $\mu^y \in \mathfrak{R}^{n_\mathfrak{H}}$ are the multipliers associated with the equality constraints and w^x, w^y are the multipliers associated with the bounds on x and y (see CN 2.2). These conditions are equivalent to

$$\nabla_x f(x, y) + \nabla_x h(x) \mu^x = 0$$

$$\nabla_y f(x, y) - \nabla_y \mathfrak{H}(y) \mu^y = 0$$

$$h(x) + s^x = 0, \quad \mathfrak{H}(y) + s^y = 0, \quad s^x \geq 0, \quad s^y \geq 0$$

$$\langle \mu^x, s^x \rangle = 0, \quad \langle \mu^y, s^y \rangle = 0, \quad \mu^x, s^x, \mu^y, s^y \geq 0.$$

The penalty formulation to ensure feasibility regarding the inequality constraints on the slack variables is realized using a barrier function such as $-\log s^{y^i}$ and $-\log s^{x^j}$ where i, j denote the corresponding elements of the slack vectors. An earlier attempt to formulate constrained minimax problems in terms of barrier, or interior penalty methods, is discussed by Sasai (1974). The approach in this section differs significantly from Sasai's in the choice of the barrier function as well as in the framework for the solution which is closely related to the interior point literature (e.g., El-Bakry et al., 1996; Akrotirianakis and Rustem, 1998). The transformed objective function and minimax problem is given by

$$\ell(x, y, \nu) \equiv f(x, y) - \nu \left[\sum_{i=1}^{n_h} \log s^{x^i} - \sum_{j=1}^{n_\mathfrak{H}} \log s^{y^j} \right] \qquad (2.12a)$$

$$\min_{x, s^x} \max_{y, s^y} \{ \ell(x, y, \nu) \mid h(x) + s^x = 0, \mathfrak{H}(y) + s^y = 0, s^x > 0, s^y > 0 \}. \qquad (2.12b)$$

Given the convexity-concavity Assumptions (1.1), (2.1), the vectors $x_*(\nu)$, $y_*(\nu)$ are said to satisfy the first order necessary and sufficient conditions (see CN 2.2) for (2.12) if

$$\nabla_x f(x_*(\nu), y_*(\nu)) + \nabla_x h(x_*(\nu)) \mu^x = 0 \qquad (2.13a)$$

$$\nabla_y f(x_*(\nu), y_*(\nu)) - \nabla_y \mathfrak{H}(y_*(\nu)) \mu^y \geq 0 \qquad (2.13b)$$

$$\mu^x - \nu [S^x]^{-1} 1^x = 0, \quad -\mu^y + [S^y]^{-1} 1^y = 0 \qquad (2.13c)$$

$$h(x_*(\nu)) + s^x = 0, \quad \mathfrak{H}(y_*(\nu)) + s^y = 0 \qquad (2.13d)$$

$$s^x > 0 \quad s^y > 0 \quad \mu^x, \mu^y \geq 0$$

where

$$S^x = \mathrm{diag}(s^x) \equiv \mathrm{diag}(s_1^x, \ldots, s_{n_h}^x) \equiv \begin{bmatrix} s_1^x & 0 & 0 \\ 0 & \ddots & 0 \\ 0 & 0 & s_{n_h}^x \end{bmatrix}$$

$$S^y = \mathrm{diag}(s^y) \equiv \mathrm{diag}(s_1^y, \ldots, s_{n_\mathfrak{H}}^y) \equiv \begin{bmatrix} s_1^y & 0 & 0 \\ 0 & \ddots & 0 \\ 0 & 0 & s_{n_\mathfrak{H}}^y \end{bmatrix}$$

$$1^x = \begin{bmatrix} 1 \\ \vdots \\ 1 \end{bmatrix} \in \Re^{n_h}, \quad 1^y = \begin{bmatrix} 1 \\ \vdots \\ 1 \end{bmatrix} \in \Re^{n_\mathfrak{H}}.$$

If we let

$$M^x = \mathrm{diag}(\mu^x) \equiv \mathrm{diag}(\mu_1^x, \ldots, \mu_{n_h}^x) \equiv \begin{bmatrix} \mu_1^x & 0 & 0 \\ 0 & \ddots & 0 \\ 0 & 0 & \mu_{n_h}^x \end{bmatrix}$$

$$M^y = \mathrm{diag}(\mu^y) \equiv \mathrm{diag}(\mu_1^y, \ldots, \mu_{n_\mathfrak{H}}^y) \equiv \begin{bmatrix} \mu_1^y & 0 & 0 \\ 0 & \ddots & 0 \\ 0 & 0 & \mu_{n_\mathfrak{H}}^y \end{bmatrix}$$

we can express these conditions as the perturbed version of the optimality conditions given by

$$\nabla_x f(x_*(\nu), y_*(\nu)) + \nabla_x h(x_*(\nu))\mu^x = 0 \qquad (2.14a)$$

$$\nabla_y f(x_*(\nu), y_*(\nu)) - \nabla_y \mathfrak{H}(y_*(\nu))\mu^y = 0 \qquad (2.14b)$$

$$S^x M^x = \nu 1^x, \quad S^y M^y = \nu 1^y \qquad (2.14c)$$

$$h(x_*(\nu)) + s^x = 0, \quad \mathfrak{H}(y_*(\nu)) + s^y = 0 \qquad (2.14d)$$

$$s^x > 0, \quad s^y > 0, \quad \mu^x, \mu^y \geq 0.$$

Equations (2.14a–d) are the perturbed version of conditions (2.6)–(2.9). It can be seen from (2.14c) that, with $\nu \to 0$, the sequence of solutions to (2.14a–d) converges to the solution of the original saddle point problem (2.4), given by (2.6)–(2.9), and hence $\{(x_*(\nu), \; y_*(\nu)\} \to (x_*, y_*)$. The formal discussion of this is given by Fiacco and McCormick (1968). We illustrate the property with an example.

Example Consider the problem

$$\min_{x^1, x^2} \max_{y^1, y^2} \left\{ (x^1 - 1)^2 + (x^2 - 1)^2 - (y^1 - 2)^2 - (y^2 - 2)^2 \mid y^1 + y^2 \leq 2, \right.$$

$$\left. x^1 + x^2 \geq 1 \right\}.$$

Without slack variables, we apply the barrier function directly to the inequality constraints

$$\min_{x^1, x^2} \max_{y^1, y^2} \left\{ (x^1 - 1)^2 + (x^2 - 1)^2 - (y^1 - 2)^2 - (y^2 - 2)^2 \right.$$

$$\left. + \nu \Big[\log (2 - y^1 - y^2) - \log x^1 + x^2 - 1) \Big] \right\}.$$

The first order conditions for a saddle point of the barrier function yield

$$(x^1 - 1) - \frac{\nu/2}{x^1 + x^2 - 1} = 0, \quad (x^2 - 1) - \frac{\nu/2}{x^1 + x^2 - 1} = 0$$

$$(y^1 - 2) + \frac{\nu/2}{2 - y^1 - y^2} = 0, \quad (y^2 - 2) + \frac{\nu/2}{2 - y^1 - y^2} = 0.$$

The solution of the above system of nonlinear equations is

$$x^1(\nu) = x^2(\nu) = \frac{3 \pm \sqrt{9 - 8(1 - (\nu/2))}}{4}$$

$$y^1(\nu) = y^2(\nu) = \frac{3 \pm \sqrt{9 - 2(4 - (\nu/2))}}{2}.$$

The negative term in both cases correspond to the minimum with respect to x and maximum with respect to y. As $\nu \to 0$, the constrained minimax solution is given by

$$x^1 = x^2 = \frac{1}{2}$$

$$y^1 = y^2 = 1.$$

The perturbed conditions (2.14) of (2.4) are equivalent to the saddle point conditions of the barrier function problem (2.12), given by (2.13). Following the discussion in El-Bakry et al. (1996), we note that the perturbed conditions (2.14) are not the saddle point conditions (2.13) of the barrier function problem. Furthermore, the iterates of the Newton algorithm, discussed below, applied to the perturbed conditions (2.14) are not the same as the iterates of the Newton algorithm applied to conditions (2.13). In other words, (2.14) and (2.13) have the same solutions, and are consequently equivalent. However, the sequence of iterates generated by the Newton algorithm for solving (2.14) or (2.13) are not equivalent.

In the spirit of the development in Section 1, we redefine the vector ζ as

$$\zeta \equiv \begin{bmatrix} x \\ y \\ \mu^x \\ \mu^y \\ s^x \\ s^y \end{bmatrix}.$$

Let

$$\mathcal{F}^\nu(\zeta) = 0$$

denote the system (2.14a–d). The method for solving $\mathcal{F}^\nu(\zeta) = 0$ is a Newton-type algorithm for each value of ν. In addition, a stepsize strategy is required to ensure sufficient progress with respect to x, y, for fixed ν, and this is discussed in relation to the quasi-Newton algorithm below. Another strategy is required to reduce ν, such that $\{\nu\} \rightarrow 0$, and finally a stepsize strategy is required to ensure s^x, s^y, μ^x, $\mu^y > 0$. The latter two strategies are discussed further by El-Bakry et al. (1996), Zakovic et al. (2000) and Akrotirianakis and Rustem (1998). Most of the analysis below is directed at the property of Newton-type algorithms for solving $\mathcal{F}^\nu(\zeta) = 0$ for fixed ν. Hence, the super-script ν is omitted and we refer to (2.14a–d) as $\mathcal{F}(\zeta) = 0$. The analysis related to convergence of the overall algorithm, while reducing ν, is a consequence of the optimization algorithm in Akrotirianakis and Rustem (1998).

2.4 Quasi-Newton Algorithm for Nonlinear Systems

The final algorithm we consider approximates the Jacobian $\nabla \mathcal{F}_j$ with a quasi-Newton scheme. There are numerous schemes for this purpose, especially in view of the underlying problem structure. We discuss the general case. The basic algorithm aims to solve a system of equations such as (1.4) or (2.14a–d).

Let the system be denoted by $\mathcal{F}(\zeta) = 0$. Starting from an initial point ζ_0, the algorithm generates a sequence $\{\zeta_k\}$ by using (1.16) where d_k is given by

$$\nabla\hat{\mathcal{F}}_k d_k = -\mathcal{F}_k \tag{2.15}$$

and α_k is determined using (2.2). The matrix $\nabla\hat{\mathcal{F}}_k$ is an approximation to $\nabla\mathcal{F}_k$. The principal approximation procedure is the Broyden update given by

$$\nabla\hat{\mathcal{F}}_k = \nabla\hat{\mathcal{F}}_{k-1} + \frac{[\mathcal{F}_k - \mathcal{F}_{k-1} - \nabla\hat{\mathcal{F}}_{k-1}(\zeta_k - \zeta_{k-1})](\zeta_k - \zeta_{k-1})^{\mathrm{T}}}{\langle(\zeta_k - \zeta_{k-1}), (\zeta_k - \zeta_{k-1})\rangle}. \tag{2.16}$$

In Becker and Rustem (1993), this approximation is augmented with periodic re-evaluation of columns of $\nabla\mathcal{F}_k$ by numerical differentiation. The corresponding approximate columns of $\nabla\hat{\mathcal{F}}_k$ are replaced with these re-evaluated columns. The full re-evaluation of all columns at every iteration would yield the discrete Newton algorithm (Ortega and Rheinboldt, 1970). The Broyden update may not always converge elementwise to the Jacobian at the solution. Dennis and Schnabel (1983) give an example in which the update converges to a different matrix. However, this does not alter the desirable theoretical properties of the method, especially when it is coupled with periodic numerical reevaluation of the Jacobian (Moré and Trangenstein, 1976). Furthermore, numerical experience with the method is generally reported to be good.

3 GLOBAL CONVERGENCE OF NEWTON-TYPE ALGORITHMS

In this section, we consider the convergence properties of the quasi-Newton algorithm based on (1.16), (2.2), (2.15), (2.16). We establish global convergence with relaxed descent directions and relate the achievement of unit stepsizes to the difference of the Jacobian approximation and the actual Jacobian at the solution. In Section 4, we discuss the local convergence rate by providing a simple derivation of the Dennis–Moré characterization of Q-superlinear convergence.

The direction d_k employed by the algorithm need not necessarily satisfy (2.2). The global convergence of the algorithm requires that d_k is a descent direction for the sum of squares merit function such that

$$\langle[\nabla\mathcal{F}_k]^{\mathrm{T}}\mathcal{F}_k, d_k\rangle \leq -m\|d_k\|^2 \tag{3.1a}$$

for some $m > 0$. This is clearly satisfied by the steepest descent and, with a nonsingular Jacobian, the Newton directions. The quasi-Newton direction satisfies

$$\langle[\nabla\hat{\mathcal{F}}_k]^{\mathrm{T}}\mathcal{F}_k, d_k\rangle \leq -m\|d_k\|^2. \tag{3.1b}$$

The stepsize is adjusted to ensure the inequality

$$\frac{1}{2}\|\mathcal{F}(\zeta_k + \alpha_k d_k)\|_2^2 - \frac{1}{2}\|\mathcal{F}_k\|_2^2 \le \alpha_k \rho \langle [\nabla\mathcal{F}_k]^{\mathrm{T}}\mathcal{F}_k, d_k \rangle \qquad (3.2)$$

where $\langle [\nabla\mathcal{F}_k]^{\mathrm{T}}\mathcal{F}_k, d_k \rangle \le 0$. If, even for small stepsizes, the Newton or the quasi-Newton direction (2.2) does not satisfy (3.2), then the steepest descent direction $d_k = -[\nabla\mathcal{F}_k]^{\mathrm{T}}\mathcal{F}_k$ in Section 1, $-\mathcal{F}_k$, as discussed in Section 2, or even $-d_k$ may be tried (Li, 1989).

When an approximate Jacobian $\nabla\hat{\mathcal{F}}_k$ is used and (3.2) is not satisfied for $\alpha_k = 1$, with this approximation, then the algorithm computes columns of the Jacobian numerically before attempting to reduce the stepsize in order to satisfy the stepsize criterion (3.2).

Assumption 3.1

(i) There exists a solution, ζ_*, to $\mathcal{F}(\zeta) = 0$.

(ii) $\mathcal{F} \in \mathbb{C}^1(\mathfrak{R}^n)$.

(iii) The direction d_k satisfies (3.1).

Assumption 3.1 (i) is required to ensure that the algorithm is well defined. We also require that $\nabla\mathcal{F}$ is Lipschitz continuous to establish the monotonic decrease of the sequence $\{\|\mathcal{F}_k\|_2^2\}$ discussed in Theorem 3.1 and also to establish global convergence. The monotonic decrease may also be obtained without Lipschitz continuity if it is assumed that $\|\mathcal{F}_k\| = \phi\|d_k\|$, for some $\phi \in (0, \infty)$. The latter is satisfied both by the steepest descent with a non-singular Jacobian approximation and with the Newton and quasi-Newton directions.

In Theorems 4.1–4.3, we further require $\mathcal{F} \in \mathbb{C}^2(\mathfrak{R}^n)$ to establish the achievement of unit stepsizes (i.e., $\alpha_k = 1$). The descent condition in Assumption 3.1 (iii) is satisfied by the Newton and steepest descent directions. For the quasi-Newton direction, the algorithm ensures satisfaction by refining the Jacobian approximation. If the descent condition is not satisfied, even after a full numerical evaluation of the Jacobian, then the algorithm fails. In Theorem 3.1, we have allowed for a general descent direction including the Newton, quasi-Newton or steepest descent steps. For example, one reasonable way of establishing the quasi-Newton case is by assuming $\|\mathcal{F}_k\| = \phi\|d_k\|$ as in Theorem 3.1 (ii).

Theorem 3.1 (Monotonic Decrease) *Let Assumptions 3.1 be satisfied and furthermore let either*

(i) $\nabla\mathcal{F}$ *be Lipschitz continuous:* $\|\nabla\mathcal{F}(\zeta) - \nabla\mathcal{F}(\hat{\zeta})\| \le \ell\|\zeta - \hat{\zeta}\|,$ *for* $\ell > 0;$ *or*

(ii) d_k *satisfies* $\|\mathcal{F}_k\| = \phi\|d_k\|,$ *for some* $\phi \in (0, \infty).$

(iii) d_k *satisfies (3.1).*

Then, the stepsize computed in the N-SCE and N-B-SCE algorithms is such that $\alpha_k \in (0, 1]$ and hence the sequence $\{\zeta_k\}$ generates a corresponding monotonically decreasing sequence $\{\|\mathcal{F}_k\|_2^2\}$.

Remark In Theorems 3.1–3.2, the bound $\rho \in (0, 1]$ is sufficient. However, we have adopted the tighter bound $\rho \leq \frac{1}{2}$ since the latter is required to establish the convergence to unit stepsizes.

Proof. Using the first order expansion of $\|\mathcal{F}\|_2^2$, we have

$$\frac{1}{2}\|\mathcal{F}_{k+1}\|_2^2 - \frac{1}{2}\|\mathcal{F}_k\|_2^2 = \alpha_k \langle [\nabla \mathcal{F}_k]^{\mathrm{T}} \mathcal{F}_k, d_k \rangle$$

$$+ \alpha_k \int_0^1 \left\langle [\nabla \mathcal{F}(\zeta(t))]^{\mathrm{T}} \{\mathcal{F}(\zeta(t)) - \mathcal{F}_k\} + \left\{[\nabla \mathcal{F}(\zeta(t))]^{\mathrm{T}} - [\nabla \mathcal{F}_k]^{\mathrm{T}}\right\} \mathcal{F}_k, d_k \right\rangle dt$$

$$(3.3)$$

where $\zeta(t) = \zeta_k + t\alpha_k d_k$. For (i), given the Lipschitz continuity and (3.1a) we have

$$\frac{1}{2}\|\mathcal{F}_{k+1}\|_2^2 - \frac{1}{2}\|\mathcal{F}_k\|_2^2 \leq \alpha_k \langle [\nabla \mathcal{F}_k]^{\mathrm{T}} \mathcal{F}_k, d_k \rangle \left[1 - \frac{\alpha_k}{2m}\{\psi^2 + \ell\|\mathcal{F}_k\|\}\right] \quad (3.4)$$

where $\|\nabla \mathcal{F}\| \leq \psi$. The scalar $\rho \in (0, 1)$ in the stepsize strategy (3.2) clearly determines α_k such that

$$\rho \leq 1 - \frac{\alpha_k}{2m}\{\psi^2 + \ell\|\mathcal{F}_k\|\} \leq \frac{1}{2}.$$

By (3.1), there exists $\alpha_k \in [0, 1]$ satisfying the inequalities (3.2) and (3.3). Suppose α^0 is the largest $\alpha \in [0, 1]$ satisfying these inequalities. Thus, all $\alpha \leq \alpha^0$ also satisfy these conditions and that the stepsize strategy selects a $\alpha_k \in [\overline{\alpha}\alpha^0, \alpha^0]$, where $\overline{\alpha} \in (0, 1)$. By (3.1), it follows that $\{\|\mathcal{F}_k\|_2^2\}$ is a monotonically decreasing sequence. For (ii), we can use (3.3) to derive a relationship similar to (3.4) by invoking $\|\mathcal{F}_k\| = \phi\|d_k\|$. $\square$

We discuss the global convergence of the basic Newton and quasi-Newton algorithms. A review of convergence concepts is given in CN 1.7.

Lemma 3.1 *Let the assumptions of Theorem 3.1 be satisfied. We then have*

$$\lim_{k\to\infty} \langle [\nabla \mathcal{F}_k]^{\mathrm{T}} \mathcal{F}_k, d_k \rangle = 0. \qquad 3.5)$$

Proof. The level set

$$\mathbb{F} = \left\{\zeta \in \mathfrak{R}^n \mid \|\mathcal{F}(\zeta)\|_2^2 \leq \|\mathcal{F}(\zeta_{k_0})\|_2^2\right\}$$

is bounded (see CN 2) for some $k_0 \geq 0$. We state the proof for the Lipschitz continuous case (i) in Theorem 3.1. Case (ii) can also be similarly established using the compactness of $\mathbb{F}$. Given $\rho \in (0, \frac{1}{2})$, by (3.4), the choice

$$\alpha_0 = \min\left\{1, \frac{1-\rho}{\frac{1}{2m}(\psi^2 + \ell\|\mathcal{F}_k\|)}\right\}$$

always satisfies the stepsize strategy (3.2). Clearly, α_k, chosen in the algorithms is in the range $\alpha_k \in [\bar{\alpha}\alpha_0, \alpha_0]$ and thereby also satisfies (3.2). As $\mathcal{F}$ is Lipschitz continuous and $\mathbb{F}$ is compact, there exists a scalar $M < \infty$ and $\|\mathcal{F}_k\| < M$. Thus, as $\ell \in (0, \infty)$ and $m > 0$, we have $\alpha_k \geq \hat{\alpha} > 0, \forall j$, for some $\hat{\alpha}$ and stepsize strategy (3.2). The boundedness of $\|\mathcal{F}(\zeta)\|_2^2$ on $\mathbb{F}$ and $\langle [\nabla \mathcal{F}_k]^{\mathrm{T}} \mathcal{F}_k, d_k \rangle \leq 0$ imply

$$0 \leq \rho \sum_j \alpha_k |\langle [\nabla \mathcal{F}_k]^{\mathrm{T}} \mathcal{F}_k, d_k \rangle| \leq \sum_j \left(\frac{1}{2}\|\mathcal{F}_k\|_2^2 - \frac{1}{2}\|\mathcal{F}_{k+1}\|_2^2\right) < \infty$$

which yields (3.5). $\square$

Lemma 3.2 *Inequality (3.1) and Lemma 3.1 imply* $\lim_{k\to\infty}\|d_k\| = 0$.

Proof. The result follows from (3.1) and (3.5). $\square$

We can hence show the global convergence for any algorithm satisfying (3.1) as well as $\|\mathcal{F}_k\| = \phi\|d_k\|$, for some $\phi \in (0, \infty)$. As mentioned earlier, this is satisfied by the Newton, quasi-Newton directions and the steepest descent direction with a nonsingular Jacobian.

Theorem 3.2 (Global Convergence) *Let Assumptions 3.1 be satisfied and let d_k satisfy $\|\mathcal{F}_k\| = \phi\|d_k\|$, for some $\phi \in (0, \infty)$. Then the algorithms either terminate at a solution of the system $\mathcal{F}(\zeta) = 0$ or they generate an infinite sequence $\{\zeta_k\}$ with a subsequence $k \in K \subset \{0, 1, ...\}$ such that $\{\|d_k\|\} \to 0$ and thus every accumulation point ζ_* of the infinite sequence $\{\zeta_k\}$ is a solution of $\mathcal{F}(\zeta) = 0$.*

Proof. By Lemmas 3.1 and 3.2, there exists a subsequence $\{\zeta_k\}, k \in K$, such that $\|d_k\| \to 0$. Let there exist ζ_* such that $\{\zeta_k\} \to \zeta_*$. The existence of such points is ensured since the algorithms decrease $\|\mathcal{F}(\zeta_k)\|_2^2$ at each iteration, thereby ensuring $\zeta_k \in \mathbb{F}$, with $\mathbb{F}$ compact (see CN 2). The result then follows by letting $k \to \infty, k \in K$, since $\|\mathcal{F}(\zeta_*)\| = \lim_{k\to\infty}\|\mathcal{F}(\zeta_k)\| = \lim_{k\to\infty}\phi\|d_k\| = 0$. $\square$

4 ACHIEVEMENT OF UNIT STEPSIZES AND SUPERLINEAR CONVERGENCE

In this section, we demonstrate convergence to unit stepsizes in terms of a condition on the Jacobian or its approximation. To establish these results, we need to further assume that d_k is a Newton or quasi-(or approximate) Newton direction and strengthen the assumption on differentiability. We show that the convergence to unit stepsizes depends on the Jacobian. We first establish the result for an exact Jacobian and in Theorem 4.2 discuss the case for an approximate Jacobian such as a quasi-Newton approximation. *Let $\mathcal{F}_* \equiv \mathcal{F}(\zeta_*)$ and $\nabla \mathcal{F}_* \equiv \nabla \mathcal{F}(\zeta_*)$.*

Theorem 4.1 (Convergence to Unit Stepsizes – Exact Jacobian) *Let Assumptions 3.1 be satisfied. Also let $\mathcal{F} \in \mathbb{C}^2(\mathfrak{R}^n)$ and*

$$m\|v\|^2 \le \langle v, [\nabla \mathcal{F}_k]^{\mathrm{T}}[\nabla \mathcal{F}_k]v \rangle \le M\|v\|^2, \quad m > 0, \quad \forall v \ne 0.$$

Then there is a stage K such that strategy (3.2) is satisfied with $\alpha_k = 1$, $\forall k \ge K$.

Proof. Premultiplying the relationship $\nabla \mathcal{F}_k d_k = -\mathcal{F}_k$ by $\mathcal{F}_k$, we obtain

$$\langle ([\nabla \mathcal{F}_k]^{\mathrm{T}} \mathcal{F}_k, d_k \rangle = -\langle \mathcal{F}_k, \mathcal{F}_k \rangle = -\langle d_k, [\nabla \mathcal{F}_k]^{\mathrm{T}}[\nabla \mathcal{F}_k]d_k \rangle. \quad (4.1)$$

Consider the second order expansion of $\|\mathcal{F}\|_2^2$

$$\frac{1}{2}\|\mathcal{F}_{k+1}\|_2^2 - \frac{1}{2}\|\mathcal{F}_k\|_2^2 = \langle [\nabla \mathcal{F}_k]^{\mathrm{T}} \mathcal{F}_k, d_k \rangle + \frac{1}{2}\langle d_k, [\nabla \mathcal{F}_k]^{\mathrm{T}}[\nabla \mathcal{F}_k]d_k \rangle$$

$$+ \int_0^1 (1-t)\langle d_k, \Big\{ \mathcal{Q}(\zeta_k + td_k) - [\nabla \mathcal{F}_*]^{\mathrm{T}}[\nabla \mathcal{F}_*]$$

$$+ [\nabla \mathcal{F}_*]^{\mathrm{T}}[\nabla \mathcal{F}_*] - [\nabla \mathcal{F}_k]^{\mathrm{T}}[\nabla \mathcal{F}_k] \Big\} d_k \rangle \, dt \quad (4.2)$$

where $\mathcal{Q}(\cdot)$ is the Hessian of $\frac{1}{2}\|\mathcal{F}(\cdot)\|_2^2$, given by $[\nabla \mathcal{F}(\cdot)]^{\mathrm{T}}[\nabla \mathcal{F}(\cdot)] + \sum_i \nabla^2 \mathcal{F}^i(\cdot)\mathcal{F}^i(\cdot)$, with the last term vanishing at the solution $\mathcal{F}(\zeta_*) = 0$. Using (4.1), (4.2) becomes

$$\frac{1}{2}\|\mathcal{F}_{k+1}\|_2^2 - \frac{1}{2}\|\mathcal{F}_k\|_2^2 \le \langle [\nabla \mathcal{F}_k]^{\mathrm{T}} \mathcal{F}_k, d_k \rangle \left[\frac{1}{2} - \left(\eta_k + \frac{\xi_k}{2} \right) \right]$$

where

$$\eta_k = \int_0^1 (1-t)\|\mathcal{Q}(\zeta_k + td_k) - [\nabla \mathcal{F}_*]^{\mathrm{T}}[\nabla \mathcal{F}_*]\| \, dt$$

$$\xi_k = \|[\nabla \mathcal{F}_*]^{\mathrm{T}}[\nabla \mathcal{F}_*] - [\nabla \mathcal{F}_k]^{\mathrm{T}}[\nabla \mathcal{F}_k]\|.$$

Since, by Theorem 3.2, $\{\zeta_k\} \to \zeta_*$, we have $\{\eta_k\}, \{\xi_k\} \to 0$. For $\alpha_k = 1$, the scalar $\rho \in [0, \frac{1}{2})$ of the stepsize strategy is bounded by

$$\rho \leq \frac{1}{2} - \left(\eta_k + \frac{\xi_k}{2}\right).$$

Thus, when $\frac{1}{2} - \rho \geq (\eta_k + (\xi_k/2))$, the stepsize strategy is satisfied with $\alpha_k = 1$. $\square$

The main difficulty about the use of approximate Jacobians is that d_k does not always satisfy the descent condition $\langle [\nabla \mathcal{F}_k]^{\mathrm{T}} \mathcal{F}_k, d_k \rangle \leq 0$, and that $\langle [\nabla \hat{\mathcal{F}}_k]^{\mathrm{T}} \mathcal{F}_k, d_k \rangle \leq 0$ does not necessarily imply descent. If descent is not ensured, then global convergence cannot be established. A better Jacobian approximation is required and the algorithms in the previous section aim to achieve this.

Theorem 4.2 (Convergence to Unit Stepsizes – Approximate Jacobian)
Let Assumptions 3.1 be satisfied. Also let $\mathcal{F} \in \mathbb{C}^2(\mathfrak{R}^n)$ and

$$m\|v\|^2 \leq \langle v, [\nabla \hat{\mathcal{F}}_k]^{\mathrm{T}} [\nabla \hat{\mathcal{F}}_k] v \rangle \leq M\|v\|^2, \quad m > 0, \quad \forall v \neq 0$$

and let d_k that solves

$$\nabla \hat{\mathcal{F}}_k d = -\mathcal{F}_k \tag{4.3}$$

satisfy the descent condition

$$\langle [\nabla \mathcal{F}_k]^{\mathrm{T}} \mathcal{F}_k, d_k \rangle \leq -\langle \mathcal{F}_k, \mathcal{F}_k \rangle. \tag{4.4}$$

Then the stepsize strategy is satisfied for some $\alpha_k \in (0, 1]$ and the monotonic decrease of the sequence $\{\|\mathcal{F}_k\|_2^2\}$ is ensured. Also, there is a number χ such that if

$$\|[\nabla \mathcal{F}_*]^{\mathrm{T}} [\nabla \mathcal{F}_*] - [\nabla \hat{\mathcal{F}}_k]^{\mathrm{T}} [\nabla \hat{\mathcal{F}}_k]\| \leq \chi$$

then the stepsize strategy is satisfied for $\alpha_k = 1$.

Proof. Consider the second order expansion (4.2)

$$\frac{1}{2}\|\mathcal{F}_{k+1}\|_2^2 - \frac{1}{2}\|\mathcal{F}_k\|_2^2 \leq \alpha_k \langle [\nabla \hat{\mathcal{F}}_k]^{\mathrm{T}} \mathcal{F}_k, d_k \rangle + \alpha_k \langle [\nabla \mathcal{F}_k - \nabla \hat{\mathcal{F}}_k]^{\mathrm{T}} \mathcal{F}_k, d_k \rangle$$

$$+ \frac{1}{2}(\alpha_k)^2 \langle d_k, [\nabla \hat{\mathcal{F}}_k]^{\mathrm{T}} [\nabla \hat{\mathcal{F}}_k] d_k \rangle$$

$$+ (\alpha_k)^2 \left\{ \eta_k + \frac{1}{2}\|[\nabla \mathcal{F}_*]^{\mathrm{T}} [\nabla \mathcal{F}_*] - [\nabla \hat{\mathcal{F}}_k]^{\mathrm{T}} [\nabla \hat{\mathcal{F}}_k]\| \right\} \|d_k\|^2 > dt.$$

$$\tag{4.6}$$

From (4.3) and (4.4), we note that $\langle [\nabla \mathcal{F}_k - \nabla \hat{\mathcal{F}}_k]^{\mathrm{T}} \mathcal{F}_k, d_k \rangle \leq 0$. Also, from

(4.3), we have

$$\langle ([\nabla \hat{\mathcal{F}}_k]^{\mathrm{T}} \mathcal{F}_k, d_k \rangle = -\langle \mathcal{F}_k, \mathcal{F}_k \rangle = -\langle d_k, [\nabla \hat{\mathcal{F}}_k]^{\mathrm{T}} [\nabla \hat{\mathcal{F}}_k] d_k \rangle. \quad (4.7)$$

Using (4.7), (4.6) becomes

$$\frac{1}{2} \|\mathcal{F}_{k+1}\|_2^2 - \frac{1}{2} \|\mathcal{F}_k\|_2^2 \le \alpha_k \langle [\nabla \hat{\mathcal{F}}_k]^{\mathrm{T}} \mathcal{F}_k, d_k \rangle$$

$$\times \left[1 - \frac{1}{2} \alpha_k \left[1 + \frac{1}{m} \left\{ 2\eta_k + \|[\nabla \mathcal{F}_*]^{\mathrm{T}} [\nabla \mathcal{F}_*] - [\nabla \hat{\mathcal{F}}_k]^{\mathrm{T}} [\nabla \hat{\mathcal{F}}_k] \| \right\} \right] \right].$$

Thus, there exists an $\alpha_k \in (0, 1]$ such that

$$\rho \le 1 - \frac{1}{2} \alpha_k \left[1 + \frac{1}{m} \left\{ 2\eta_k + \|[\nabla \mathcal{F}_*]^{\mathrm{T}} [\nabla \mathcal{F}_*] - [\nabla \hat{\mathcal{F}}_k]^{\mathrm{T}} [\nabla \hat{\mathcal{F}}_k] \| \right\} \right] \le \frac{1}{2} \quad (4.8)$$

which satisfies the stepsize strategy. The monotonic decrease of the sequence $\{\|\mathcal{F}_k\|_2^2\}$ follows from this property. Global convergence follows from Theorem 3.2. Thus, $\{\zeta_k\} \to \zeta_*$ and $\{\eta_k\} \to 0$.

We can use (4.8) to conclude that if $\chi > 0$ is such that

$$\frac{1}{2} - \rho \ge \left[\frac{1}{m} \{ 2\eta_k + \chi \} \right]$$

(in view of $\{\eta_k\} \to 0$, this defines the number χ) then the stepsize strategy holds with $\alpha_k = 1$. $\square$

The local superlinear convergence rate of the quasi-Newton algorithm is established by the Dennis and Moré (1974) characterization (see CN 3). The following is a simple derivation of this characterization which only requires the once continuous differentiability of $\mathcal{F}$ and does not require the Lipschitz continuity of the Jacobian.

Lemma 4.1 *Let* $\{\zeta^j \to \zeta^*$, *then* $\{\zeta^j\}$ *is Q-superlinearly convergent if and only if* $\|d^j\| \le \gamma^j \|d^{j-1}\|$, $\lim_{j \to \infty} \gamma^j = 0$.

Proof. Suppose $\|d_k\| \le \gamma_k \|d_{k-1}\|$, $\lim_{j \to \infty} \gamma_k = 0$, holds. We have

$$\|\zeta_* - \zeta_k\| \le \lim_{t \to \infty} \sum_{i=k}^{t-1} \|\zeta_{i+1} - \zeta_i\|$$

$$\le \gamma_k \|d_{k-1}\| (1 + \omega + \omega^2 + \omega^3 + \cdots)$$

$$\le \frac{\gamma_k}{1 - \omega} \{ \|\zeta_k - \zeta_*\| + \|\zeta_* - \zeta_{k-1}\| \}$$

for some $\omega \in [0, 1)$. As $\{\gamma_k\} \to 0$, ω is chosen such that $\gamma_k + \omega < 1$,

$\forall k \geq K_0$. K_0 is an integer and is such that $\gamma_k < 1$, $\forall k \geq K_0$. Rearranging the above expression,

$$\frac{\|\zeta_k - \zeta_*\|}{\|\zeta_k - \zeta_*\| + \|\zeta_{k-1} - \zeta_*\|} \leq \frac{\gamma_k}{1 - \omega}$$

$$\|\zeta_* - \zeta_k\| \leq \frac{\gamma_k}{1 - \omega - \gamma_k} \|\zeta_* - \zeta_{k-1}\|$$

which establishes the Q-superlinear convergence of $\{\zeta_k\}$.

Suppose that $\|\zeta_* - \zeta_k\| \leq \beta_k \|\zeta_* - \zeta_{k-1}\|$, $\lim_{j \to \infty} \beta_k = 0$, with $\beta_k < 1$. This yields the inequality

$$\|\zeta_* - \zeta_k\| \leq \beta_k \|\zeta_* - \zeta_k\| + \beta_k \|d_{k-1}\|$$

$$\leq \left(\frac{\beta_k}{1 - \beta_k}\right) \|d_{k-1}\|. \tag{4.9}$$

Next, consider

$$-\nabla \hat{\mathcal{F}}_d k = \mathcal{F}_k = \mathcal{F}_* + \int_0^1 \nabla \mathcal{F}(\zeta_k + t(\zeta_* - \zeta_k))(\zeta_* - \zeta_k)\, dt$$

which yields, for $c_1 \in [0, \infty)$,

$$\|d_k\| \leq c_1 \|\zeta_* - \zeta_k\|. \tag{4.10}$$

Using (4.10) in (4.9), leads to the required result

$$\|d_k\| \leq c_1 \left(\frac{\beta_k}{1 - \beta_k}\right) \|d_{k-1}\|. \quad \square$$

We give a simple proof of the Dennis and Moré (1974) characterization of the local Q-superlinear convergence rate of the algorithms.

Theorem 4.3 *Let $D \subset \mathfrak{R}^n$ be an open convex set, Assumptions 3.1 be satisfied and let $\{\nabla \hat{\mathcal{F}}_k\}$ be a sequence of nonsingular matrices and suppose that for some $\zeta_0 \in D$ the sequence generated by*

$$\zeta_{k+1} = \zeta_k + d_k \quad \text{with} \quad \nabla \hat{\mathcal{F}}_k d_k = -\mathcal{F}_k \tag{4.11}$$

remains in D, and $\lim_{j \to \infty} \zeta_k = \zeta_*$. Then $\{\zeta_k\}$ satisfies $\|d_k\| \leq \gamma_k \|d_{k-1}\|$, $\lim_{k \to \infty} \gamma_k = 0$, and thence converges Q-superlinearly to ζ_*, in some norm $\|\cdot\|$ and $\mathcal{F}(\zeta_*) = 0$ if and only if

$$\lim_{j \to \infty} \frac{\left\| \left[\nabla \hat{\mathcal{F}}_k - \nabla \mathcal{F}_*\right] d_k \right\|}{\|d_k\|} = 0.$$

Proof. In order to establish this result in both directions, we need only to consider (4.11) and the first order expansion of $\mathcal{F}(x)$

$$-\nabla\hat{\mathcal{F}}_k d_k = \mathcal{F}_k = \mathcal{F}_{k-1} + \nabla\hat{\mathcal{F}}_{k-1}d_{k-1} + \Psi_k + \left[\nabla\mathcal{F}_* - \nabla\hat{\mathcal{F}}_{k-1}\right]d_{k-1}$$

where

$$\Psi_k = \int_0^1 [\nabla\mathcal{F}(\zeta(t)) - \nabla\mathcal{F}_*]d_{k-1}\, dt$$

with $\{\Psi_k\} \to 0$ and $\mathcal{F}_{k-1} + \nabla\hat{\mathcal{F}}_{k-1}d_{k-1} = 0$. As $\nabla\hat{\mathcal{F}}_k$ is nonsingular, dividing the above expression by $\|d_{k-1}\|$ and invoking Lemma 4.1 yields the required result. $\square$

5 CONCLUDING REMARKS

The computation of saddle points can be done using numerous approaches. A number of algorithms are developed and discussed for this purpose. A general quasi-Newton algorithm is discussed and its properties are established. It is shown that the stepsize α is guaranteed to converge to unity as the Jacobian approximation is sufficiently close to its value at the solution. A novel short proof of Q-superlinear convergence is given.

The convergence and accuracy of the solution can be regulated in Newton-type algorithms. One advantage of Newton-type algorithms, is that they solve for the zero of a system $\mathcal{F}(\zeta) = 0$ and it is possible to apply them to general structures.

References

Akrotirianakis, I. and B. Rustem (1998). "A Globally Convergent Interior Point Algorithm for General Non-Linear Programming Problems", Technical Report 98-14, Department of Computing, Imperial College.

Apostol, T. (1981). *Mathematical Analysis*, 2nd Edition, Addison Wesley, Massachusetts.

Becker, R. and B. Rustem (1993). "Algorithms for Solving Nonlinear Dynamic Decision Models", *Annals of Operations Research*, 44, 117–142.

Danilin, Y.M. and V.M. Panin (1974). "Methods of Searching Saddle Points", *Kibernetika*, 3, 119–124.

Demyanov, V.F. and A.B. Pevnyi (1972). "Numerical Methods for Finding Saddle Points", *USSR Computational Mathematics and Mathematical Physics*, 12, 1099–1127.

Demyanov, V.F. and A.B. Pevnyi (1972). "Some Estimates in Minmax Problems", Kibernetika, 1, 107-112.

Dennis, Jr., J.E. and J.J. Moré (1974). "A Characterization of Superlinear Convergence and its Application to Quasi-Newton Methods", *Mathematics of Computation*, 28, 549–560.

Dennis, Jr., J.E. and R.B. Schnabel (1983). *Numerical Methods for Unconstrained Optimization and Nonlinear Equations*, Prentice Hall, Englewood Cliffs, NJ.

El-Bakry, A.S., R.A. Tapia, T. Tsuchiya and Y. Zhang (1996). "On the Formulation and Theory of the Newton Interior Point method for Nonlinear Programming", *Journal of Optimization Theory and Applications*, 89, 507–541.

Fiacco, A.V. and G.P. McCormick (1968). *Nonlinear Programming: Sequential Unconstrained Minimization Techniques*, John Wiley, New York.

Li, G. (1989). "Successive Column Correction Algorithms for Solving Sparse Nonlinear Systems of Equations", *Mathematical Programming*, 43, 187–208.

Moré, J.J. and J.A. Trangenstein (1976). "On the Global Convergence of Broyden's Method", *Mathematics of Computation*, 30, 523–540.

Ortega, J. and W.C. Rheinboldt (1970). *Iterative Solution of Nonlinear Equations in Several Variables*, Academic Press, New York.

Qi, L. and W. Sun (1995). "An Iterative Method for the Minimax Problem", in: D.-Z. Du and P.M. Pardalos (editors), *Minimax and Applications*, Kluwer, Dordrecht, 55–67.

Rustem, B. (1998). *Algorithms for Nonlinear Programming and Multiple Objective Decisions*, J. Wiley & Sons, Chichester.

Sasai, H. (1974). "An Interior Penalty Method for Minimax for Problems with Constraints", *SIAM Journal on Control*, 12, 643–649.

Wright, S.J. (1997). *Primal-dual Interior Point Methods*, SIAM, Philadelphia, PA.

Zakovic, S., C. Pantelides, B. Rustem (2000). "An Interior Point Algorithm for Computing a Saddle Point to a Constrained Continuous Minimax Problem", *Annals of Operations Research*, 99, 59–77.

COMMENTS AND NOTES

CN 1: A Mean Value Theorem for $f : \mathbb{F} \subset \Re^n \to \Re^m$, $m > 1$

We note that mean value theorems usually apply to mappings $f : \mathbb{F} \subset \Re^n \to \Re^1$ and do not hold in general for mappings $: \mathbb{F} \subset \Re^n \to \Re^m$, $m > 1$. We construct the result by treating each element of the vector individually (see Ortega and Rheinboldt, 1970, Theorem 3.2.2).

CN 2: Notes on Convergence

A sequence of vectors $x_0, x_1, \ldots, x_k, \ldots$, in $S \subseteq \Re^n$, denoted by $\{x_k\}$, is said to *converge* to the limit $x_* \in S$ if

$$\lim_{k \to \infty} \|x_k - x_*\| = 0$$

(i.e., given $\epsilon > 0$, there is a N such that $k \geq N$ implies $\|x_k - x_*\| < \epsilon$). If $\{x_k\}$ converges to x, we write $\{x_k\} \rightarrow x_*$.

The *open ball* around a point x is a set of the form

$$\{y \in \mathfrak{R}^n \mid \|y - x\| < \epsilon\}$$

for some $\epsilon > 0$. A subset $S \subseteq \mathfrak{R}^n$ is *open* if around every point $x \in S$, there is an open ball contained in S (i.e., for $x \in S$, there is an $\epsilon > 0$ such that $\|y - x\| < \epsilon$, with $y \in S$). For example, the ball $\{y \in \mathfrak{R}^n \mid \|y\| < \epsilon\}$ is open. The *interior* of any set $S \subseteq \mathfrak{R}^n$ is the set of points $x \in S$ which are at the center of some ball contained in S. A set is said to be open if all its points are all interior points. Thus, the interior of a set is always open. The interior of the ball

$$\{y \in \mathfrak{R}^n \mid \|y\| \leq \epsilon\}$$

is the open ball

$$\{y \in \mathfrak{R}^n \mid \|y\| < \epsilon\}.$$

A point x is a *limit* of $\{x_k\}$ if there is a subsequence of $\{x_k\}$ convergent to x. The subsequence is denoted in terms of the subset $\mathcal{K}$ of the positive integers such that $\{x_k\} \rightarrow x$ for $k \in \mathcal{K}$. A related concept is an *accumulation point*: if $S \subseteq \mathfrak{R}^n$ and $x \in \mathfrak{R}^n$, then x is called an accumulation point of S if every ball around x contains at least one point of S distinct from x. For example, the set of numbers $1/n, n = 1, 2, 3, \ldots$, has zero as an accumulation point.

A set S is *closed* if its complement $\mathfrak{R}^n - S$ is open. Alternatively, a set S is *closed* if every point arbitrarily close to S is a member of S (i.e., S is closed if $\{x_k\} \in S$ and $\{x_k\} \rightarrow x$ imply that $x \in S$). For example, the ball $\{y \in \mathfrak{R}^n \mid \|y\| \leq \epsilon\}$ is closed. The *closure* of any set $S \subseteq \mathfrak{R}^n$ is the smallest closed set containing S. The boundary of a set is that part of the set that is not its interior.

A set is *bounded* if it lies entirely within a ball of some radius $\epsilon > 0$,

$$\{y \in \mathfrak{R}^n \mid \|y - x\| \leq \epsilon\}.$$

A set is *compact* if and only if it is closed and bounded.

For a compact set S, the Bolzano-Weierstrass theorem (Apostol, 1981) can be used to show that every infinite subset of S has an accumulation point in S. Thus, if $\{x_k\} \subset S$ (i.e., each member of the sequence is in S), then $\{x_k\}$ has a *limit point in* S. This establishes that there is a subsequence of $\{x_k\}$ converging to a point in S. (For example, the sequence defined by $x_{k+1} = -x_k, x_0 = 1$, has a subsequence $\{x_{2k}\}$ with limit point 1 and a subsequence $\{x_{2k+1}\}$ with limit point -1. Both sequences and their limits are clearly in the compact set $\{x \mid \|x\| \leq 1\}$.)

The convergence of the algorithms in this chapter and in Chapters 7, 8, 9 and 12 depends on the Bolzano–Weierstrass theorem. We are concerned with algorithms that aim to solve for a fixed optimal point while generating a sequence $\{x_k\}$ within a compact set S such that a certain descent property is satisfied for a merit function (e.g., $f(x_{k+1}) < f(x_k)$ and for x_* solving the problem, $f(x_*) \leq f(x_k), \forall k$). There need not be a unique accumulation point for each subsequence. However, each accumulation point would need to satisfy the optimality condition of the problem.

CN 3: Q-Rates of Convergence of a Sequence $\{x_k\} \to x_*$

If there exists a constant $\gamma \in [0, 1)$, and an integer $\hat{k} \geq 0$ such that $\forall k \geq \hat{k}$

$$\|x_* - x_{k+1}\| \leq \gamma \|x_* - x_k\|$$

then $\{x_k\}$ is said to be convergent to x_* at a *Q-linear rate*. If for some sequence $\{\gamma_k\}$,

$$\lim_{k \to \infty} \gamma_k = 0$$

and

$$\|x_* - x_{k+1}\| \leq \gamma_k \|x_* - x_k\|$$

then $\{x_k\}$ is said to be convergent to x_* at a *Q-superlinear rate*. If, instead of this rate, for some integer j, we have

$$\|x_* - x_{k+j}\| \leq \gamma_k \|x_* - x_k\|$$

then $\{x_j\}$ is said to be convergent to x_* at a *j-step Q-superlinear rate*. If there exist constant values $i > 1$, $\gamma \geq 0$ and an integer $\hat{k} \geq 0$ such that $\forall k \geq \hat{k}$

$$\|x_* - x_{k+1}\| \leq \gamma \|x_* - x_k\|^i$$

then $\{x_k\}$ is said to be converge to x_* with *a Q-order at least i*. For $i = 2$, the rate is said to be *Q-quadratic*. An authoritative discussion of general rates of convergence is given by Ortega and Rheinboldt (1970, Chapter 9).

Chapter 4

A quasi-Newton algorithm for continuous minimax

In this chapter, we develop the algorithm models considered in Chapter 2 to consider a fast algorithm for the continuous minimax problem. This extends the steepest descent approach of Panin (1981) and the convex combination rule for subgradients of Kiwiel (1987) to quasi-Newton search directions, conditional on an approximate maximizer.

In effect, we evaluate the choice between two alternative directions. The first is relatively easy to compute and is based on an augmented maximization to ensure that the multiplicity of maximizers does not result in an inferior search direction. The second involves a quadratic suproblem to determine the minimum norm subgradient. In Chapter 5, we discuss the relative merits of an algorithm based only on the former, simpler, direction.

We establish the descent property of the direction chosen by the algorithm. A step is taken using an Armijo-type stepsize strategy consistent with the direction. This ensures the monotonic decrease of the maximizing function and the convergence of the algorithm. We show that the stepsize selected by the algorithm converges to unity and that the local convergence rate is Q-superlinear.

1 INTRODUCTION

As in Chapters 1 and 2, we consider the problem

$$\min_{x \in \Re^n} \max_{y \in \mathcal{Y}} f(x, y) \tag{1.1}$$

where $\mathcal{Y} \subset \Re^m$ and $f : \Re^n \times \mathcal{Y} \mapsto \Re^1$. Let

$$\Phi(x) = \max_{y \in \mathcal{Y}} f(x, y) \tag{1.2}$$

for all $x \in \Re^n$. We call $\Phi(x)$ the max-function. Hence, (1.1) can be written as

$$\min_{x \in \Re^n} \Phi(x). \tag{1.3}$$

In this chapter, we discuss a quasi-Newton algorithm to solve (1.3). The salient features of the algorithm are the generation of a descent direction based on a subgradient of $f(x, \cdot)$ and an approximate Hessian, in the presence of possible multiple maximizers of (1.2), and a stepsize strategy that ensures sufficient decrease in $\Phi(x)$ at each iteration.

Assumption 1.1 $\mathcal{Y} \subset \mathfrak{R}^m$ *is a convex and compact infinite set.*

Remarks *Convexity* is required to ensure a *global maximum* (see CN 1.11) for y. This is important for the algorithm below. If an algorithm for computing global maxima is assumed to exist for feasible sets which may not necessarily be convex, this assumption would not be needed. It should be noted that a similar assumption regarding the concavity of $f(\cdot, y)$ is not made as it is assumed the global maximization of $f(\cdot, y)$ over a convex feasible set is possible.

Compactness ensures that, by Weierstrass' theorem, the sequence generated by the algorithm has a limit point in this set (see CN 3.2, Apostol, 1981; Luenberger, 1984)

Assumption 1.2 $f(x, y)$ *is continuous in x and y; twice continuously differentiable in x*

Assumption 1.3 *In the neighborhood of the solution* x_* *of (1.3), there exists a scalar* $b > 0$ *such that*

$$\forall x \in \{x \in \mathfrak{R}^n \mid \|x - x_*\| < b\}$$

the Hessian with respect to x of $f(x, y)$ *for all* $y \in \mathcal{Y}$ *is positive definite.*

Remark *Positive definiteness* is required for the demonstration of convergence to unit stepsize and local superlinear convergence.

As we discuss in Chapter 1, if $\mathcal{Y}$ is a finite set, (1.3) becomes the discrete minimax problem

$$\min_{x \in \mathfrak{R}^n} \max_{j \in 1,\dots,J} \left\{ f^j(x) \right\} \tag{1.4}$$

and this is discussed in Chapters 6 and 7. Most algorithms for solving (1.4) involve the transformation of (1.4) into the nonlinear programming problem

$$\min_{x,z \in \mathfrak{R}^{n+1}} \left\{ z \mid f^j(x) \leq z, j = 1, \dots, J \right\}. \tag{1.5}$$

In the continuous case for $\mathcal{Y}$, the formulation of (1.5) is the semi-infinite optimization problem

$$\min_{x,z \in \mathfrak{R}^{n+1}} \{ z \mid f(x, y) \leq z, \forall y \in \mathcal{Y} \}, \tag{1.6}$$

with an infinite number of constraints corresponding to the elements in $\mathcal{Y}$.

Problem (1.3) poses several difficulties. First, $\Phi(x)$ is, in general, continuous but may have kinks. Therefore, it may not be straightforwardly differentiable. The presence of kinks makes the optimization problem difficult to solve. At a kink, the maximizer is not unique and the choice of subgradient to generate a search direction is not simple compared to smooth functions. Furthermore, the Hessian of the Lagrangian of (1.6), is viewed in the context of multiple maximizers. The existence of the Hessian of the *max-function* for finite $\mathcal{Y}$ (i.e., the discrete minimax case) is discussed by Wierzbicki (1982). The approach can be used to establish that the Hessian of (1.6) represents the Hessian of the *max-function at the solution*. The results below require the existence of the Hessian of (1.6) and the algorithm utilizes an approximation to this. Second, $\Phi(x)$ may not be computed accurately as it would require infinitely many iterations of an algorithm to maximize $f(x, y)$, with respect to $y \in \mathcal{Y}$. In practice, the maximization algorithm is terminated when a sufficiently good maximum is attained. (We assume that $\Phi(x)$ is the exact value of the *max-function*. In the numerical experiments reported in Chapter 5, $\Phi(x)$ is computed with an accuracy of at least 10^{-8}.) Third, (1.3) requires a global maximum, in view of possible multiple maximizers, such as corner solutions (see Remark 3.1a). The use of nonglobal maxima cannot guarantee a monotonic decrease in $\Phi(x)$.

In this chapter, we discuss an algorithm for solving (1.1) which extends the first order approach of Panin (1981) and Kiwiel (1987) to quasi-Newton descent directions, conditional on the maximizer, and attempts to deal with the problem of multiple maximizers. Two alternative directions are considered. The first is based on the maximizer at x, corresponding to the minimum norm subgradient of $\nabla_x f(x, \cdot)$, and a Hessian approximation. It involves an augmented maximization subproblem to ensure that any multiplicity of maximizers does not result in an inferior search direction. For this purpose, the quasi-Newton descent direction is based on a quadratic approximation of $f(x, y)$ and on y_{k+1} which minimizes (with respect to y) simultaneously

$$\left[\Phi(x_k) - f(x_k, y)\right]^2 \quad \text{and} \quad \|\nabla_x f(x_k, y)\|^2_{\mathfrak{C}_k^{-1}}.$$

Following Lemma 3.1 below, whenever possible, the maximizer with the minimum-norm subgradient is chosen. The second direction involves more computation. This is a quasi-Newton descent direction based on the combination of the gradients corresponding to the multiple maximizers. The evaluation of the combination entails the solution of a quadratic programming problem. The descent property is established in Lemma 3.2 below. In the discrete minimax case, it can be shown that the algorithm based on the second direction only is equivalent to the discrete minimax algorithm in Rustem (1992). An Armijo-type stepsize strategy, consistent with the search direction, is used to determine the stepsize. The algorithm is shown to be globally convergent.

The algorithm also attains unit stepsizes and a Q-superlinear convergence rate, depending on the accuracy of the Hessian approximation used.

We introduce useful basic concepts in Section 2 and present the algorithm and its basic descent property in Section 3. The monotonic decrease of the sequence $\{\Phi(x_k)\}$ is discussed in Section 4. Convergence to unit stepsizes, global and local convergence results are discussed in Section 5.

Numerical results for test problems with $f(x, y)$ convex-concave in $x - y$; convex in both x, y; generally nonlinear in x and linear in y are reported in Chapter 5 using the quasi-Newton algorithm below, a simplified version, based on the direction $\bar{d}$ only, and Kiwiel's (1987) algorithm.

2 BASIC CONCEPTS AND DEFINITIONS

At any point x, we define the set of maximizers by

$$\mathcal{Y}(x) \equiv \left\{ y \mid y = \arg\max_{\eta \in \mathcal{Y}} f(x, \eta) \right\}. \tag{2.1}$$

Assumption 2.1 *At a point x, all members of the set $\mathcal{Y}(x)$ are computable.*

We define the directional derivative of $\Phi(x)$, along the direction $d \in \mathfrak{R}^n$, at x_k, as

$$\frac{\partial \Phi(x_k)}{\partial d} \equiv \max_{y \in \mathcal{Y}(x_k)} \langle \nabla_x f(x_k, y), d \rangle$$

(Demyanov and Malozemov, 1974). The proposed algorithm generates quasi-Newton directions that ensure descent. To this end, let $f_k(d, y)$ denote an augmented quadratic approximation to $f(x, y)$

$$f_k(d, y) = f(x_k, y) + \langle \nabla_x f(x_k, y), d \rangle + \frac{1}{2} \|d\|_{\mathfrak{C}_k}^2 - C[\Phi(x_k) - f(x_k, y)]^2 \tag{2.2}$$

where $C \geq 0$ is a penalty parameter for deviations from the maximizer at x_k (see Appendix B); $\mathfrak{C}_k$ is a positive definite approximation to the Hessian, with respect to x, of the Lagrangian of (1.6) at the kth iteration; and $\|d\|_{\mathfrak{C}_k}^2 = \langle d, \mathfrak{C}_k d \rangle$. The max-function corresponding to this approximation is given by

$$\Phi_k(d) = \max_{y \in \mathcal{Y}} f_k(d, y). \tag{2.3}$$

Assumption 2.2 For the Hessian approximation $\mathfrak{C}_k$, $k = 1, 2, \ldots$, there exist numbers $\bar{\zeta}, \underline{\zeta} > 0$ such that

$$\underline{\zeta}\|x\|^2 \leq \|x\|_{\mathfrak{C}_k}^2 \leq \bar{\zeta}\|x\|^2, \quad \forall x \in \mathfrak{R}^n.$$

The choice of Hessian, discussed below, is consistent with the convex duality theory in Wierzbicki (1982; Lemma 1), for determining the minimum-norm subgradient in the subdifferential

$$\partial\Phi(x) = \text{conv}\{\nabla_x f(x,y) \mid y = \arg\max f(x,y)\}$$

$$= \{\text{conv}\nabla_x f(x,y) \mid f(x,y) = \Phi(x)\}$$

of the *max-function*. By Wierzbicki (1982), this is equivalent to the problem of minimizing the quadratic approximation to the Lagrangian of (1.6). The equivalence holds for $y = \arg\max f(x,y)$, or $f(x,y) = \Phi(x)$, ensured for sufficiently large C in (2.2)–(2.3). Hence we can express the above subdifferential equivalently as

$$\partial\Phi(x) = \text{conv}\{\nabla_x f(x,y) \mid y \in \mathcal{Y}(x)\}.$$

Furthermore, we note that

$$\max_{y \in \mathcal{Y}_{x_k}} \langle \nabla_x f(x_k,y), d_k \rangle = \max_{\nabla_x f(x_k,y) \in \partial\Phi(x_k)} \langle \nabla_x f(x_k,y), d_k \rangle \qquad (2.4)$$

(Demyanov and Malozemov, 1974, p. 195, proof of Theorem 3.1). This follows from the equivalence of the maximum over the convex combination of $|\mathcal{Y}(x_k)|$ numbers (i.e., the inner product of the gradient (2.12) below with d_k) to the maximum of these $|\mathcal{Y}(x_k)|$ numbers, corresponding to the individual inner products $\langle \nabla_x f(x_k,y), d_k \rangle$ arising from each maximizer.

Let $\overline{d}$ be the direction that minimizes the approximation to the max-function, that is,

$$\overline{d} = \arg\min_{d \in \mathfrak{R}^n} \Phi_k(d). \qquad (2.5)$$

To evaluate $\overline{d}$, we note that (2.3) is minimized by $d(y) = -\mathfrak{C}_k^{-1}\nabla_x f(x_k,y)$, for $y \in \mathcal{Y}$. Using this d in (2.3) determines the maximizer y_{k+1} given by

$$y_{k+1} = \arg\max_{y \in \mathcal{Y}} \left\{ f(x_k,y) - \frac{1}{2}\|\nabla_x f(x_k,y)\|^2_{\mathfrak{C}_k^{-1}} - C[\Phi(x_k) - f(x_k,y)]^2 \right\}. \quad (2.6)$$

It is shown in Lemma 3.1, that the set of $y \in \mathcal{Y}$ that solve (2.3) (or (2.6)) is a subset of $\mathcal{Y}(x_k)$.

We define the set of maximizers of $f_k(\overline{d},y)$ as

$$\mathcal{Y}_{k+1} \equiv \{ y_{k+1} \in \mathcal{Y}(x_k) \mid y_{k+1} = \arg\max f_k(\overline{d},y)\}. \qquad (2.7)$$

The resulting direction

$$\overline{d} = -\mathfrak{C}_k^{-1}\nabla_x f(x_k, y_{k+1}) \qquad (2.8)$$

is a descent direction for $\langle \overline{d}, \nabla_x f(x_k, y_{k+1})\rangle$ and, if $\overline{d}$ is a descent direction for all the other maximizers, it is a descent direction for the *max-function*. If y_{k+1} is nonunique (i.e., $|\mathcal{Y}_{k+1}| > 1$), we choose an arbitrary element of $\mathcal{Y}_{k+1}$ and compute

$$\bar{y} = \arg\max_{y \in \mathcal{Y}(x_k)} \langle \nabla_x f(x_k, y), \bar{d} \rangle. \tag{2.9}$$

Remark   The computation in (2.9) can be realized by evaluating the inner product for the maximizers $y \in \mathcal{Y}(x_k)$, or by solving

$$\max_{y \in \mathcal{Y}} \langle \nabla_x f(x_k, y), \bar{d} \rangle - C[\Phi(x_k) - f(x_k, y)]^2.$$

To solve (1.3), the quasi-Newton algorithm constructs the sequence

$$x_{k+1} = x_k + \alpha_k d_k \tag{2.10}$$

where α_k is calculated according to a rule discussed below while d_k is given by

$$d_k = \begin{cases} -\mathbb{C}_k^{-1} \nabla \Phi_k & \text{if } \langle \nabla_x f(x_k, \bar{y}), \bar{d} \rangle \geq -\xi \\ \bar{d} & \text{otherwise.} \end{cases} \tag{2.11}$$

ξ is the termination accuracy and $\nabla \Phi_k$ is a convex combination of $\nabla_x f(x_k, y), y \in \mathcal{Y}(x_k)$, given by (2.12)–(2.13) below.

For nonunique $y \in \mathcal{Y}(x_k)$, by Caratheodory's Theorem (Theorem 1.1.1, see also Rockafeller, 1972), a vector $\nabla \Phi_k \in \partial \Phi(x_k)$ can be characterized by at most $(n + 1)$ vectors

$$\nabla_x f(x_k, y) \in \partial \Phi(x_k)$$

such that

$$\nabla \Phi_k \equiv \sum_{y \in \mathcal{Y}(x_k)} \lambda_{k+1}^y \nabla_x f(x_k, y), \quad \lambda_{k+1}^y \geq 0, \quad \sum_{y \in \mathcal{Y}(x_k)} \lambda_{k+1}^y = 1. \tag{2.12}$$

As in Wierzbicki (1982, Lemmas 1 and 3), λ_{k+1}^y are chosen to ensure that $\nabla \Phi_k$ is the minimum-norm subgradient in $\partial \Phi(x_*)$ and hence

$$\lambda_{k+1}^y = \arg\min_{\lambda^y} \left\{ \left\| \sum_{y \in \mathcal{Y}(x_k)} \lambda^y \nabla_x f(x_k, y) \right\|_{\mathbb{C}_k^{-1}}^2 \,\middle|\, \lambda^y \geq 0, \sum_y \lambda^y = 1 \right\}. \tag{2.13}$$

As all $y \in \mathcal{Y}(x_k)$ correspond to the same function value, $\nabla \Phi_k$ and the Hessian using λ_{k+1}^y, are consistent. The solution to the minimum-norm problem is unique when $\nabla_x f(x_k, y), y \in \mathcal{Y}(x_k)$, are linearly independent. Otherwise, a minimum length λ_{k+1}^y is determined.

We note that d_k based on $\arg\min_{d \in \mathbb{R}^n} \Phi_k(d)$ corresponds to the selection of the direction based on the maximizer that yields the least steep gradient. This is straightforward to compute. On the other hand d_k based on (2.12)–(2.13) entails the solution of a quadratic programming problem. The former is an overall descent direction if it is also a descent direction for the other maximizers. The latter is always a descent direction.

In Proposition 1 below, we discuss the restriction that ensures either way of computing d_k yields the same direction. The equivalence is considered further in Assumption A.1 (see Appendix A).

Proposition 1 (Equivalence of Both Choices for d_k) *Directions $\overline{d}$ and $-\mathbb{C}_k^{-1}\nabla\Phi_k$ are equivalent if the minimization in (2.13) is solved by*

$$\lambda_{k+1}^y = 1$$

corresponding to a unique y_{k+1}

$$y_{k+1} = \underset{y\in\mathcal{Y}(x_k)}{\arg\min}\left\{\frac{1}{2}\|\nabla_x f(x_k, y)\|_{\mathbb{C}_k^{-1}}^2\right\}. \qquad (2.14)$$

Proof. The condition that ensures the equivalence of both choices for d_k in (2.11),

$$-\mathbb{C}_k^{-1}\nabla_x f(x_k, y_{k+1}) = -\mathbb{C}_k^{-1}\sum_{y\in\mathcal{Y}(x_k)}\lambda_{k+1}^y\nabla_x f(x_k, y)$$

is given by

$$\nabla_x f(x_k, y_{k+1}) = \sum_{y\in\mathcal{Y}(x_k)}\lambda_{k+1}^y\nabla_x f(x_k, y). \qquad \square \qquad (2.15)$$

The Lagrangian of (1.6) is considered given the maximizers at x_k,

$$\mathcal{L}(x, v, \lambda) = v + \sum_{y\in\mathcal{Y}(x_k)}(f(x, y) - v)\lambda^y. \qquad (2.16)$$

Thus, at x_k, the Hessian of $\mathcal{L}$, with respect to x, is given by

$$\sum_{y\in\mathcal{Y}(x_k)}\nabla_x^2 f(x_k, y)\lambda_{k+1}^y \qquad (2.17)$$

where, consistent with the above discussion, λ_{k+1}^y is the solution of (2.13). We also define the difference

$$\Psi_k \equiv \begin{cases} -\frac{1}{2}\|\nabla\Phi_k\|_{\mathbb{C}_k^{-1}}^2 & \text{if } \langle\nabla_x f(x_k, \overline{y}), \overline{d}\rangle \geq -\xi \\ \underset{d\in\Re^n}{\min}\{\Phi_k(d) - \Phi(x_k)\} & \text{otherwise} \end{cases} \qquad (2.18)$$

which is used in the algorithm.

Let x_* be the solution of the minimax problem (1.1). We note that, by Theorem 1.3.1, at x_* the following variational inequality is satisfied as the *necessary condition for an extremum (nce)*:

$$\frac{\partial \Phi(x_*)}{\partial d} = \max_{y \in \mathcal{Y}(x_*)} \langle \nabla_x f(x_*, y), d \rangle \geq 0, \forall d \in \Re^n \qquad (2.19)$$

(see also Demyanov and Malozemov, 1974, p. 191, Theorem 2.2).

3 THE QUASI-NEWTON ALGORITHM

The algorithm utilizes the direction $\overline{d}$ whenever possible as the evaluation of $\nabla \Phi_k$ entails the solution of a quadratic programming problem. If $\overline{d}$ is a descent direction, as discussed above, it is used to determine d_k in Step 2, and subsequently Ψ_k in Step 3. It then utilizes a stepsize strategy to determine the progress along d_k and updates the Hessian approximation d_k.

The full quasi-Newton algorithm (QN1) is discussed below. An alternative, simplified algorithm (QN2), based on direction $\overline{d}$ alone, is also evaluated empirically in Chapter 5. The default parameter values are those used in the numerical experiments in Chapter 5.

Quasi-Newton Algorithm (QN1)

Step 0: Given x_0, y_0, $\mathfrak{C}_0$, set: $k = 0$;

termination accuracy: $1 \gg \xi \geq 0$, $(\xi = 10^{-8})$;

line search parameter $c \in (0, 1)$, $(c = 10^{-4})$;

stepsize factor $\sigma \in (0, 1)$, $(\sigma = 0.5)$;

penalty coefficient $C \in [0, \infty)$, $(C = 10^6)$

Step 1: Maximization at x_k: compute the global solution to the nonlinear programming problem

$$\Phi(x_k) = \max_{y \in \mathcal{Y}} \{f(x_k, y)\}. \qquad (3.1)$$

Step 2: Direction-finding subproblem:

(a) compute y_{k+1} given by (2.6): if y_{k+1} is nonunique (i.e., $|\mathcal{Y}_{k+1}| > 1$), choose an arbitrary element of $\mathcal{Y}_{k+1}$.

(b) Compute $\overline{d}$ and $\overline{y}$ given by (2.8) and (2.9), respectively. Set d_k given by (2.11):

if

$$\langle \nabla_x f(x_k, \overline{y}), \overline{d} \rangle < -\xi \qquad (3.2)$$

go to Step 3 (b). Else, compute $\nabla \Phi_k$ given by (2.12)–(2.13) and go to Step 3 (a).

Step 3: (a) (if $\langle \nabla_x f(x_k, \bar{y}), \bar{d} \rangle \geq -\xi$) compute

$$\Psi_k = -\frac{1}{2} \|\nabla \Phi_k\|^2_{\mathfrak{C}_k^{-1}}. \tag{3.3}$$

(b) (if $\langle \nabla_x f(x_k, \bar{y}), \bar{d} \rangle < -\xi$) compute

$$\Psi_k = f(x_k, y_{k+1}) + \langle \nabla_x f(x_k, y_{k+1}), d_k \rangle$$

$$+ \frac{1}{2} \|d_k\|^2_{\mathfrak{C}_k} - C[\Phi(x_k) - f(x_k, y_{k+1})]^2 - \Phi(x_k). \tag{3.4}$$

Stop if

$$\Psi_k \geq -\xi. \tag{3.5}$$

Else, perform the line search

$$\alpha_k = \max\left\{\alpha \mid \Phi(x_k + \alpha d_k) - \Phi(x_k) \leq c\alpha\Psi_k; \alpha = (\sigma)^i, i = 0, 1, 2, ...\right\}. \tag{3.6}$$

Set x_{k+1} using (2.10), update the Hessian and set $k = k + 1$, go to Step 1.

In Step 3, while updating the Hessian, although x_{k+1} is known, λ_{k+2} in (2.13) cannot be computed as $\mathfrak{C}_{k+1}$ is not yet known. To overcome this difficulty, an approximate λ_{k+2} is evaluated using $\mathfrak{C}_k$ in (2.13). With this $\lambda_{k+2}, \nabla \Phi_{k+1}$ in (2.12) is approximated and the matrix $\mathfrak{C}_k$ is updated to $\mathfrak{C}_{k+1}$. Given this $\mathfrak{C}_{k+1}$, the algorithm proceeds to Steps 1 and 2 to compute y_{k+2} and a more accurate $\nabla \Phi_{k+1}$.

The condition

$$\max_{y \in \mathcal{Y}(x_k)} \langle \nabla_x f(x_k, y), \bar{d} \rangle < -\xi \tag{3.7}$$

ensures that the direction d_k chosen by the algorithm is a descent direction for the max-function. This property is established in Lemma 3.2.

In the case of the discrete minimax problem (1.4), with Assumption 2.1 and the direction of search given by $d_k = -\mathfrak{C}_k^{-1}\nabla\Phi_k$, the above algorithm is equivalent to the discrete minimax algorithm discussed in Rustem (1992). In particular, this equivalence applies to the computation of $\nabla\Phi_k$ in (2.12)–(2.13) and the quadratic subproblem in Rustem (1992) as well as the stepsize strategies of either algorithm.

Remark 3.1 (Choice of Maximizer) (a) If the maximum in (3.1) is attained by more than one y, these constitute $\mathcal{Y}(x_k)$. In the case of $\mathcal{Y}$ given by upper and lower bounds as in some examples in Chapter 5, a potential subset of $\mathcal{Y}(x_k)$ is evaluated by considering local solutions of the nonlinear program and the value of $f(x_k, y)$ at every vertex of the hypercube $\mathcal{Y}$. This practical approach can be refined, with increased algorithmic complexity, by adopting a global optimization procedure, based on branch-and-bound, with greater assurance to reach the global maximum (e.g., Pardalos and Rosen, 1987; Floudas and Pardalos, 1992, 1995).

If the minimum in (2.6) is attained by more than one y, these constitute $\mathcal{Y}_{k+1}$. *By Lemma 3.2 below, d_k is a descent direction, for all* $y_{k+1} \in \mathcal{Y}_{k+1}, |\mathcal{Y}_{k+1}| > 1$. The algorithm is shown to be convergent for any choice of $y_{k+1} \in \mathcal{Y}_{k+1}$. (As in Step 1, in numerical experiments with problems with only upper and lower bounds on y, the objective in (2.6) is checked at every vertex.)

(b) Let (3.7) be satisfied. We have, $y_{k+1} \in \mathcal{Y}_{k+1} \subseteq \mathcal{Y}(x_k), d_k = \bar{d}$ in (2.8) and

$$\nabla_x f(x_k, y_{k+1}) + \mathfrak{C}_k d_k = 0. \quad \square$$

Remark 3.2 (The BFGS Hessian Approximation Formula) The approximate Hessian $\mathfrak{C}_k$ is computed using the BFGS quasi-Newton formula (Broyden, 1969, 1970; Fletcher, 1970; Goldfarb, 1970; Shanno, 1970). We have

$$\delta_k = x_{k+1} - x_k$$

$$\gamma_k = \sum_{y \in \mathcal{Y}(x_{k+1})} \lambda_{k+2}^y \nabla_x f(x_{k+1}, y) - \sum_{y \in \mathcal{Y}(x_k)} \lambda_{k+1}^y \nabla_x f(x_k, y) \quad (3.8a)$$

and

$$\mathfrak{C}_{k+1} = \mathfrak{C}_k - \frac{\mathfrak{C}_k \delta_k \delta_k^T \mathfrak{C}_k}{\langle \delta_k, \mathfrak{C}_k \delta_k \rangle} + \frac{\gamma_k \gamma_k^T}{\langle \gamma_k, \delta_k \rangle} \quad (3.8b)$$

The use of this formula in the case of a unique maximizer is discussed in CN 1. Following Theorem 5.1 below, the algorithm converges for any positive definite matrix $\mathfrak{C}_k$. The BFGS formula is used to approximate the Hessian for achieving unit stepsizes and superlinear convergence.

It is possible for a maximizing algorithm to terminate at a solution for (2.6), satisfying

$$0 \leq \Phi(x_k) - f(x_k, y_{k+1}) \leq \varepsilon(C) \quad (3.9a)$$

for a small number $\varepsilon(C) \geq 0$. In Lemma 3.2, d_k is a descent direction even when Assumption 3.1 below is not satisfied. This is due to the fact that

$$-[\Phi(x_k) - f(x_k, y_{k+1})] - C[\Phi(x_k) - f(x_k, y_{k+1})]^2 \leq 0 \quad (3.9b)$$

(see (3.4) and Lemma 3.1). Nevertheless, the satisfaction of equality (3.10) below is required by subsequent results, and enforced in the algorithm by appropriate choice of C.

Assumption 3.1 *There exists a $C \geq 0$ such that equality*

$$\Phi(x_k) - f(x_k, y_{k+1}) = 0 \tag{3.10}$$

is satisfied for all k.

Remark 3.3 Assumption 3.1 can be replaced by a strategy for adjusting C in the algorithm to satisfy (3.10) (Fiacco and McCormick, 1968). The default value of C has, in practice, been sufficient to ensure (3.10).

Subproblem (2.6) ensures $y_{k+1} \in \mathcal{Y}(x_k)$ through the penalty term. By Remark 3.3, Assumption 3.1 can be relaxed using a strategy for increasing C to ensure (3.10). This is discussed in Lemma 3.1 along with the possible nonuniqueness of y_{k+1}. In the latter case, the descent property and subsequent convergence results apply to all members of $\mathcal{Y}_{k+1}$.

Lemma 3.1 *Let*

 (i)   *Assumptions (1.1)–(1.3), (2.1), (2.2) and (3.1) hold,*

 (ii) *$f(x, y)$ be continuous in y and once continuously differentiable in x; and*

 (iii) *the scalar C be chosen from the range $0 \leq C < \infty$.*

Then
 (a) *in (2.6), we have $y_{k+1} \in \mathcal{Y}(x_k)$, and*
 (b)

$$y_{k+1} \in \mathcal{Y}_{k+1} = \left\{ y \in \mathcal{Y}(x_k) \mid y = \arg\min_{y \in \mathcal{Y}(x_k)} \left\{ \frac{1}{2} \| \nabla_x f(x_k, y) \|^2_{\mathbb{C}_k^{-1}} \right\} \right\}.$$

Proof. For a given $C, y_{k+1} \in \mathcal{Y} \setminus \mathcal{Y}(x_k)$ in (2.6) implies that

$$\Phi(x_k) - f(x_k, y_{k+1}) > 0.$$

By Assumption 3.1, appropriately increasing $C \in [0, \infty)$, ensures (3.10), and consequently $y_{k+1} \in \mathcal{Y}(x_k)$.

To show (b), we use the definition (2.7) and since $\Phi(x_k) - f(x_k, y) = 0$, for all $y \in \mathcal{Y}(x_k)$, we have

$$y_{k+1} = \arg\min_{y \in \mathcal{Y}(x_k)} \left\{ -f(x_k, y) + \frac{1}{2} \| \nabla_x f(x_k, y) \|^2_{\mathbb{C}_k^{-1}} + C[\Phi(x_k) - f(x_k, y)]^2 \right\}$$

$$= \arg\min_{y \in \mathcal{Y}(x_k)} \left\{ -\Phi(x_k) + \frac{1}{2} \| \nabla_x f(x_k, y) \|^2_{\mathbb{C}_k^{-1}} + C[\Phi(x_k) - \Phi(x_k)]^2 \right\}$$

$$= \arg\min_{y \in \mathcal{Y}(x_k)} \left\{ -\Phi(x_k) + \frac{1}{2} \| \nabla_x f(x_k, y) \|^2_{\mathbb{C}_k^{-1}} \right\}.$$

Thus, $y_{k+1} \in \mathcal{Y}(x_k)$ is chosen to satisfy (b). $\quad \square$

Corollary 3.1 *Let Assumptions (2.1) and (2.2) be satisfied. Then,*
(a) for $\max_{y \in \mathcal{Y}(x_k)} \langle \nabla_x f(x_k, y), \overline{d} \rangle < -\xi$, *we have* $\overline{d} = d_k$, *and*

$$-\xi > \langle \nabla_x f(x_k, \overline{y}), \overline{d} \rangle \geq -\|d_k\|^2_{\mathbb{C}_k}$$

(b) for $d_k = -\mathbb{C}_k^{-1} \nabla \Phi_k$, *we have*

$$\max_{y \in \mathcal{Y}(x_k)} -\langle \nabla_x f(x_k, y), \mathbb{C}_k^{-1} \nabla \Phi_k \rangle = -\|d_k\|^2_{\mathbb{C}_k}.$$

Proof. By (2.11) and Lemma 3.1 (b), we have

$$\max_{y \in \mathcal{Y}(x_k)} \langle \nabla_x f(x_k, y), \overline{d} \rangle = \max_{y \in \mathcal{Y}(x_k)} -\langle \nabla_x f(x_k, y), \mathbb{C}_k^{-1} \nabla_x f(x_k, y_{k+1}) \rangle$$

$$\geq -\langle \nabla_x f(x_k, y_{k+1}), \mathbb{C}_k^{-1} \nabla_x f(x_k, y_{k+1}) \rangle$$

$$= \max_{y \in \mathcal{Y}(x_k)} -\langle \nabla_x f(x_k, y), \mathbb{C}_k^{-1} \nabla_x f(x_k, y) \rangle$$

$$= -\|d_k\|^2_{\mathbb{C}_k}$$

which yields (a).

For (b), consider

$$\max_{y \in \mathcal{Y}(x_k)} -\left\langle \nabla_x f(x_k, y), \mathbb{C}_k^{-1} \sum_{y \in \mathcal{Y}(x_k)} \lambda^y_{k+1} \nabla_x f(x_k, y) \right\rangle$$

$$= \max_{\lambda^y} \left\{ -\left\langle \sum_{y \in \mathcal{Y}(x_k)} \lambda^y \nabla_x f(x_k, y), \mathbb{C}_k^{-1} \sum_{y \in \mathcal{Y}(x_k)} \lambda^y_{k+1} \nabla_x f(x_k, y) \right\rangle \middle| \lambda^y \geq 0, \sum_{y \in \mathcal{Y}(x_k)} \lambda^y = 1 \right\}$$

$$= \max_{\lambda^y} \left\{ -\left\langle \sum_{y \in \mathcal{Y}(x_k)} \lambda^y \nabla_x f(x_k, y), \mathbb{C}_k^{-1} \sum_{y \in \mathcal{Y}(x_k)} \lambda^y \nabla_x f(x_k, y) \right\rangle \middle| \lambda^y \geq 0, \sum_{y \in \mathcal{Y}(x_k)} \lambda^y = 1 \right\}$$

$$\equiv -\left\langle \left[\sum_{y \in \mathcal{Y}(x_k)} \lambda^y_{k+1} \nabla_x f(x_k, y) \right], \mathbb{C}_k^{-1} \left[\sum_{y \in \mathcal{Y}(x_k)} \lambda^y_{k+1} \nabla_x f(x_k, y) \right] \right\rangle$$

$$= -\|d_k\|^2_{\mathbb{C}_k}$$

where the first equality is due to the equivalence of the maximum over $y \in \mathcal{Y}(x_k)$ to the maximum over the convex combination. The second equality is due to the symmetry of the two convex combinations which leads to equivalent maximizers. □

Corollary 3.2 *Let* λ_{k+1}^y *be computed by (2.12)–(2.13) and* y_{k+1} *be computed using (2.6). We have*

$$\left\| \sum_{y \in \mathcal{Y}(x_k)} \lambda_{k+1}^y \nabla_x f(x_k, y) \right\|_{\mathbb{C}_k^{-1}}^2 \leq \| \nabla_x f(x_k, y_{k+1}) \|_{\mathbb{C}_k^{-1}}^2 \leq \| \nabla_x f(x_k, y) \|_{\mathbb{C}_k^{-1}}^2, y \in \mathcal{Y}(x_k).$$

Proof. The first inequality follows from (2.12)–(2.13) and the second follows from Lemma 3.1 (b). □

Lemma 3.2 (Descent Property of d_k) *Let*

(i) *Assumptions (1.1), (2.1), (2.2) and (3.1) hold; and,*

(ii) $f(x, y)$ *be continuous in* y *and once continuously differentiable in* x.

Then,

(a) *we have*

$$\Psi_k = -\frac{1}{2} \| d_k \|_{\mathbb{C}_k}^2 \leq -\frac{\zeta}{2} \| d_k \|^2 \leq 0 \qquad (3.11)$$

(b) *if, furthermore,* $\max_{y \in \mathcal{Y}(x_k)} \langle \nabla_x f(x_k, y), \overline{d} \rangle < -\xi$, *we have*

$$-\frac{1}{2} \| \overline{d} \|_{\mathbb{C}_k}^2 \leq -\frac{1}{2} \| \mathbb{C}_k^{-1} \nabla \Phi_k \|_{\mathbb{C}_k}^2$$

and

$$\Psi_k \geq \max_{y \in \mathcal{Y}(x_k)} \langle \nabla_x f(x_k, y), d_k \rangle.$$

Remark 3.4 In (a), we establish the relation of Ψ_k with the directional derivative, and d_k as a descent direction. Using (3.9, b), the descent property can also be demonstrated for approximate solutions satisfying (3.9, a). In (b), the relation of $\overline{d}$ to $-\mathbb{C}^{-1} \nabla \Phi_k$ is shown and the directional derivative along d_k is related to Ψ_k.

Proof of Lemma 3.2. Inequality (3.11) is immediate for (3.3) and (3.4). For (b), we note that

$$\langle \nabla_x f(x_k, y_{k+1}), \bar{d} \rangle + \frac{1}{2} \left\| \bar{d} \right\|_{\mathbb{C}_k}^2 = -\frac{1}{2} \| \nabla_x f(x_k, y_{k+1}) \|_{\mathbb{C}_k^{-1}}^2$$

$$\leq -\frac{1}{2} \left\| \sum_{y \in \mathcal{Y}(x_k)} \lambda_{k+1}^y \nabla_x f(x_k, y) \right\|_{\mathbb{C}_k^{-1}}^2$$

$$= -\frac{1}{2} \langle \nabla \Phi_k, \mathbb{C}_k^{-1} \nabla \Phi_k \rangle.$$

Using (2.11) and Assumption 3.1, this yields

$$\Psi_k = \langle \nabla_x f(x_k, y_{k+1}), \bar{d} \rangle + \frac{1}{2} \left\| \bar{d} \right\|_{\mathbb{C}_k}^2 = -\frac{1}{2} \left\| \bar{d} \right\|_{\mathbb{C}_k}^2 \leq -\frac{1}{2} \| \nabla \Phi_k \|_{\mathbb{C}_k^{-1}}^2. \quad (3.12)$$

To show the second inequality in (b), we note that if $\bar{d}$ is a descent direction for the *max-function*, $d_k = \bar{d}$, and, with Assumption 3.1

$$\Phi_k(d_k) = \max_{y \in \mathcal{Y}} \left\{ f(x_k, y) + \langle \nabla_x f(x_k, y), d_k \rangle + \frac{1}{2} \| d_k \|_{\mathbb{C}_k}^2 - C[\Phi(x_k) - f(x_k, y)]^2 \right\}$$

$$\geq \Phi(x_k) + \max_{y \in \mathcal{Y}} \left\{ \langle \nabla_x f(x_k, y), d_k \rangle + \frac{1}{2} \| d_k \|_{\mathbb{C}_k}^2 - C[\Phi(x_k) - f(x_k, y)]^2 \right\}$$

$$\geq \Phi(x_k) + \max_{y \in \mathcal{Y}} \left\{ \langle \nabla_x f(x_k, y), d_k \rangle \right\}$$

$$\geq \Phi(x_k) + \max_{y \in \mathcal{Y}(x_k)} \left\{ \langle \nabla_x f(x_k, y), d_k \rangle \right\}$$

where the last inequality is due to $\mathcal{Y}(x_k) \subseteq \mathcal{Y}$. Thus, Ψ_k can be expressed as

$$\Psi_k = \Phi_k(d_k) - \Phi(x_k) \geq \max_{y \in \mathcal{Y}(x_k)} \langle \nabla_x f(x_k, y), d_k \rangle. \quad \square$$

Lemma 3.3 *Let*

(i) *Assumptions (1.1), (2.1), (2.2) and (3.1) hold; and,*

(ii) *$f(x, y)$ be continuous in y and once continuously differentiable in x.*

Then, $\{d_k\} \to 0$ if and only if $\{\Psi_k\} \to 0$.

Proof. The result follows from (3.11). $\quad \square$

4 BASIC CONVERGENCE RESULTS

In this section, we establish the monotonic decrease of the sequence of *max-function* values, $\{\Phi(x_k)\}$, generated by the algorithm.

Assumption 4.1 *For some* x_{k_0}, *the set* $\mathbb{F} \equiv \{x \in \Re^n | \Phi(x) \leq \Phi(x_{k_0})\}$ *is bounded.*

Remark An implication of $\mathbb{F}$ being *bounded* is that the Hessian of the $f(x, y)$, with respect to x, has an upper bound $\forall x, y$ and consequently, for bounded λ_{k+1}^y, the Hessian of the Lagrangian with respect to x, given by (2.17), has an upper bound:

$$\left\| \sum_{y \in \mathcal{Y}(x_k)} \nabla_x^2 f(x_k, y) \lambda_{k+1}^y \right\| \leq M, \quad \forall x_k \in \mathbb{F}$$

(see Theorem 5.1).

We note that (1.2) has a solution since $f(x, y)$ is continuous on the compact set $\mathcal{Y}$.

Lemma 4.1 *Let*

(i) Assumptions (1.1), (2.1), (2.2) and (3.1) hold; and,

(ii) $f(x, y)$ be continuous in y and once continuously differentiable in x.

Then, condition $d_k = 0$ is necessary and sufficient for point x_k to satisfy the nce of (1.1).

Proof. For $d_k = -\mathbb{C}_k^{-1} \sum_{y \in \mathcal{Y}(x_k)} \lambda_{k+1}^y \nabla_x f(x_k, y)$ as $\mathbb{C}_k$ is nonsingular, we note that $d_k = 0$ is equivalent to *nce* (2.19). For $d_k = \bar{d}$, consider subproblem (2.5) expressed as

$$\min_{d \in \Re^n} \Phi_k(d) = \min_{d \in \Re^n} \max_{y \in \mathcal{Y}_{k+1}} \{f(x_k, y) + \langle \nabla_x f(x_k, y), d \rangle$$

$$+ \frac{1}{2} \|d\|_{\mathbb{C}_k}^2 - C[\Phi(x_k) - f(x_k, y)]^2\} \tag{4.1}$$

where d_k is the minimizer of $\Phi_k(d)$. With $d = d_k$, (4.1) becomes

$$\Phi_k(d_k) = \max_{y \in \mathcal{Y}_{k+1}} \left\{ f(x_k, y) + \langle \nabla_x f(x_k, y), d_k \rangle + \frac{1}{2} \|d_k\|_{\mathbb{C}_k}^2 - C[\Phi(x_k) - f(x_k, y)]^2 \right\}.$$

$$\tag{4.2}$$

Since $\mathcal{Y}_{k+1} \subseteq \mathcal{Y}(x_k)$, the last term in (4.2) vanishes. Using (2.19), the *nce* of (4.1) is thus given by

$$\max_{y \in \mathcal{Y}_{k+1}} \{\langle \nabla_x f(x_k, y) + \mathbb{C}_k d_k, d \rangle\} \geq 0, \quad \forall d \in \Re^n. \tag{4.3}$$

For $d_k = 0$, by (2.2), we have $f_k(0, y) = f(x_k, y)$, $\mathcal{Y}_{k+1} = \mathcal{Y}(x_k)$ and (4.3) coincides with the *nce* of the original problem.

Conversely, let the *nce* be satisfied at point $x_k = x_*$ and let

$$f_*(x - x_*, y) = f(x_*, y) + \langle \nabla_x f(x_*, y), x - x_* \rangle$$

$$+ \frac{1}{2} \|x - x_*\|_{\mathfrak{C}}^2 - C[\Phi(x_*) - f(x_*, y)]^2$$

$$\Phi_*(x - x_*) = \max_{y \in \mathcal{Y}} \{f_*(x - x_*, y)\} \tag{4.4}$$

where $\mathfrak{C}$ is a symmetric positive definite matrix satisfying Assumption 2.2. At point $x_k = x_*$, subproblem in (2.5) has the form

$$\min_{x \in \mathfrak{R}^n} \Phi_*(x - x_*). \tag{4.5}$$

Let $\bar{x}$ be the solution of (4.5) and let $\overline{\mathcal{Y}} \equiv \{y \in \mathcal{Y} \mid y = \arg\max_\eta f_*(\bar{x} - x_*, \eta)\}$. The *nce* of (4.5), is given by

$$\max_{y \in \mathcal{Y}} \langle \nabla_x f(x_*, y) + \mathfrak{C}(\bar{x} - x_*), x - x_* \rangle \geq 0, \quad \forall x \in \mathfrak{R}^n. \tag{4.6}$$

The functions $f_k(d, y)$ and $f_*(x - x_*, y)$ are strictly convex in d, x, for all $y \in \mathcal{Y}$. Thus, d_k and $\bar{x}$ are unique solutions to problems (2.3) and (4.5), respectively. Since the *nce* of the original problem is satisfied at x_*, (4.6) also holds for $\bar{x} = x_*$ and, as $x_k = x_*$, we have $\overline{\mathcal{Y}} = \mathcal{Y}(x_*) = \mathcal{Y}(x_k)$ and hence $d_k = 0$. $\quad\square$

Corollary 4.1 *Let*

(i) Assumptions (1.1), (2.1), (2.2) and (3.1) hold; and,

(ii) $f(x, y)$ be continuous in y and once continuously differentiable in x.

Condition $\Psi_k = 0$ is necessary and sufficient for the nce of (1.1) to be satisfied at x_k.

Proof. The proof follows from Lemma 3.2 and Lemma 4.1. $\quad\square$

Lemma 4.2 *Let Assumptions (1.1), (1.2), (2.1), (2.2), (3.1) and (4.1) hold. Then, the stepsize computed in Step 3 of the algorithm is such that $\alpha_k \in (0, 1]$ and the sequence $\{x_k, y_k\}$ computed by the algorithm satisfies the stepsize strategy*

$$\Phi(x_{k+1}) - \Phi(x_k) \leq c\alpha_k \Psi_k \tag{4.7}$$

and generates a corresponding monotonically decreasing sequence $\{\Phi(x_k)\}$.

Proof. Consider the case $d_k = \bar{d}$. Since the maximum of a set of numbers is equal to the maximum of their convex combination, we have

$$\Phi(x_{k+1}) = \max_{\lambda^y} \left\{ \sum_{y \in \mathcal{Y}} \lambda^y f(x_{k+1}, y) \,\middle|\, \lambda^y \geq 0, \sum_{y \in \mathcal{Y}} \lambda^y = 1 \right\}$$

$$= \max_{\lambda^y} \left\{ \sum_{y \in \mathcal{Y}(x_{k+1})} \lambda^y f(x_{k+1}, y) \,\middle|\, \lambda^y \geq 0, \sum_{y \in \mathcal{Y}(x_{k+1})} \lambda^y = 1 \right\}.$$

Let λ_{k+2}^y be the solution of the above problem. The second order expansion of $f(x, \cdot)$ yields

$$\Phi(x_{k+1}) = \max_{\lambda^y} \left\{ \sum_{y \in \mathcal{Y}(x_{k+1})} \lambda^y \Big(f(x_k, y) + \alpha_k \langle \nabla_x f(x_k, y), d_k \rangle + \frac{1}{2} \alpha_k^2 \|d_k\|_{\mathfrak{C}_k}^2 \right.$$

$$\left. + \alpha_k^2 \int_0^1 (1 - t) \langle d_k, [\nabla_x^2 f(x_k + t\alpha_k d_k, y) - \mathfrak{C}_k] d_k \rangle \, dt \right)$$

$$\lambda^y \geq 0, \sum_{y \in \mathcal{Y}(x_{k+1})} \lambda^y = 1 \right\}$$

$$\leq \max_{\lambda^y} \left\{ \sum_{y \in \mathcal{Y}(x_{k+1})} \lambda^y \Big(f(x_k, y) + \alpha_k \langle \nabla_x f(x_k, y), d_k \rangle + \frac{1}{2} \alpha_k^2 \|d_k\|_{\mathfrak{C}_k}^2 \Big) \right.$$

$$\left. \lambda^y \geq 0; \sum_{y \in \mathcal{Y}(x_{k+1})} \lambda^y = 1 \right\} + \rho_k \alpha_k^2 \|d_k\|^2 \qquad (4.8)$$

$$\leq \max_{y \in \mathcal{Y}} \left\{ f(x_k, y) + \alpha_k \langle \nabla_x f(x_k, y), d_k \rangle + \frac{1}{2} \alpha_k^2 \|d_k\|_{\mathfrak{C}_k}^2 \right\} + \rho_k \alpha_k^2 \|d_k\|^2$$

where the last inequality is due to $\mathcal{Y}(x_k) \subseteq \mathcal{Y}$ and

$$\rho_k = \int_0^1 (1 - t) \left\| \sum_{y \in \mathcal{Y}(x_{k+1})} \lambda_{k+2}^y \nabla_x^2 f(x_k + t\alpha_k d_k, y) - \mathfrak{C}_k \right\| dt. \qquad (4.9)$$

We also have the relation

$$\max_{y \in \mathcal{Y}} - (\Phi(x_k) - f(x_k, y))^2 + \min_{y \in \mathcal{Y}} (\Phi(x_k) - f(x_k, y))^2 = 0$$

Hence

$$\Phi(x_{k+1}) \le \max_{y \in \mathcal{Y}} \left\{ f(x_k, y) + \alpha_k \langle \nabla_x f(x_k, y), d_k \rangle + \frac{1}{2} \alpha_k^2 \|d_k\|_{\mathbb{C}_k}^2 \right\} + \rho_k \alpha_k^2 \|d_k\|^2$$

$$+ \max_{y \in \mathcal{Y}} -C(\Phi(x_k) - f(x_k, y))^2 + \min_{y \in \mathcal{Y}} C(\Phi(x_k) - f(x_k, y))^2$$

$$\le \max_{y \in \mathcal{Y}} \left\{ f(x_k, y) + \alpha_k \langle \nabla_x f(x_k, y), d_k \rangle \right.$$

$$\left. + \frac{1}{2} \alpha_k^2 \|d_k\|_{\mathbb{C}_k}^2 - C(\Phi(x_k) - f(x_k, y))^2 \right\}$$

$$+ \min_{y \in \mathcal{Y}} C(\Phi(x_k) - f(x_k, y))^2 + \rho_k \alpha_k^2 \|d_k\|^2.$$

In view of Assumption 3.1, we have $y_{k+1} \in \mathcal{Y}(x_k)$ and

$$\Phi(x_{k+1}) = \Phi(x_k + \alpha_k d_k) \le \Phi_k(\alpha_k d_k) + \alpha_k^2 \rho_k \|d_k\|^2 \qquad (4.10)$$

Since $\Phi_k(d)$ is convex, and in view of (4.2), we can write the first term on the right of (4.10) as

$$\Phi_k(\alpha_k d_k) \le \alpha_k \Phi_k(d_k) + (1 - \alpha_k)\Phi_k(0)$$

$$\le \alpha_k \Phi_k(d_k) + (1 - \alpha_k)\Phi(x_k)$$

$$= \Phi(x_k) + \alpha_k [\Phi_k(d_k) - \Phi(x_k)]$$

$$= \Phi(x_k) + \alpha_k \Psi_k. \qquad (4.11)$$

Thus, (4.10) may be expressed as

$$\Phi(x_{k+1}) \le \Phi(x_k) + \alpha_k \Psi_k + \alpha_k^2 \rho_k \|d_k\|^2. \qquad (4.12a)$$

For $d_k = -\mathbb{C}_k^{-1} \sum_{y \in \mathcal{Y}(x_k)} \lambda_{k+1}^y \nabla_x f(x_k, y)$, we invoke Corollary 3.1 (b) to write (4.8) as

$$\Phi(x_{k+1}) \le \Phi(x_k) - \frac{1}{2} \alpha_k \|d_k\|_{\mathbb{C}_k}^2 + \alpha_k^2 \rho_k \|d_k\|^2 \qquad (4.12b)$$

where we have also used the equality $\Phi(x_k) = \max_{y \in \mathcal{Y}} f(x_k, y)$. Using Lemma 3.2 and (3.11), (4.12) can be written as

$$\Phi(x_k + \alpha_k d_k) - \Phi(x_k) \le \alpha_k \Psi_k \left[1 - \frac{2\alpha_k \rho_k}{\underline{\zeta}} \right]. \qquad (4.13)$$

The scalar $c \in (0, 1)$ in (4.7) determines α_k such that

$$0 < c \le 1 - \frac{2\alpha_k \rho_k}{\underline{\zeta}} \le 1. \qquad (4.14)$$

Since $\Psi_k \leq -\frac{\zeta}{2}\|d_k\|^2 \leq 0$, there exists a $\alpha_k \in (0, 1]$ to ensure (4.14) and (4.7). Suppose α^0 is the largest $\alpha \in (0, 1]$ satisfying (4.13). It follows that all $\alpha \leq \alpha^0$ also satisfy this condition. Thus, the strategy in Step 3 selects $\alpha_k \in [\sigma\alpha^0, \alpha^0]$. It follows that $\{\Phi(x_k)\}$ is a monotonically decreasing sequence. $\square$

5 GLOBAL CONVERGENCE AND LOCAL CONVERGENCE RATES

In this section, we discuss the global convergence of $\{x_k\}$ generated by the quasi-Newton algorithm, the convergence of the stepsize α_k to unity and the convergence rate of $\{x_k\}$.

Theorem 5.1 [Global Convergence] *Let*

(i) *Assumptions (1.1), (1.2), (2.1), (2.2), (3.1) and (4.1) hold; and,*

(ii) *the nce of (1.1) be satisfied at x_*.*

Then:

(a) *the algorithm either terminates with $\Psi_k = 0$ or it generates an infinite sequence $\{\Psi_k\}$ with $\lim_{k \to \infty} \Psi_k = 0$;*

(b) *the algorithm either terminates with $d_k = 0$ or it generates an infinite sequence $\{d_k\}$ with $\lim_{k \to \infty} d_k = 0$; and,*

(c) *the algorithm either terminates at x_* or it generates an infinite sequence $\{x_k\}$ in which there exists a subsequence $\{x_k\}$ with $k \in K \subset \{1, 2, ...\}$ such that $\{d_k\} \to 0$ and every accumulation point x_* of the infinite sequence $\{x_k\}$ satisfies the nce of (1.1).*

Proof. To show (a), note that, by (2.11), Lemma 3.3 and Lemma 4.1, the algorithm terminates if $\Psi_k = 0$. Alternatively, given $c \in (0, 1)$, by (4.13) the choice

$$\alpha^0 = \min\left\{1, \frac{(1-c)\underline{\zeta}}{2\rho_k}\right\} \qquad (5.1)$$

always satisfies the stepsize strategy $\Phi(x_k + \alpha_k d_k) - \Phi(x_k) \leq c\alpha_k\Psi_k$. Clearly, $\alpha_k = (\sigma)^i$ chosen in Step 3 is in the range $\alpha_k \in [\sigma\alpha^0, \alpha^0]$ and thereby also satisfies this strategy.

Since $f(x, y)$ is twice continuously differentiable with respect to x and $\mathbb{F}$ is compact, there is a scalar $\overline{M} < \infty$ such that $\|\nabla_x^2 f(x, y)\| \leq \overline{M}$. As the Hessian in (4.9) is a convex combination of $\nabla_x^2 f(x, y)$, we have $\rho_k \leq \overline{M} < \infty$.

As $\underline{\zeta} > 0$, we have established that there is an $\overline{\alpha} > 0$ such that the stepsize $\alpha_k \geq \overline{\alpha} > 0$, $\forall k$. The boundedness of $\Phi(x)$ on $\mathbb{F}$ and $\Psi_k \leq 0$ imply

$$0 \le c \sum_k \alpha_k \mid \Psi_{k'} \mid \le \sum_k [\Phi(x_k) - \Phi(x_{k+1})] < \infty \qquad (5.2)$$

which yields (a). Results (b) and (c) follow from Lemma 3.3 and Lemma 4.1.□

Remark 5.1   When $\mid y_{k+1} \mid > 1$, there is more than one minimum-norm subgradient $\nabla_x f(x_k, y)$ with corresponding Hessian $\nabla_x^2 f(x_k, y)$. It is then possible for each of these subgradient and Hessian pairs to generate a descent direction leading to a local minimax solution. For $f(x, y)$ convex in x, this would imply the existence of more than one direction leading to a global minimax solution. For $f(x, y)$ nonconvex in x, it would indicate the existence of possibly more than one direction leading to possibly more than one local minimax solution.

We note that Theorems 2 and 3 apply to a local solution within a neighborhood such that Assumption 1.3 (the local convexity of $f(x, y)$ in x) holds.

Theorem 5.2 (Unit Stepsize Achievement)   *Let*

(i)   *Assumptions (1.1)–(1.3), (2.1), (2.2), (3.1) and (4.1) hold;*

(ii) $\mathfrak{C}_*$ *be the Hessian (2.17) at* x_**; and,*

(iii)   *the sequence* $\{x_k\}$ *converges to a local solution,* x_**, satisfying the nce of (1.1).*

Then there exist a number $\tau > 0$*, and an integer* $K_0 > 0$ *such that if*

$$\|\mathfrak{C}_k - \mathfrak{C}_*\| \le \tau, \quad k \ge K_0, \qquad (5.3)$$

the stepsize strategy in Step 3 is satisfied for $\alpha_k = 1, k \ge K_0$*.*

Proof. Writing (4.9) as

$$\rho_k = \int_0^1 (1 - t) \left\| \sum_{y \in y(x_{k+1})} \lambda_{k+2}^y \nabla_x^2 f(x_k + t\alpha_k d_k, y) - \mathfrak{C}_* + \mathfrak{C}_* - \mathfrak{C}_k \right\| dt \quad (5.4)$$

(4.12) becomes

$$\Phi(x_k + \alpha_k d_k) \le \Phi(x_k) + \alpha_k \Psi_k + \alpha_k^2 \rho_k^* \|d_k\|^2 + \frac{1}{2} \alpha_k^2 \langle d_k, \{\mathfrak{C}_* - \mathfrak{C}_k\} d_k \rangle \quad (5.5)$$

where

$$\rho_k^* = \int_0^1 (1 - t) \left\| \sum_{y \in y(x_{k+1})} \lambda_{k+2}^y \nabla_x^2 f(x_k + t\alpha_k d_k, y) - \mathfrak{C}_* \right\| dt. \qquad (5.6)$$

From Lemma 3.2, we have $\Psi_k \le -\frac{1}{2} \underline{\zeta} \|d_k\|^2 \le 0$. Thus, (5.5) can be written as

$$\Phi(x_k + \alpha_k d_k) - \Phi(x_k) \le \alpha_k \Psi_k \left[1 - \frac{\alpha_k}{\underline{\zeta}} [2\rho_k^* + \|\mathbb{C}_* - \mathbb{C}_k\|] \right] \quad (5.7)$$

The scalar $c \in (0, 1)$ in Step 3 requires α_k to satisfy

$$0 < c \le 1 - \frac{\alpha_k}{\underline{\zeta}} [2\rho_k^* + \|\mathbb{C}_* - \mathbb{C}_k\|] \le 1. \quad (5.8)$$

From Lemma 4.2, (4.13)–(4.14), there exists an $\alpha_k \in (0, 1]$ satisfying (5.7) and hence (4.7). If τ in (5.3) is such that

$$\frac{1}{\underline{\zeta}} [2\rho_k^* + \tau] \le 1 - c \quad (5.9)$$

(in view of $\{x_k\} \to x_*, \{\rho_k^*\} \to 0$ this defines the number τ) then (5.8) holds with $\alpha_k = 1$, and therefore, because $\Psi_k \le 0$, the stepsize computed in (5.7) and Step 3 is $\alpha_k = 1$. $\square$

Lemma 5.1 *Let $\{x_k\} \to x_*$. Then, $\{x_k\}$ is Q-superlinearly convergent, that is,*

$$\lim_{k \to \infty} \frac{\| x_* - x_{k+1} \|}{\| x_* - x_k \|} = 0 \quad (5.10)$$

if

$$\| d_k \| \le \pi_k \| d_{k-1} \|, \lim_{k \to \infty} \pi_k = 0. \quad (5.11)$$

Proof. Suppose (5.11) holds. We have

$$\| x_* - x_k \| \le \lim_{t \to \infty} \sum_{i=k}^{t-1} \| x_{i+1} - x_i \| \le \pi_k \| d_{k-1} \| (1 + w + w^2 + w^3 + \cdots)$$

$$\le \frac{\pi_k}{1 - w} \{ \| x_k - x_* \| + \| x_* - x_{k-1} \| \}$$

for some $w \in [0, 1)$. As $\{\pi_k\} \to 0$, w is chosen such that $\pi_k + w < 1$, $\forall k \ge K_0$. K_0 is an integer and is such that $\pi_k < 1, \forall k \ge K_0$. Rearranging the above expression,

$$\| x_* - x_k \| \le \frac{\pi_k}{1 - w - \pi_k} \| x_* - x_{k-1} \|$$

which establishes the Q-superlinear convergence of $\{x_k\}$. $\square$

We give a proof of the Dennis and Moré (1974) characterization of the local Q-superlinear convergence rate of the algorithm.

Theorem 5.3 (Q-Superlinear Convergence) *Let*

(i) *Assumptions of Theorem 5.2 be satisfied, $\mathbb{C}_*$ be the Hessian (2.17) at x_*, and*

(ii) *for some $x_0 \in \mathcal{D}$, the sequence $\{x_k\}$ generated by $x_{k+1} = x_k + d_k$ remains in $\mathcal{D}$, and $\lim_{k \to \infty} x_k = x_*$.*

Then, $\{x_k\}$ satisfies (5.11) and thence converges Q-superlinearly to x_*, in some norm $\| \cdot \|$ if

$$\lim_{k \to \infty} \frac{\|(\mathbb{C}_k - \mathbb{C}_*)(x_{k+1} - x_k)\|}{\|x_{k+1} - x_k\|} = 0. \tag{5.12}$$

Proof. Consider the first order expansion

$$\langle d, \nabla_x f(x_k, y) \rangle = \Big\langle d, \Big[\nabla_x f(x_{k-1}, y) + [\mathbb{C}_{k-1} + [\mathbb{C}_* - \mathbb{C}_{k-1}]$$

$$+ \int_0^1 [\nabla_x^2 f(x_{k-1} + td_{k-1}, y) - \mathbb{C}_*] \, dt] d_{k-1} \Big] \Big\rangle.$$

Evaluating the maximum of both sides with respect to $y \in \mathcal{Y}(x_k)$ yields

$$\max_{y \in \mathcal{Y}(x_k)} \langle d, \nabla_x f(x_k, y) \rangle$$

$$= \max_{y \in \mathcal{Y}} \Big\{ \langle d, \nabla_x f(x_k, y) \rangle - C(\Phi(x_k) - f(x_k, y))^2 \Big\}$$

$$= \max_{y \in \mathcal{Y}} \Big\{ \langle d, \nabla_x f(x_{k-1}, y) + \mathbb{C}_{k-1} d_{k-1} \rangle - C(\Phi(x_k) - f(x_k, y))^2$$

$$+ \Big\langle d, \Big[[\mathbb{C}_* - \mathbb{C}_{k-1}] + \int_0^1 \big[\nabla_x^2 f(x_{k-1} + td_{k-1}, y) - \mathbb{C}_* \big] dt \Big] d_{k-1} \Big\rangle \Big\}$$

$$\geq \max_{y \in \mathcal{Y}} \{ \langle d, [\nabla_x f(x_{k-1}, y) + \mathbb{C}_{k-1} d_{k-1}] \rangle \}$$

$$+ \max_{y \in \mathcal{Y}} \Big\{ \Big\langle d, \Big[[\mathbb{C}_* - \mathbb{C}_{k-1}] + \int_0^1 \big[\nabla_x^2 f(x_{k-1} + td_{k-1}, y) - \mathbb{C}_* \big] dt \Big] d_{k-1} \Big\rangle$$

$$- C(\Phi(x_k) - f(x_k, y))^2 \Big\}. \tag{5.13}$$

For $d_k = -\mathbb{C}_k^{-1} \sum_{y \in \mathcal{Y}(x_k)} \lambda_{k+1}^y \nabla_x f(x_k, y)$, we have the inequality

$$\max_{y \in \mathcal{Y}} \{\langle d, [\nabla_x f(x_{k-1}, y) + \mathfrak{C}_{k-1} d_{k-1}] \rangle\}$$

$$\geq \max_{y \in \mathcal{Y}(x_{k-1})} \{\langle d, [\nabla_x f(x_{k-1}, y) + \mathfrak{C}_{k-1} d_{k-1}] \rangle\}$$

$$\geq \left\langle d, \left[\sum_{y \in \mathcal{Y}(x_k)} \lambda_{k+1}^y \nabla_x f(x_k, y) + \mathfrak{C}_{k-1} d_{k-1} \right] \right\rangle$$

$$= 0.$$

For $d_k = \bar{d}$, in view of the *nce* (4.3), we have

$$\max_{y \in \mathcal{Y}} \{\langle d, [\nabla_x f(x_{k-1}, y) + \mathfrak{C}_{k-1} d_{k-1}] \rangle\}$$

$$\geq \max_{y \in \mathcal{Y}_{k+1}} \{\langle d, [\nabla_x f(x_{k-1}, y) + \mathfrak{C}_{k-1} d_{k-1}] \rangle\}$$

$$\geq 0.$$

Hence, the following inequality applies for d_k in general

$$\max_{y \in \mathcal{Y}(x_k)} \langle d, \nabla_x f(x_k, y) \rangle$$

$$\geq \max_{y \in \mathcal{Y}} \left\{ \left\langle d, \left[[\mathfrak{C}_* - \mathfrak{C}_{k-1}] + \int_0^1 [\nabla_x^2 f(x_{k-1} + t d_{k-1}, y) - \mathfrak{C}_*] dt \right] d_{k-1} \right\rangle \right.$$

$$\left. - C(\Phi(x_k) - f(x_k, y))^2 \right\}$$

$$= \max_{y \in \mathcal{Y}(x_k)} \left\{ \left\langle d, \left[[\mathfrak{C}_* - \mathfrak{C}_{k-1}] + \int_0^1 [\nabla_x^2 f(x_{k-1} + t d_{k-1}, y) - \mathfrak{C}_*] dt \right] d_{k-1} \right\rangle \right\}$$

$$= \max_{\lambda_y} \left\{ \left\langle d, \left[[\mathfrak{C}_* - \mathfrak{C}_{k-1}] \right. \right. \right.$$

$$\left. \left. \left. + \int_0^1 \sum_{y \in \mathcal{Y}(x_k)} [\lambda^y \nabla_x^2 f(x_{k-1} + t d_{k-1}, y) - \mathfrak{C}_*] dt \right] d_{k-1} \right\rangle \right| \lambda^y \geq 0;$$

$$\left. \sum_{y \in \mathcal{Y}(x_{k+1})} \lambda^y = 1 \right\}$$

$$\geq \{\langle d, [[\mathfrak{C}_* - \mathfrak{C}_{k-1}] + \Lambda_{k-1}] d_{k-1} \rangle\}$$

where

$$\Lambda_{k-1} = \int_0^1 \left[\left[\sum_{y \in \mathcal{Y}(x_k)} \lambda_{k+1}^y \nabla_x^2 f(x_{k-1} + td_{k-1}, y) \right] - \mathfrak{C}_* \right] dt$$

and we have used the fact that λ_{k+1}^y is not necessarily the maximizer of the last expression. Choosing $d = d_k$, using Lemma 3.2, Corollary 3.1, Assumption 2.2 and (2.11) yields

$$\underline{\zeta} \|d_k\|^2 \leq \langle d_k, \mathfrak{C}_k d_k \rangle \leq [\|\Lambda_{k-1}\| \|d_{k-1}\| + \|[\mathfrak{C}_* - \mathfrak{C}_{k-1}]d_{k-1}\|] \|d_k\|$$

and therefore

$$\|d_k\| \leq \frac{1}{\underline{\zeta}} \left[\|\Lambda_{k-1}\| + \frac{\|[\mathfrak{C}_* - \mathfrak{C}_{k-1}]d_{k-1}\|}{\|d_{k-1}\|} \right] \|d_{k-1}\|.$$

As $\{x_k\} \to x_*, \{\|\Lambda_k\|\} \to 0$. Using Lemma 5.1, we have the desired result. $\square$

References

Apostol, T.M. (1981). *Mathematical Analysis*, Addison Wesley, Reading, MA.

Broyden, C.G. (1969). "A New Method for Solving Nonlinear Simultaneous Equations", *Computer Journal*, 12, 95–100.

Broyden, C.G. (1970). "The Convergence of a Class of Double-Rank Minimisation Algorithms 2. The New Algorithm", *Journal of the Institute of Mathematics and its Applications*, 6, 222–231.

Byrd, R.H., R.A. Tapia and Y. Zhang (1992). "An SQP Augmented Lagrangian BFGS Algorithm for Constrained Optimization", *SIAM Journal on Optimization*, 2, 210–241.

Chaney, R.W. (1982) "A Method of Centers Algorithm for Certain Minimax Problems", *Mathematical Programming*, 22, 206–226.

Coleman, T.F (1978). "A Note on New Algorithms for Constrained Minimax Optimization", *Mathematical Programming*, 15, 239–242.

Conn, A.R. (1979). "An Efficient Second Order Method to Solve the Constrained Minmax Problem", Report, Department of Combinatorics and Optimization, University of Waterloo.

Conn, A.R. and Y. Li (1992). "A Structure Exploiting Algorithm for Nonlinear Minimax Problems", *SIAM Journal on Optimization*, 2, 242–263.

Demyanov, V.F. and V.N. Malozemov (1974). *Introduction to Minimax*, John Wiley, New York.

Dennis Jr., J.E. and J.J. Moré (1974). "A Characterization of Superlinear Convergence and its Application to Quasi-Newton Methods", *Mathematics of Computation*, 28, 549–560.

Fiacco, A.V. and G.P. McCormick (1968). *Nonlinear Programming: Sequential Unconstrained Minimization Techniques*, John Wiley, New York.

Fletcher, R. (1970). "A New Approach to Variable Metric Algorithms", *Computer Journal*, 13, 317–322.

Floudas, C.A. and P.M. Pardalos (1992). *Recent Advances in Global Optimization*, Princeton Series in Computer Science, Princeton University Press, Princeton, NJ.

Floudas, C.A. and P.M. Padralos (1995). *State of the Art in Global Optimization: Computational Methods and Applications, Nonconvex Optimization and Applications*, Kluwer, Dordrecht.

Goldfarb D. (1970). "A Family of Variable Metric Algorithms Derived by Variational Means", *Mathematics of Computation*, 24, 23–26.

Kiwiel, K.C. (1987). "A Direct Method of Linearization for Continuous Minimax Problems", *Journal of Optimization Theory and Applications*, 55, 271–287.

Klessig, R. and E. Polak (1973). "A Method of Feasible Directions Using Function Approximations with Applications to Minimax Problems", *Journal of Mathematical Analysis and Applications*, 41, 583–602.

Luenberger, D. (1984). *Linear and Nonlinear Programming*, Addison-Wesley, Reading, MA.

Makela, M. and P. Neittaanmaki (1992). *Nonsmooth Optimization: Analysis and Algorithms with Applications to Optimal Control*, World Scientific, Singapore.

Panin, V.M. (1981). "Linearization Method for Continuous Min-Max Problems", *Kibernetika*, 2, 75–78.

Pardalos, P.M. and J.B. Rosen (1987). "Constrained Global Optimization: Algorithms and Applications", *Lecture Notes in Computer Science No: 268*, Springer-Verlag, Berlin.

Polak, E. (1989). "Basics of Minimax Algorithms", in: F.H. Clarke, V.F. Demyanov and F. Giannessi (editors), *Nonsmooth Optimization and Related Topics*, Plenum Press, New York, 343–369.

Rockafeller, R.T. (1972). *Convex Analysis*, Princeton University Press, Princeton, NJ.

Rustem, B. (1992). "A Constrained Min-Max Algorithm for Rival Models of the Same Economic System", *Mathematical Programming*, 53, 279–295.

Shanno, D.F. (1970). "Conditioning of Quasi-Newton Methods for Function Minimization", *Mathematics of Computation*, 24, 647–654.

Wierzbicki, A.P. (1982). "Lagrangian Functions and Nondifferentiable Optimization", in: E.A. Nurminski (editor), *Progress in Nondifferentiable Optimization*, Publication CP-82-58, IIASA, 2361 Laxenburg, Austria.

APPENDIX A: IMPLEMENTATION ISSUES

In Proposition 1, the condition that ensures the equivalence of $\bar{d}$ and $-\mathfrak{C}^{-1}\nabla\Phi_k$ is discussed. In some problems, this condition may be naturally satisfied in which case the algorithm simplifies considerably. Assumption A.1 below addresses such situations.

Assumption A.1 *Let $\mathcal{D} \subset \mathfrak{R}^n$ be an open convex neighborhood of x_*. For $x_k \in \mathcal{D}, y \in \mathcal{Y}(x_k)$, the minimization in (2.13) is solved by $\lambda^y_{k+1} = 1 \in \mathfrak{R}^1$ for a unique y_{k+1} given by (2.14).*

Assumption A.1 can be seen as a *straightforward extension of the unique maximizer case to multiple maximizers*. We note that the maximizer is unique for $f(\cdot, y)$ concave in y. However, the requirement in Assumption A.1 is that the maximizer with minimum norm subgradient to be unique. This ensures a simplified gradient representation and consistent Hessian approximation when there is more than one maximizer.

Justification of Assumption A.1   For y_{k+1} given by (2.14), Corollary 3.1(a) and 1(b) coincide such that

$$y_{k+1} = \arg\max_{y \in \mathcal{Y}(x_k)} \langle \nabla_x f(x_k, y), d_k \rangle$$

and

$$\max_{y \in \mathcal{Y}(x_k)} -\langle \nabla_x f(x_k, y), d_k \rangle = -\|d_k\|_{\mathbb{C}_k}^2.$$

The expression on the left is the directional derivative along d_k (see also Lemma 3.1). In view of (2.4), $\nabla_x f(x_k, y_{k+1})$ corresponds to the subgradient in $\partial \Phi(x_k)$ that yields the maximal directional derivative for d_k. The same $\nabla_x f(x_k, y_{k+1})$ also corresponds to the minimum-norm subgradient. On the other hand, the failure of Assumption A.1 is significant when, at the solution, $\nabla_x f(x_k, y_{k+1})$ may not necessarily correspond to the minimum-norm subgradient $0 \in \partial \Phi(x_*)$. Hence, a linear combination of at most $n + 1$ elements of $\mathcal{Y}(x_*)$ may be required to find $\in \partial \Phi(x_*)$.

Approximation implied by Assumption A.1   (i) If $\lambda_{k+1}^y = 1$ in (2.12)–(2.13), then strict equality holds for

$$\left\| \sum_{y \in \mathcal{Y}(x_k)} \lambda^y \nabla_x f(x_k, y) \right\|_{\mathbb{C}_k^{-1}}^2 \leq \sum_{y \in \mathcal{Y}(x_k)} \lambda^y \| \nabla_x f(x_k, y) \|_{\mathbb{C}_k^{-1}}^2.$$

(ii) For

$$\| \nabla_x f(x_k, y_{k+1}) \|_{\mathbb{C}_k^{-1}}^2 = \min_{y \in \mathcal{Y}(x_k)} \| \nabla_x f(x_k, y) \|_{\mathbb{C}_k^{-1}}^2$$

the problem

$$\min_{\lambda^y} \left\{ \sum_{y \in \mathcal{Y}(x_k)} \lambda^y \| \nabla_x f(x_k, y) \|_{\mathbb{C}_k^{-1}}^2 \;\middle|\; \lambda^y \geq 0, \sum_{y \in \mathcal{Y}(x_k)} \lambda^y = 1 \right\}$$

is solved by $\lambda_{k+1}^y = 1$. Thus, y_{k+1} provides an approximate solution to the minimization in (2.13) even when Assumption A.1 is not satisfied.

(iii) When y_{k+1} in (2.14) is nonunique, as the subgradient norm is the same for all $_{k+1}$, the inequality

$$\left\|\sum_{y\in \mathcal{Y}(x_k)} \lambda^y \nabla_x f(x_k,y)\right\|_{\mathfrak{C}_k^{-1}}^2 \leq \left[\sum_{y\in \mathcal{Y}(x_k)} \lambda^y\right]\|\nabla_x f(x_k,y_{k+1})\|_{\mathfrak{C}_k^{-1}}^2 = \|\nabla_x f(x_k,y_{k+1})\|_{\mathfrak{C}_k^{-1}}^2.$$

holds for all y_{k+1}.

Hessian approximation in problems with unique and multiple maximizers

In general, the *max-function* is characterized by regions of smoothness and the boundaries between these smooth regions are characterized by kinks. When x_k is in a smooth region, $\partial\Phi(x_k)$ has only one subgradient, with a corresponding Hessian. When x_k coincides with a kink, the Hessian corresponding to the subgradient of minimum-norm, $\nabla_x f(x_k,y_{k+1})$, is used to define the curvature of the *max-function* at that kink. The Hessian update may then be corrupted by the sudden change of Hessian at the kink and the algorithm may exhibit linear convergence, rather than the superlinear rate discussed in Theorem 5.3. For this reason, the Hessian approximation is computed using (3.8). If, however, $\bar{d}$ is used in (2.11) consistently throughout the algorithm (such as cases in which Assumption A.1 is satisfied) we can replace γ_k in (3.8a) with the simplified version

$$\gamma_k = \nabla_x f(x_{k+1},y_{k+2}) - \nabla_x f(x_k,y_{k+1})$$

where y_{k+1} is given by (2.6). Thus, with Assumption A.1, the BFGS formula uses the fact that y_{k+1} in Step 2 is such that $y_{k+1} \in \mathcal{Y}_{k+1} \subseteq \mathcal{Y}(x_k)$ and $\nabla_x f(x_k,y_{k+1}) \in \partial\Phi(x_k)$ (see Lemma 3.1). Thus, $\nabla_x f(x_k,y_{k+1})$ is used in the quasi-Newton approximation as a subgradient at x_k. Hence, the subgradient corresponding to the maximizer solving (2.6) is used to represent the subgradient of the *max-function*. We mention this explicitly because numerical experience in Chapter 5 indicates that the choice in (2.11) is almost invariably $\bar{d}$ except in the very last two or three iterations and restricting the choice exclusively to $\bar{d}$ does not alter the performance of the algorithm for low levels of termination accuracy at $\xi = 10^{-8}$.

The idealized instance, therefore, for the use of (3.8a) is the general case of multiple maximizers, such as $\mathcal{Y}$ defined by upper and lower bounds on y, with multiple maximizers arising at the vertices (although even in this case, when $\bar{d}$ is used in (2.11), the simplified γ_k is still justified). The corresponding case for the simplified γ_k is $f(\cdot,y)$, concave in y.

Another source of poor convergence is ill-conditioning in $\mathfrak{C}_k$. As discussed in Byrd et al. (1992), the condition number of $\mathfrak{C}_k$ may become large if $\langle\gamma_k,\delta_k\rangle$ is small in relation to $\|\gamma_k\|\|\delta_k\|$. Even when we have unique maximizers at each iteration, the subgradient is a function of y_{k+1} and a sequence $\{y_k\}$ that successively produces an ill-conditioned $\mathfrak{C}_k$ will result in linear convergence.

As an illustration, consider $f(x, y)$ maximized alternately by y_1 and y_2, where $\nabla_x f(x, y_1)$ is a steep subgradient and $\nabla_x f(x, y_2)$ is a flat subgradient. These subgradients may result in an ill-conditioned Hessian, similar to the situation in nonlinear programming where successive iterates alternate between steep and flat gradients. Superlinear convergence may be achieved when there is a stable sequence $\{y_k\}$ that does not produce an ill-conditioned Hessian. For example, this may arise when $\mathcal{Y}$ is given by upper and lower bounds on y, with the maximizer due to one vertex of the hypercube dominating the other vertices. It may also arise when there is a unique maximizer for each x, with $\nabla\Phi(x_k) = \nabla_x f(x_k, y_{k+1})$, $\mathfrak{C}[x_k] = \nabla_x^2 f(x_k, y_{k+1})$ and $\nabla\Phi(x_*) = \nabla_x f(x_*, y_*) = 0$. Provided $\{y_k\}$ does not generate an ill-conditioned Hessian, then the Hessian approximation may increasingly approach the true Hessian and superlinear convergence can be achieved as discussed in Theorem 5.3. Nevertheless, an ill-conditioned Hessian may also be generated in the minimization of $\Phi(x)$ for the same reasons as in nonlinear programming problems.

APPENDIX B: MOTIVATION FOR THE SEARCH DIRECTION $\bar{d}$

If the difference $\Phi(x_k + \bar{d}) - \Phi(x_k)$ can be approximated by $\Phi_k(\bar{d}) - \Phi(x_k)$, then a descent direction for $\Phi(x)$, at x_k, can be generated by solving (2.5). The quasi-Newton direction $\bar{d}$ is the minimizer of $\Phi_k(d)$, while y_{k+1} is determined by (2.6).

To motivate the penalty term in (2.6), consider the nonlinear program $\min_x \{\mathcal{F}(x)\}$ which can be solved by successively minimizing a quadratic approximation to $\mathcal{F}(x)$:

$$\min_d \{\mathcal{F}_k(d)\}, \quad \mathcal{F}_k(d) = \mathcal{F}(x_k) + \langle \nabla_x \mathcal{F}(x_k), d \rangle + \frac{1}{2} \|d\|_{\mathfrak{C}_k}^2. \quad \text{(B1)}$$

For $d = 0$, (B1) implies that $\mathcal{F}_k(0) = \mathcal{F}(x_k)$. In other words, the graph of $\mathcal{F}_k(d)$ touches the graph of $\mathcal{F}(x)$ at x_k. The same consideration motivates the search direction of the minimax algorithm: (1.1) is solved by considering the sequential quadratic approximation (2.2) and minimizing (2.3). We note that in Makela and Neittaanmaki (1992, Theorem 5.2.8, p. 81), (1.1) is solved by considering (2.1.5) and which results in a descent direction $-\bar{g}$ where

$$\bar{g} = \arg \min_{g \in \partial\Phi(x_k)} \{\| g \|\}.$$

We note from Kiwiel (1987), and also Polak (1989), that the auxiliary algorithm to solve (2.1.5) aims to find the descent direction of minimum norm and at the same time aims to minimize the difference $\Phi(x_k) - f(x_k, y_{k+1})$. Extending Kiwiel (1987, p. 274), we consider a quadratic approximation to the change in objective function value $\Phi(x_k + \bar{d}) - \Phi(x_k)$, given by $\Phi_k(\bar{d}) - \Phi(x_k)$. When $\bar{d} = 0$, (2.3) is reduced to

$$\Phi_k(0) = \max_{y \in \mathcal{Y}} \left\{ f(x_k, y) - C[\Phi(x_k) - f(x_k, y)]^2 \right\} \leq \Phi(x_k).$$

In other words, the graph of $\Phi_k(d)$ touches the graph of $\Phi(x)$ at x_k if

$$\Phi(x_k) - f(x_k, y) = 0.$$

When $\bar{d} \neq 0$, a maximizer $y_{k+1} \in \mathcal{Y}$ is determined such that $f(x_k, y_{k+1})$ is as close as possible to the *max-function* $\Phi(x_k)$. This ensures that the quadratic approximation is close to the *max-function*: hence, (3.9, a) is satisfied for

$$\Phi_k(\bar{d}) = f(x_k, y_{k+1}) + \langle \nabla_x f(x_k, y_{k+1}), \bar{d} \rangle + \frac{1}{2} \| \bar{d} \|^2_{\mathfrak{C}_k} - C[\Phi(x_k) - f(x_k, y_{k+1})]^2.$$

To compute y_{k+1}, using the quasi-Newton direction in $d(y) = -\mathfrak{C}_k^{-1} \nabla_x f(x_k, y)$ yields (2.6). While maximizing the quadratic approximation, given $\bar{d}$, (2.6) simultaneously aims to determine y to minimize the difference $\Phi(x_k) - f(x_k, y)$. This ensures that the quadratic approximation is close to the *max-function* $\Phi(x_k)$.

For a given x_k, there may be a set of maximizers for (1.1), defined as $\mathcal{Y}(x_k)$. We solve the maximization problem at x_k and denote its value by $\Phi(x_k)$. With Assumption 3.1, for C sufficiently large, $0 \leq C < \infty$, the solution y_{k+1} ensures (3.10). If (3.7) is satisfied, it is shown in Lemma 3.2 that for y_{k+1} solving (2.6), d_k is a descent direction since (3.4) reduces to

$$\Psi_k = -\frac{1}{2} \| \nabla_x f(x_k, y_{k+1}) \|^2_{\mathfrak{C}_k^{-1}} = -\frac{1}{2} \| \bar{d} \|^2_{\mathfrak{C}_k}.$$

COMMENTS AND NOTES

CN 1: $\nabla \Phi(x_k)$ for a unique maximizer

Consider the case with a unique maximizer: the set $\mathcal{Y}(x)$ is thus a singleton,

$$\Phi(x) = f(x, \mathcal{Y}(x)).$$

$y_{k+1} = \mathcal{Y}(x_k)$ solves (3.1) and (2.6). Let

$$\mathcal{Y} \equiv \{ y \in \Re^m | \mathcal{H}(y) \leq 0 \}$$

and, in addition to Assumptions 1.1 and 1.2, assume further that $\mathcal{H}(y)$, $f(\cdot, y) \in \mathbb{C}^1$. The Lagrangian of (3.1) is given by

$$L^{\mathcal{Y}(x)}(y, \theta) = f(x, y) + \langle \mathcal{H}(y), \theta \rangle$$

where θ is the vector of multipliers. At the solution of (3.1), y, $\theta^{\mathcal{Y}(x)}$, we have $y = \mathcal{Y}(x)$ and the first order optimality conditions

$$\langle \mathcal{H}(\mathcal{Y}(x)), \theta^{\mathcal{Y}(x)} \rangle = 0$$

$$\nabla_y L^{\mathcal{Y}(x)}\Big(\mathcal{Y}(x), \theta^{\mathcal{Y}(x)}\Big) = 0.$$

We thus have

$$f(x, \mathcal{Y}(x)) = L^{\mathcal{Y}(x)}\Big(\mathcal{Y}(x), \theta^{\mathcal{Y}(x)}\Big)$$

$$\Phi(x_k) = f(x_k, y_{k+1}) = L^{\mathcal{Y}(x_k)}\Big(y_{k+1}, \theta^{\mathcal{Y}(x_k)}\Big).$$

The gradient of the *max-function*, at x_k, y_{k+1}, is obtained by

$$\nabla\Phi(x_k) = \frac{\partial f(x_k, y_{k+1})}{\partial x} + \left[\frac{\partial y}{\partial x}\right]^{\mathrm{T}}_{x=x_k\, y=y_{k+1}} \frac{\partial L^{\mathcal{Y}(x_k)}(y_{k+1}, \theta^{\mathcal{Y}(x_k)})}{\partial y}$$

$$= \nabla_x f(x_k, y_{k+1}) + \left[\frac{\partial y}{\partial x}\right]^{\mathrm{T}}_{x=x_k\, y=y_{k+1}} \nabla_y L^{\mathcal{Y}(x_k)}\Big(\mathcal{Y}(x_k), \theta^{\mathcal{Y}(x_k)}\Big)$$

$$= \nabla_x f(x_k, y_{k+1}).$$

Thus, in the case of a unique maximizer, we have $\nabla\Phi(x_k) = \nabla_x f(x_k, y_{k+1})$ (Chaney, 1982, Proposition 1.3, p. 203).

Numerical experiments with continuous minimax
algorithms

In this chapter, we consider the implementation of continuous minimax algorithms. The first is the gradient-based algorithm due to Kiwiel (1987),
discussed in Section 2.4, and the second is the quasi-Newton algorithm in
Chapter 4, and a simplified variant derived from this algorithm.

An important complication of continuous minimax is the presence of multiple maximizers. The contribution of these maximizers to the definition of the
direction of search of an algorithm is often difficult to evaluate. We consider
several strategies suggested by these algorithms for this purpose. The utility of
adopting a simplified direction is evaluated. In addition, the general performance of the algorithms are considered in terms of termination criteria, accuracy of the solution and convergence rates.

We observe a superior performance of the full quasi-Newton algorithm and
promising results obtained by adopting the variant with a simplified search
direction, not requiring convex combination of subgradients.

1 INTRODUCTION

As in Chapters 3 and 4, we consider the continuous minimax problem,
constrained in y, but unconstrained in x

$$\min_{x \in \Re^n} \max_{y \in \mathcal{Y}} f(x, y)$$

and express it as

$$\min_{x \in \Re^n} \Phi(x)$$

where $\Phi(x)$ is the *max-function*

$$\Phi(x) = \max_{y \in \mathcal{Y}} f(x, y).$$

The vector x is defined on the n-dimensional Euclidian space, $\mathcal{Y}$ is a convex
compact set, $f(x, y)$ is continuous in x, y and twice continuously differentiable
in x. In this chapter, we discuss numerical experience using two versions of a
quasi-Newton algorithm and Kiwiel's (1987) algorithm for continuous mini-

max. The aim is to test the performance of the algorithms as well as the effectiveness of a simplified strategy. This is intended to address an important difficulty in continuous minimax algorithms when, at a given x, there are multiple maximizers y. The computation of the search direction at such points is difficult and complex. We consider strategies for computing the direction and compare these with the choice of direction based on only one of the multiple maximizers.

Test problems of specific categories are considered. When the function $f(x, y)$ is convex in both x and y, with the elements of y bounded above and below, the continuous minimax problem can be expressed as a discrete minimax problem. We have tested the algorithms in such convex-convex problems as the behavior of the algorithms can be clearly observed in the presence of multiple maximizers which can be straightforwardly identified. In addition, the optima obtained in these cases are checked using a nonlinear programming formulation of the equivalent discrete minimax problem using a NAG library optimization subroutine.

Section 2 provides a brief description of the algorithms, implementation issues are considered in Section 3, Section 4 presents the test problems and the results of the experiments and Section 5 provides a summary discussion of the results. The test problems are grouped according to the convexity in the x-space and concavity in the y-space of the objective function.

2 THE ALGORITHMS

2.1 Kiwiel's Algorithm (Kiw)

Based on Panin's (1981) approach, as discussed in Section 2.4, Kiwiel (1987) has proposed the linear approximation to the *max-function* (2.4.2). A descent direction is computed at x_k by an auxiliary algorithm which solves (2.4.4), in finite number of iterations. In addition to the set of maximizers at x_k, given by $\mathcal{Y}(x_k)$ where

$$\mathcal{Y}(x) \equiv \left\{ y(x) \in \mathcal{Y} \mid y(x) = \arg\max_{y \in \mathcal{Y}} f(x, y) \right\}$$

the algorithm utilizes the set of maximizers of $f_k^{\ell}(d, y)$, defined as

$$\mathcal{Y}_{k+1} \equiv \left\{ y_{k+1} \in \mathcal{Y} \mid y_{k+1} = \arg\max_{y \in \mathcal{Y}} f_k^{\ell}(d, y) \right\}$$

and the termination criterion for the algorithm determined in terms of the function

$$\Psi_k^{\ell} = -\left\{ \|d_k\|^2 + \Phi(x_k) - f(x_k, y_{k+1}) \right\}.$$

2.2 Quasi-Newton Methods

The main quasi-Newton Algorithm (QN1) is discussed in Section 4.3. An important aspect of both quasi-Newton algorithms is the following:

Assumption *At a point x, all members of the set $\mathcal{Y}(x)$ are computable.*

The algorithm proceeds using the search direction

$$d_k = \begin{cases} -\mathbb{C}_k^{-1}\nabla\Phi_k & \text{if } \langle\nabla_x f(x_k,\bar{y},\bar{d}\rangle \geq -\xi \\ \bar{d} & \text{otherwise.} \end{cases}$$

Thus, whenever possible, $\bar{d}$ is used as the evaluation of $\nabla\Phi_k$ entails the solution of a quadratic programming problem. If $\bar{d}$ is a descent direction, as discussed above, it is used to determine d_k in Step 2, and subsequently Ψ_k in Step 3. It then utilizes a stepsize strategy to determine the progress along d_k and updates the Hessian approximation d_k.

The condition $\max_{y\in\mathcal{Y}(x_k)}\langle\nabla_x f(x_k,y),\bar{d}\rangle < -\xi$ ensures that d_k chosen by the algorithm is a descent direction for $\Phi(x_k)$. This is established in Lemma 4.3.2.

The condition for the equivalence of both choices, $\bar{d}$ and $-\mathbb{C}^{-1}\nabla\Phi_k$, for d_k is discussed in Chapter 4. In some problems, this condition may be naturally satisfied in which case the algorithm simplifies considerably. The following assumption addresses such situations.

Assumption *Let $\mathcal{D} \subset \mathfrak{R}^n$ be an open convex neighborhood of x_*. For $x_k \in \mathcal{D}, y \in \mathcal{Y}(x_k)$, the minimization in (4.2.13) is solved by $\lambda_{k+1}^y = 1 \in \mathfrak{R}^1$ for a unique y_{k+1} given by (4.2.14).*

The above assumption can be seen as a *straightforward extension of the unique maximizer case to multiple maximizers.* We note that the maximizer is unique for $f(\cdot,y)$ concave in y. However, the requirement in the assumption is that the maximizer with minimum norm subgradient to be unique. This ensures a simplified gradient representation and consistent Hessian approximation when there is more than one maximizer.

The simplified quasi-Newton algorithm (QN2) below is evaluated to investigate the effectiveness of the search direction $\bar{d}$ and an algorithm based only on this direction.

Quasi-Newton Algorithm 2 (QN2)

In this version, we assume $d_k = \bar{d}$. The alternative direction, $-\mathbb{C}_k^{-1}\nabla\Phi_k$, is ignored.

3 IMPLEMENTATION

The quasi-Newton algorithms (QN1 and QN2) and Kiwiel's algorithm (Kiw) both require a maximization subproblem to be solved in Step 1. Because this is an optimization problem within the main body of the algorithms, the minimax solution is computationally more expensive compared to most nonlinear programming algorithms. For maximization subproblems, the NAG nonlinear programming optimization subroutine E04UCC is used which can handle both linear and nonlinear constraints.

As a comprehensive gradient-based method, Kiw is used as an independent check for the solutions computed by QN1, QN2 as well as the various experiments with stopping criteria choices and directions of search. All the test problems given in this section were solved by using all the algorithms. Problems 8–16 can be formulated as, and problems 17–21 are originally, discrete minimax problems. These discrete minimax formulations are further reformulated as nonlinear programming problems and solved using NAG E04UCC as an independent check for the solutions found by the minimax algorithms.

3.1 Terminology

We specify the parameters, tests and the general output of the experiments together with a short description of their meaning. In the implementation of Kiw, the setting of the linear approximation parameter is given by $m = 2.0 \times 10^{-3}$ for all the test problems. QN1, QN2 were solved using the parameter values specified in Section 2. Any deviation from these values are specified explicitly in the results discussed in Section 4. In addition, the data corresponding to each problem are represented as follows:

x_0, y_0	initial values of x, y
$x_*^{i|j}, y_*^{i|j}$	solution values obtained by algorithm $i =$ QN1, QN2, Kiw for solution accuracy $j =$ A, B
$\Phi(x_*)$	objective function value at computed solution
x_*, y_*	computed solution x, y
$k_g \mid \langle \nabla_x f(x_k g, \overline{y}), \overline{d} \rangle \geq -\xi$	iterations in QN1 where direction $\overline{d}$ is not chosen by QN1 (i.e., the directional derivative is greater than $-\xi$). $k_g =$ none indicates that the condition is not satisfied

	at any iteration and $\bar{d}$ is chosen throughout that run.
$k_\alpha \mid \alpha = 1; \forall k \geq k_\alpha$	iteration where stepsize $\alpha = 1$ is achieved and maintained for all subsequent iterations. $k_\alpha =$ not attained indicates that $\alpha_k = 1$ is not achieved during any iteration.
No. of iterations	total number of iterations taken to solve the problem
Time	total computer time taken using a 200 megaHertz Pentium processor.
Failure: $\alpha_* = 0$ and $\Phi(x_*) =$	*max-function* value Φ attained, stepsize $\alpha_* = 0$ caused the algorithm to terminate due to insufficient progress in the direction generated.

3.2 The Stopping Criterion

The termination criterion (4.3.5) is the condition, $\Psi_k \geq -\xi$ and $\Psi_k^\ell \geq -\xi$ for Kiw. The values used in the test problems are in the range $\xi \in [10^{-07}, 10^{-11}]$. Every test problem is solved twice using all algorithms: first, using a higher value such as $\xi = 10^{-08}$, and second with a lower value such as $\xi = 1.0^{-10}$. The purpose of solving the problems twice is to investigate the benefits of evaluating the minimum-norm subgradient $\nabla\Phi_k$. Some test problems demonstrate that without the evaluation of the directional derivative, and of $\nabla\Phi_k$, QN2 is able to attain the solution, generally with superlinear convergence, for an accuracy of say $\xi = 10^{-08}$, but cannot maintain superlinear convergence for a higher accuracy of say $\xi = 10^{-10}$. Whereas the quasi-Newton algorithm without the evaluation of the directional derivative, QN2, may deteriorate in performance at such high accuracy, the quasi-Newton with the evaluation of the directional derivative, QN1, generally maintains superlinear convergence, even for very high accuracies. It is therefore possible to observe the trade-off between attaining very high accuracy with QN1 and saving computation time with QN2. In general, the performance of QN2 with high accuracy, together with the savings in computation time, make it a practical alternative to QN1.

3.3 Evaluation of the Direction of Descent

In QN1, d_k is either determined by $-\mathbb{C}_k^{-1}\nabla\Phi_k$ or $\bar{d}$. If the condition

$$\max_{y \in \mathcal{Y}(x_k)} \langle \nabla_x f(x_k, y), \bar{d} \rangle < -\xi$$

is satisfied, $\nabla\Phi_k$ is not computed and d_k is determined by $\bar{d}$ which in turn depends on y_{k+1} that maximizes the penalty function. However, if the above condition is not satisfied, d_k is determined by the value that corresponds to the minimum-norm subgradient $\nabla\Phi_k$. In the test problems, the evaluation of $\nabla\Phi_k$ depends on the type of problem.

In convex-convex problems (i.e., $f(x, y)$ convex in x and convex in y) where $\mathcal{Y}$ consists of upper and lower bounds on y, all candidate solutions at each vertex of the hypercube $\mathcal{Y}$ are explicitly evaluated. (This is because maxima of a convex function are at extreme points of the feasible region $\mathcal{Y}$.) The calculation of the minimum-norm subgradient $\nabla\Phi_k$ is based on all corner point y-values that yield a function value equal to the *max-function* at that iteration. For example, in Problem 15, there are 16 corner points and 8 of these are maximizers at the solution. In this case $\nabla\Phi_k$ is obtained by a linear combination of 8 subgradients based on 8 corner maximizers. At any iteration, only corner point maximizers are considered when evaluating $\nabla\Phi_k$. So, at early iterations, where there may only be 2 or 4 corner point maximizers, only these points are used for the computation. A criterion was implemented to test whether a corner point is a maximizer: if close to the *max-function* by 10^{-08}, then a corner point qualifies as a corner maximizer.

For the general case of nonconvex-nonconcave problems, the calculation of the minimum-norm subgradient $\nabla\Phi_k$ should be based on y_{k+1} and $\bar{y}$. Problems in this general case are not included in the test problems as they are not helpful in exhibiting the properties of the algorithms. These are more readily explored with the convex maximization problems discussed above and the continuous minimax formulations of equality constrained general nonlinear discrete minimax problems (Rustem and Nguyen, 1998).

4 TEST PROBLEMS

Tests on 21 problems are reported below. The objectives of problems 1–7 are convex-concave. These examples illustrate the algorithms when the *max-function* has a unique y maximizer for each fixed x. Although these initial problems do not test algorithm behavior for multiple maximizers, they provide insight into the work done by algorithms intended for more general problems.

Problems 8–16 have convex-convex objectives and illustrate the performance of the algorithms when the *max-function* may have multiple maximizers for a fixed x. The set of corner points comprises the extreme points in the y-space defined by the upper and lower bounds on y. At any iteration, the set of maximizers comprises the subset of corner points whose function value equals the *max-function* at that iteration.

Problems 8–16 which are convex-convex, subject to bounds constraints in y, can also be reformulated as discrete minimax to obtain an independent confirmation of the solution and consequently of the computations of continuous minimax algorithms. Consider the continuous minimax problem

$$\min_{x\in\Re^n} \max_{y\in\Re^m} \left\{ f(x,y) \mid y_{\ell_i} \le y_i \le y_{u_i}, \ i = 1,...,m \right\}$$

where $f(x,y)$ is convex-convex. The maxima are situated at the vertices of the hypercube. Let the vertices be given by $y_t, t \in T \equiv \{1,...,2^m\}$. Discrete minimax formulation is then given by

$$\min_{x\in\Re^n} \max_{t\in T} \{f(x,y_t)\}.$$

The discrete minimax formulation is equivalent to the following nonlinear programming problem in $n + 1$ variables

$$\min_{x\in\Re^n, \nu\in\Re^1} \{\nu \mid f_t(x,y_t) \le \nu, t \in T\}$$

where $f_t(x,y_t)$ are the objective values at vertex y_t. As an independent check on the computed solution, the nonlinear programming formulation is solved using the NAG nonlinear programming subroutine E04UCC.

Problems 17–21 are published constrained discrete minimax problems which have been reformulated as continuous minimax, unconstrained in $x \in \Re^n$. In this case, the maximizer may or may not be at a vertex. To obtain the continuous minimax formulations, consider the discrete minimax problem

$$\min_{x\in\Re^n} \max_i \left\{ f_i(x) \mid i = 1,...,m, \ g_j(x) = 0, \ j = 1,...,J \right\}$$

where f_i is one of m objective functions, and g_j is one of J constraints. This is reformulated as the continuous minimax problem

$$\min_{x\in\Re^n} \max_{y\in\Re^m} \left\{ \sum_{i=1}^{m} y_i f_i(x) + \mu \sum_{j=1}^{J} \left(g_j(x)\right)^2 \;\middle|\; \sum_{i=1}^{m} y_i = 1, \ 0 \le y_i \le 1, \ i = 1,...,m \right\}.$$

where $\mu \ge 0$ is a penalty parameter to ensure that the constraints on x are satisfied. The equivalence of these discrete and continuous minimax problems are discussed in Lemma 6.2.1. In the computations, the value $\mu = 10^8$ has been used. A similar penalty approach is adopted to enforce the equality constraint on y. For Problems 17–21, the objective function values f_t at the solution are reported. Although these problems do not appear to have multiple maximizers, they are difficult test problems and thus test the robustness of the algorithms.

Problem 1 [A: $\xi = 10^{-8}$; B: $\xi = 10^{-10}$]

$$f(x, y) = 5\sum_{i=1}^{2} x_i^2 - \sum_{i=1}^{2} y_i^2 + x_1(-y_1 + y_2 + 5) + x_2(y_1 - y_2 + 3) \mid -5 \le y_i \le 5; i = 1, 2$$

$x_0 = [10.0, -10.0]; \quad y_0 = [5.0, -5.0]$

$x_*^{i|A} = [-.4833563, -.3166436]; \ y_*^{i|A} = [.08335633, -.08335633]; \ i = \text{QN1}, \text{QN2}$

$x_*^{\text{Kiw}|A} = [-.4833371, -.3166634]; \ y_*^{\text{Kiw}|A} = [.08333694, -.08333694]$

$x_*^{\text{QN1}|B} = [-.4833563, -.3166436]; \ y_*^{\text{QN1}|B} = [.08335633, -.08335633]$

$x_*^{\text{QN2}|B} = [-.4833340, -.3166659]; \ y_*^{\text{QN2}|B} = [.08333408, -.08333408]$

$x_*^{\text{Kiw}|B} = [-.4833332, -.3166667]; \ y_*^{\text{Kiw}|B} = [.08333322, -.08333322]$

$\Phi^{i|j}(x_*) = -1.683333; \ i = \text{QN1}, \text{QN2}, \text{Kiw}; \ j = A, B$

Problem 2 [$C = 10^{10}$]; [A: $\xi = 10^{-8}$]; B: $\xi = 10^{-10}$]

$$f(x, y) = 4(x_1 - 2)^2 - 2y_1^2 + x_1^2 y_1 - y_2^2 + 2x_2^2 y_2 \mid -5 \le y_i \le 5; i = 1, 2$$

$x_0 = [4.0, 4.0]; y_0 = [5.0, 5.0]$

$x_*^{\text{QN1}|A} = [1.695415, -.2245884 \times 10^{-1}]; y_*^{\text{QN1}|A} = [.7186081, .5043996 \times 10^{-3}]$

$x_*^{\text{QN2}|A} = [1.695415, -.1202846 \times 10^{-1}]; y_*^{\text{QN2}|A} = [.7186081, .1447486 \times 10^{-3}]$

$x_*^{\text{Kiw}|A} = [1.695410, -.2677936 \times 10^{-1}]; y_*^{\text{Kiw}|A} = [.7186039, .7171342 \times 10^{-3}]$

$x_*^{\text{QN1}|B} = [1.695415, -.3170148 \times 10^{-2}]; y_*^{\text{QN1}|B} = [.7186081, .1004984 \times 10^{-4}]$

$x_*^{\text{QN2}|B} = [1.695415, -.3178420 \times 10^{-2}]; y_*^{\text{QN2}|B} = [.7186081, .1010232 \times 10^{-4}]$

$x_*^{\text{Kiw}|B} = [1.695414, -.1257688 \times 10^{-1}]; y_*^{\text{Kiw}|B} = [.7186077, .1581781 \times 10^{-3}]$

$\Phi^{i|j}(x_*) = 1.403883; \ i = \text{QN1}, \text{QN2}, \text{Kiw}; \ j = A, B$

Problem 3 [$C = 10^{10}$]; [A: $\xi = 10^{-8}$; B: $\xi = 10^{-10}$]

$$f(x, y) = x_1^4 y_2 + 2x_1^3 y_1 - x_2^2 y_2(y_2 - 3) - 2x_2(y_1 - 3)^2 \mid 0 \le y_i \le 3; i = 1, 2$$

$x_0 = [10.0, 0.1]; \ y_0 = [3.0, 3.0]$

$\Phi(x_*) = -2.468775$

$x_*^{i|A} = [-1.180675, .9128218]; \ y_*^{i|A} = [2.098478, 2.666058]; i = \text{QN1}, \text{QN2}$

$x_*^{\text{Kiw}|A} = [-1.180675, .9128322]; \ y_*^{\text{Kiw}|A} = [2.098490, 2.666029]$

$x_*^{i|B} = [-1.180674, .9128242]; \ y_*^{i|B} = [2.098483, 2.666048]; \ i = \text{QN1}, \text{QN2}$

$x_*^{\text{Kiw}|B} = [-1.180674, .9128261]; \ y_*^{\text{Kiw}|B} = [2.098486, 2.666042]$

$\Phi^{i|j}(x_*) = -2.468775; \ i = \text{QN1}, \text{QN2}, \text{Kiw}; \ j = A, B$

Problem 4 $[C = 10^{10}]; [A : \xi = 10^{-8}; B: \xi = 10^{-10}]$

$$f(x, y) = -\sum_{i=1}^{3}(y_i - 1)^2 + \sum_{i=1}^{2}(x_i - 1)^2 + y_3(x_2 - 1) + y_1(x_1 - 1) + y_2 x_1 x_2$$

$-3 \le y_i \le 3; i = 1, 2, 3$

$x_0 = [-2.0, -1.0]; \ y_0 = [0.0, 0.0, 0.0]$

$x_*^{\text{QN1}|A} = [.4181272, .4181122]; \ y_*^{\text{QN1}|A} = [.7090636, 1.087412, .7090561]$

$x_*^{\text{QN2}|A} = [.4181271, .4181124]; \ y_*^{\text{QN2}|A} = [.7090635, 1.087412, .7090562]$

$x_*^{\text{Kiw}|A} = [.4181456, .4181456]; \ y_*^{\text{Kiw}|A} = [.7090728, 1.087422, .7090728]$

$x_*^{\text{QN1}|B} = [.4181405, .4181230]; 2y_*^{\text{QN1}|B} = [.7090702, 1.087417, .7090615]$

$x_*^{\text{QN2}|B} = [.4181376, .4181119]; \ y_*^{\text{QN2}|B} = [.7090688, 1.087414, .7090559]$

$x_*^{\text{Kiw}|B} = [.4181300, .4181300]; \ y_*^{\text{Kiw}|B} = [.7090650, 1.087416, .7090650]$

$\Phi^{i|j}(x_*) = -.1348339; \ i = \text{QN1}, \text{QN2}, \text{Kiw}; \ j = A, B$

Problem 5 $[C = 10^{10}]; [A: \xi = 10^{-8}; B: \xi = 10^{-10}]$

$$f(x, y) = -(x_1 - 1)y_1 - (x_2 - 2)y_2 - (x_3 - 1)y_3 + 2x_1^2 + 3x_2^2 + x_3^2 - \sum_{i=1}^{3} y_i^2$$

$-1 \le y_i \le 1; i = 1, 2, 3$

$x_0 = [2.0, 2.0, 2.0];\ y_0 = [1.0, 1.0, 1.0]$

$x_*^{i|A} = [.1111101, .1538398, .2000022];\ y_*^{i|A} = [.4444449, .9230800, .3999988];$

$\quad\quad i = QN1, QN2$

$x_*^{Kiw|A} = [.1111111, .1538355, .2000000];\ y_*^{Kiw|A} = [.4444444, .9230822, .4000000]$

$x_*^{i|B} = [.1111110, .1538457, .2000003];\ y_*^{i|B} = [.4444444, .9230770, .3999998];$

$\quad\quad i = QN1, QN2$

$x_*^{Kiw|B} = [.1111111, .1538471, .2000000];\ y_*^{Kiw|B} = [.4444444, .9230763, .4000000]$

$\Phi^{i|j}(x_*) = .1345299;\ i = QN1, QN2, Kiw;\ j = A, B$

Problem 6 $[C = 10^{10}]$; [A: $\xi = 10^{-8}$; B: $\xi = 10^{-10}$]

$f(x, y) = y_1(x_1^2 - x_2 + x_3 - x_4 + 2) + y_2(-x_1 + 2x_2^2 - x_3^2 + 2x_4 + 1)$

$\quad\quad + y_3(2x_1 - x_2 + 2x_3 - x_4^2 + 5) + 5x_1^2 + 4x_2^2 + 3x_3^2 + 2x_4^2$

$$-\sum_{i=1}^{3} y_i^2$$

$-2 \le y_i \le 2; i = 1, 2, 3$

$x_0 = [-2.0, 2.0, -2.0, 2.0]; y_0 = [2.0, 2.0, 2.0]$

$x_*^{QN1|A} = [-.2315507, .2228134, -.6755366, -.08376960];$

$y_*^{QN1|A} = [.6195175, .3534766, 1.477997]$

$x_*^{QN2|A} = [-.2315505, .2228133, -.6755375, -.08377083];$

$y_*^{QN2|A} = [.6195178, .3534747, 1.477996]$

$x_*^{Kiw|A} = [-.2315566, .2228125, -.6755220, -.08376166];$

$y_*^{Kiw|A} = [.6195227, .3534970, 1.478006]$

$x_*^{QN1|B} = [-.2315586, .2228107, -.6755238, -.08376519];$

$y_*^{QN1|B} = [.6195250, .3534925, 1.478003]$

$x_*^{\text{QN2}|B} = [-.2315585, .2228105, -.6755239, -.08376509];$

$y_*^{\text{QN2}|B} = [.6195249, .3534923, 1.478003]$

$x_*^{\text{Kiw}|B} = [-.2315600, .2228110, -.6755205, -.08376863];$

$y_*^{\text{Kiw}|B} = [.6195285, .3534921, 1.478005]$

$\Phi^{i|j}(x_*) = 4.542969; \quad i = \text{QN1}, \text{QN2}, \text{Kiw}; \quad j = A, B$

Problem 7 $[C = 10^{10}]; [\text{A: } \xi = 10^{-8}; \text{B: } \xi = 10^{-10}]$

$f(x, y) = 2x_5 x_1 + 3x_4 x_2 + x_5 x_3 + 5x_4^2 + 5x_5^2 - x_4(y_4 - y_5 - 5) + x_5(y_4 - y_5 + 3)$

$$+ \sum_{i=1}^{3} y_i(x_i^2 - 1) - \sum_{i=1}^{5} y_i^2$$

$-3 \le y_i \le 3; i = 1, 2, 3, 4, 5$

$x_0 = [-4.0, -4.0, -4.0, -4.0, -4.0]; \quad y_0 = [2.0, 2.0, 2.0, 2.0, 2.0]$

$x_*^{\text{QN1}|j} = [1.425210, 1.661231, 1.258522, -.9744131, -.7348520]; \quad j = A, B$

$y_*^{\text{QN1}|j} = [.5156123, .8798455, .2919393, .1197805, -.1197805]; \quad j = A, B$

$x_*^{\text{QN2}|j} = [1.425210, 1.661231, 1.258522, -.9744131, -.7348521]; \quad j = A, B$

$y_*^{\text{QN2}|j} = [.5156128, .8798452, .2919395, .1197804, -.1197804]; \quad j = A, B$

$x_*^{\text{Kiw}|A} = [1.425208, 1.661239, 1.258526, -.9744230, -.7348516];$

$y_*^{\text{Kiw}|A} = [.5156097, .8798588, .2919450, .1197857, -.1197857];$

$x_*^{\text{Kiw}|B} = [1.425208, 1.661228, 1.258529, -.9744119, -.7348506];$

$y_*^{\text{Kiw}|B} = [.5156093, .8798407, .2919481, .1197806, -.1197806];$

$\Phi^{i|j}(x_* = -6.350915; \quad i = \text{QN1}, \text{QN2}, \text{Kiw}; \quad j = A, B$

Problem 8 $[\text{A: } \xi = 10^{-8}; \text{B: } \xi = 10^{-10}]$

$f(x, y) = \dfrac{3}{2} \left(\dfrac{1}{2}x_1 + (4 + x_1)y_1 \right)^2 | -2 \le y_1 \le 2;$

$x_0 = 3.0; \; y_0 = 2.0$

$x_*^{i|A} = [-3.999999]; \; i = \text{QN1}, \text{QN2}; \; x_*^{\text{Kiw}|A} = [-4.000000]$

$x_*^{i|B} = [-4.000000]; i = \text{QN1}, \text{QN2}, \text{Kiw}$

$y_*^{i|j} = [2]; [-2]$ (two maximizers); $\forall i, j; \; i = \text{QN1}, \text{QN2}, \text{Kiw}; \; j = A, B$

$\Phi^{i|j}(x_*) = 6.000000; \; i = \text{QN1}, \text{QN2}, \text{Kiw}; \; j = A, B$

Problem 9 $[C = 10^{15}]; \; [\text{A: } \boldsymbol{\xi} = 10^{-8}; \; \text{B: } \boldsymbol{\xi} = 10^{-9}]$

$$f(x, y) = \frac{1}{2}\Big((2x_2 + 4y_1 + x_1 y_1)^2 + (x_1 + 2x_1 y_1 + x_2 y_2)^2\Big) \mid -5 \le y_i \le 5; \; i = 1, 2$$

$x_0 = [0.0, 0.0]; \; y_0 = [5.0, 5.0]$

$x_*^{\text{QN1}|j} = [-.6849296, -.6715307 \times 10^{-7}]; \; j = A, B$

$x_*^{\text{QN2}|A} = [-.6849318, -.5766055 \times 10^{-7}]; \; x_*^{\text{QN2}|B} = [-.6849442, -.1776339 \times 10^{-13}];$

$x_*^{\text{Kiw}|A} = [-.6849312, -.1574302 \times 10^{-8}]; \; x_*^{\text{Kiw}|B} = [-.6849314, -.1634144 \times 10^{-9}]$

$y_*^{i|j} = [5.0, 5.0]; [5.0, -5.0]$ (two maximizers); $i = \text{QN1}, \text{QN2}, \text{Kiw}; \; j = A, B$

$\Phi^{i|j}(x_*) = 1.657534 \times 10^2; \; i = \text{QN1}, \text{QN2}, \text{Kiw}; \; j = A, B$

Problem 10 $[C = 10^{15}]; \; [\text{A: } \boldsymbol{\xi} = 10^{-7}; \; \text{B: } \boldsymbol{\xi} = 10^{-9}]$

$$f(x, y) = \frac{1}{2}\Big((5x_2 + 5y_1 + 3x_1 y_1)^2 + (2x_1 + 5x_1 y_1 + 3x_2 y_2)^2\Big) \mid -5 \le y_i \le 5; \; i = 1, 2$$

$x_0 = [1.0, 1.0]; \; y_0 = [5.0, 5.0]$

$x_*^{i|A} = [-.3930712, -.7271695 \times 10^{-8}]; \; x_*^{i|B} = [-.3930818, -.3121502 \times 10^{-10}];$

$i = QN1, QN2$

$x_*^{\text{Kiw}|A} = [-.3930819, -.8013015 \times 10^{-9}]; \; x_*^{\text{Kiw}|B} = [-.3930817, -.8128688 \times 10^{-11}]$

$y_*^{i|j} = [5.0, 5.0]; [5.0, -5.0]$ (two maximizers); $\forall i, j; \; i = \text{QN1}, \text{QN2}, \text{Kiw}; \; j = A, B$

$\Phi^{i|j}(x_*) = 2.387971 \times 10^2; \; i = \text{QN1}, \text{QN2}, \text{Kiw}; \; j = A, B$

Problem 11 $[C = 10^{15}]$; $[A: \xi = 10^{-8}; B: \xi = 10^{-10}]$

$$f(x, y) = \frac{1}{2}\left((x_1x_2 - y_1(1 - x_3x_4))^2 + (x_2x_3 - y_2(2 + x_4x_1))^2 + \sum_{i=1}^{2} y_i^2\right) \mid -5 \le y_i \le 5;$$

$$i = 1, 2$$

$x_0 = [1.0, 1.0, 1.0, 1.0];\ y_0 = [3.0, 3.0]$

$x_*^{i|A} = [-1.541662, -.1086857 \times 10^{-4}, .7708296, 1.297309];\ i = QN1, QN2$

$x_*^{Kiw|A} = [-1.691932, .5904574 \times 10^{-5}, .8459655, 1.182080]$

$x_*^{i|B} = [-1.541658, -.2811413 \times 10^{-6}, .7708293, 1.297303];\ i = QN1, QN2$

$x_*^{Kiw|B} = [-1.691932, .6060316 \times 10^{-6}, .8459661, 1.182080]$

$y_*^{i|j} = [5.0, 5.0]; [5.0, -5.0]; [-5.0, 5.0]; [-5.0, -5.0]$(four maximizers);

$$i = QN1, QN2, Kiw;\ j = A, B$$

$\Phi^{i|j}(x_*) = 2.5 \times 10^1;\ i = QN1, QN2, Kiw;\ j = A, B$

Problem 12 $[C = 10^{15}]$; $[A: \xi = 10^{-8}; B: \xi = 10^{-10}]$

$$f(x, y) = \sum_{i=1}^{3} y_i^2 + y_1(x_1^2 - x_2 + x_3 - x_4 + 2) + y_2(-x_1 + 2x_2^2 - x_3^2 + 2x_4 - 10)$$

$$+ y_3(2x_1 - x_2 + 2x_3 - x_4^2 - 5) + 5\sum_{i=1}^{4} x_i^2$$

$-2 \le y_i \le 2;\ i = 1, 2, 3$

$x_0 = [10.0, 1.0, 1.0, 10.0];\ y_0 = [2.0, 2.0, 2.0]$

$x_*^{i|j} = [.1428570, .1232059 \times 10^{-5}, .1428569, .4285713];\ i = QN1, QN2;\ j = A, B$

$x_*^{Kiw|A} = [.1428532, -.7079020 \times 10^{-6}, .1428573, .4285667]$

$x_*^{Kiw|B} = [.1428567, -.7087013 \times 10^{-7}, .1428571, .4285709]$

$y_*^{i|j} = [2.0, -2.0, -2.0];\ i = QN1, QN2, Kiw;\ j = A, B$

$\Phi^{i|j}(x_*) = 4.442857 \times 10^1;\ i = QN1, QN2, Kiw;\ j = A, B$

Problem 13 $[C = 10^{15}];$ **[A:** $\xi = 10^{-8};$ **B:** $\xi = 10^{-10}]$

$$f(x,y) = \sum_{i=1}^{4} y_i^2 + y_1(x_1^2 - x_2 + x_3 - x_4 + 2) + y_2(-x_1 + 2x_2^2 - x_3^2 + 2x_4 - 10)$$

$$+y_3(2x_1 - x_2 + 2x_3 - x_4^2 - 5) + 5y_4(x_1^2 + x_2^2) + 5\sum_{i=3}^{4} x_i^2$$

$-2 \le y_i \le 2; i = 1, 2, 3, 4$

$x_0 = [1.0, -1.0, -1.0, 1.0];\ y_0 = [2.0, 2.0, 2.0, 2.0]$

$x_*^{i|A} = [.8333487 \times 10^{-1}, .8263346 \times 10^{-5}, .1428512, .4285743];\ i = \text{QN1, QN2}$

$x_*^{i|B} = [.8333332 \times 10^{-1}, .6406437 \times 10^{-9}, .1428571, .4285714];\ i = \text{QN1, QN2}$

$x_*^{\text{Kiw}|A} = [.8333627 \times 10^{-1}, .0000000, .1428571, .4285714];$

$x_*^{\text{Kiw}|B} = [.8333296 \times 10^{-1}, .0000000, .1428571, .4285714];$

$y_*^{i|j} = [2.0, -2.0, -2.0, 2.0];\ i = \text{QN1, QN2, Kiw};\ j = A, B$

$\Phi^{i|j}(x_*) = 4.848809 \times 10^1; \forall i, j;\ i = \text{QN1, QN2, Kiw};\ j = A, B$

Problem 14 $[C = 10^{15}];$ **[A:** $\xi = 10^{-8};$ **B:** $\xi = 10^{-10}]$

$$f(x,y) = \sum_{i=1}^{4} y_i^2 + y_1(x_1^2 - 2.2x_2 + x_3 - 10x_4 + 10) + y_2(-2x_1 + 2x_2^2 - x_3^2 + 3x_4 - 10)$$

$$+y_3(2x_1 - x_2 + 6x_3 - x_4^2 - 5) + 5y_4(x_1^2 + x_2^2) + 5\sum_{i=3}^{4} x_i^2$$

$-2 \le y_i \le 2; i = 1, 2, 3, 4$

$x_0 = [0.0, 0.0, 0.0, 0.0];\ y_0 = [2.0, 2.0, 2.0, 2.0]$

$x_*^{i|A} = [-.1151853 \times 10^{-6}, .4685527 \times 10^{-3}, .7920836, 1.079105];\ i = \text{QN1, QN2}$

$x_*^{i|B} = [.1692001 \times 10^{-6}, .4589308 \times 10^{-3}, .7920886, 1.079107];\ i = \text{QN1, QN2}$

$x_*^{\text{Kiw}|j} = [-.3438374, .4667710, .1742058, 1.178821];\ j = A, B$

$y_*^{i|j} = [2.0, -2.0, -2.0, 2.0]; [-2.0, -2.0, -2.0, 2.0]; [2.0, -2.0, -2.0, -2.0];$

$[-2.0, -2.0, -2.0, -2.0]$ (four maximizers); $\forall i, j$; $i = QN1, QN2$, $j = A, B$

$y_*^{Kiw|j} = [-2.0, -2.0, -2.0, 2.0]$(one maximizer); $j = A, B$

$\Phi^{i|j}(x_*) = 4.256435 \times 10^1$; $i = QN1, QN2$; $j = A, B$;

$\Phi^{Kiw|j}(x_*) = 5.524477 \times 10^1$; $j = A, B$

Problem 15 [$C = 10^{15}$]; [A: $\xi = 10^{-8}$; B: $\xi = 10^{-10}$]

$$f(x, y) = \sum_{i=1}^{4} y_i^2 + y_1(x_1^2 - 2.2x_2 + x_3 - 10x_4 + 10) + y_2(-2x_1 + 2x_2^2 - x_3^2 + 3x_4 - 10)$$

$$+y_3(2x_1 - x_2 + 5.91x_3 - x_4^2 - 15) + 5y_4(x_1^2 + x_2^2) + 5\sum_{i=3}^{4} x_i^2$$

$-2 \le y_i \le 2$; $i = 1, 2, 3, 4$

$x_0 = [1.0, 1.0, 1.0, 1.0]$; $y_0 = [2.0, 2.0, 2.0, 2.0]$

$x_*^{i|A} = [-.2384806 \times 10^{-5}, .7624597 \times 10^{-6}, .7793389, 1.077933]$; $i = QN1, QN2$

$x_*^{i|B} = [-.1476035 \times 10^{-5}, .1873070 \times 10^{-5}, .7793423, 1.077933]$; $i = QN1, QN2$

$x_*^{Kiw|j} = [.115145, .3474200, .8373482, .9378763]$; $j = A, B$

$y_*^{i|j} = [2.0, -2.0, -2.0, 2.0]$; $[-2.0, -2.0, -2.0, 2.0]$; $[2.0, -2.0, -2.0, -2.0]$;

$[-2.0, -2.0, -2.0, -2.0]$ (four maximizers); $i = QN1, QN2$; $j = A, B$

$y_*^{Kiw|j} = [2.0, -2.0, -2.0, 2.0]$ (one maximizer); $j = A, B$

$\Phi^{i|j}(x_*) = 6.270578 \times 10^1$; $i = QN1, QN2; j = A, B$;

$\Phi^{Kiw|j}(x_*) = 6.450730 \times 10^1; j = A, B$

Problem 16 [$C = 10^{15}$]; [A: $\xi = 10^{-8}$; B: $\xi = 10^{-10}$]

$$f(x, y) = \frac{1}{2}\sum_{i=1}^{4} y_i^2 + y_1(x_1^2 - 2x_2 + x_3 - 10x_4 + 2) + y_2(-2x_1 + 2x_2^2 - x_3^2 + 3x_4 - 5)$$

$$+y_3(2x_1 - x_2 + 5x_3 - x_4^2 + 2) + y_4(x_1^2 + x_2^2 + \sum_{i=3}^{4} x_i^2)$$

$-2 \le y_i \le 2; \; i = 1, 2, 3, 4$

$x_0 = [0.1, 0.0, 1.0, 0.0]; \; y_0 = [2.0, 2.0, 2.0, 2.0]$

$x_*^{i|j} = [-.7466470, -1.062980, -.2750698, .4408373]; \; \forall i, j; \; i = \text{QN1}, \text{QN2}; \; j = A, B$

$x_*^{\text{Kiw}|j} = [-.6159915, -1.117989, -.2436400, .4410787]; \; j = A, B$

$y_*^{i|j} = [2.0, 2.0, 2.0, 2.0]; [2.0, 2.0, -2.0, 2.0]; [2.0, -2.0, 2.0, 2.0]; [2.0, -2.0, -2.0, 2.0];$

$\qquad\qquad [-2.0, 2.0, 2.0, 2.0]; [-2.0, 2.0, -2.0, 2.0]; [-2.0, -2.0, 2.0, 2.0];$

$\qquad\qquad [-2.0, -2.0, -2.0, 2.0]; \; \text{(eight maximizers)}; \; i = \text{QN1}, \text{QN2}; \; j = A, B$

$y_*^{\text{Kiw}|j} = [-2.0, -2.0, 2.0, 2.0] \; \text{(one maximizer)}; \; j = A, B$

$\Phi^{i|j}(x_*) = 1.191482 \times 10^1; \; i = \text{QN1}, \text{QN2}; \; j = A, B;$

$\Phi^{\text{Kiw}|j}(x_*) = 1.279971 \times 10^1; \; j = A, B$

Problem 17 $[C = 10^{15}]$; [A: $\xi = 10^{-8}$; B: $\xi = 10^{-10}$] (Charalambous and Bandler, 1976)

$f_1(x) = x_1^2 + x_2^4; \; f_2(x) = (2 - x_1)^2 + (2 - x_2)^2; \; f_3(x) = 2e^{(x_2 - x_1)}$

$x_0 = [5.5, -2.0]; \; y_0 = [1.00.0, 0.0]$

$x_*^{i|A} = [1.139049, .8995505]; \; x_*^{i|B} = [1.139041, .8995568]; \; i = \text{QN1}, \text{QN2};$

$x_*^{\text{Kiw}|A} = [1.139030.8995658]; \; x_*^{\text{Kiw}|B} = [1.139038, .8995595]$

$y_*^{i|j} = [0.0, 1.0, 0.0]; \; i = \text{QN1}, \text{QN2}; \; j = A, B; \; y_*^{\text{Kiw}|j} = [1.0, 0.0, 0.0]; \; j = A, B$

$\Phi^{i|j}(x_*) = 1.952224; \; i = \text{QN1}, \text{QN2}, \text{Kiw}; \; j = A, B$

Problem 18 $[C = 10^{15}]$; [A: $\xi = 10^{-8}$; B: $\xi = 10^{-10}$] (Charalambous and Bandler, 1976)

$f_1(x) = x_1^4 + x_2^2; \; f_2(x) = (2 - x_1)^2 + (2 - x_2)^2; \; f_3(x) = 2e^{(x_2 - x_1)}$

$x_0 = [0.0, -0.1]; \; y_0 = [0.01.0, 0.0]$

$x_*^{i|A} = [1.000000, 1.000053]; \; y_*^{i|A} = [1.0, 0.0, 0.0]; \; i = \text{QN1}, \text{QN2}$

$x_*^{i|B} = [.9999999, 1.000004];$ $y_*^{i|B} = [0.0, 0.0, 1.0];$ $i = QN1, QN2$

$x_*^{Kiw|A} = [.9999999, .9999999];$ $x_*^{Kiw|B} = [1.000000, 1.000000];$

$y_*^{Kiw|j} = [0.0, 1.0, 0.0];$ $j = A, B$

$\Phi^{i|A}(x_*) = 2.000106;$ $\Phi^{i|B}(x_*) = 2.000000;$ $i = QN1, QN2;$

$\Phi^{Kiw|j}(x_*) = 2.000000;$ $j = A, B$

Problem 19 $[C = 10^{15}]$; $[A: \xi = 10^{-7}; B: \xi = 10^{-10}]$ (Demyanov and Malozemov, 1974)

$f_1(x) = 5x_1 + x_2;$ $f_2(x) = -5x_1 + x_2;$ $f_3(x) = x_1^2 + x_2^2 + 4x_2$

$x_0 = [-2.1, -0.1];$ $y_0 = [.333333, .333333, .333333]$

$x_*^{i|A} = [-.4606293 \times 10^{-7}, -2.999999];$ $y_*^{i|j} = [0.0, 1.0, 0.0];$ $i = QN1, QN2; j = A, B$

$x_*^{i|B} = [-.2133713 \times 10^{-7}, -3.000000];$ $i = QN1, QN2;$

$x_*^{Kiw|A} = [.4050685 \times 10^{-7}, -3.000000];$ $y_*^{Kiw|j} = [1.0, 0.0, 0.0];$ $j = A, B$

$x_*^{Kiw|B} = [.5556585 \times 10^{-10}, -3.000000];$

$\Phi^{i|A}(x_*) = -2.999999;$ $\Phi^{i|B}(x_*) = -3.000000;$ $i = QN1, QN2, Kiw.$

Problem 20 $[C = 10^{15}]$; $[A: \xi = 10^{-7} ; B: \xi = 10^{-10}]$ (Conn, 1979)

$f_1(x) = x_1^2 + x_2^2;$ $f_2(x) = (2 - x_1)^2 + (2 - x_2)^2;$ $f_3(x) = 2e^{(x_2 - x_1)}$

$g_1(x) = x_1 + x_2 - 2;$ $g_2(x) = -x_1^2 - x_2^2 + 2.25$

$x_0 = [1.1, 0.1];$ $y_0 = [0.0, 1.0, 0.0]$

$x_*^{i|j} = [1.353553, .6464466];$ $y_*^{i|j} = [1.0, 0.0, 0.0];$ $i = QN1, QN2, Kiw; j = A, B$

$\Phi^{i|A}(x_*) = 2.250000;$ $i = QN1, QN2;$ $\Phi^{Kiw|A}(x_*) = 2.249999;$

$\Phi^{i|B}(x_*) = 2.249999;$ $i = QN1, QN2, Kiw$

Problem 21 $[C = 10^{15}]$; $[\text{A: } \xi = 10^{-9}$; $\text{B: } \xi = 10^{-11}]$ **(Polak et al., 1991)**

$$f_1(x) = \exp\left(\frac{x_1^2}{1000} + (x_2 - 1)^2\right); \quad f_2(x) = \exp\left(\frac{x_1^2}{1000} + (x_2 + 1)^2\right)$$

$x_0 = [0.02, 0.15]; \quad y_0 = [0.0, 1.0]$

$x_*^{i|j} = [-.3082574 \times 10^{-4}, .6049532 \times 10^{-9}]; \quad i = QN1, QN2; \; j = A, B$

$x_*^{\text{Kiw}|A} = [.1007579 \times 10^{-4}, .1085434 \times 10^{-9}];$

$x_*^{\text{Kiw}|B} = [.1005085 \times 10^{-2}, .105424 \times 10^{-11}];$

$y_*^{i|j} = [0.0, 1.0]; \quad i = QN1, QN2, \text{Kiw}; \; j = A, B$

$\Phi^{i|j}(x_*) = 2.718282; \quad i = QN1, QN2, \text{Kiw}; \; j = A, B$

5 SUMMARY OF THE RESULTS

A number of test problems are used to test the performance of QN1 and QN2. In general, superlinear convergence is observed, and the solutions found are consistent with those found by Kiw. Where applicable, the problems are also solved using a nonlinear programming formulation and consistent results were confirmed.

The test problems, reported in Table 1A–E, highlight a number of issues regarding the performance of the optimizer, namely: iterations when the condition in (4.2.11),

$$\langle \nabla_x f(x_k, \bar{y}), \bar{d} \rangle \geq -\xi$$

which triggers the use of $\nabla \Phi_k$, is satisfied. This affects the calculation of $\nabla \Phi_k$; superlinear convergence; the termination criterion and accuracy of the solution. We discuss these issues below.

5.1 Iterations When $\langle \nabla_x f(x_k, \bar{y}), \bar{d} \rangle \geq -\xi$ is Satisfied

In all test problems, the satisfaction of the above condition is observed in general during the last few iterations. Mostly, this occurs in the very last iteration. Thus, QN1 and QN2 performed comparably through all iterations except at the very last iterations where the algorithm may attempt to satisfy a possibly stringent termination criterion. This result also suggests that the algorithm without the evaluation of $\nabla \Phi_k$ (i.e., QN2) is a good approximation to QN1 which includes the evaluation of $\nabla \Phi_k$.

The parameter ξ represents the accuracy of both the solution in the termination criterion (4.3.5) and the descent test for $\bar{d}$ in (4.2.11). The advantage of higher solution accuracy (i.e., a smaller value of ξ) is clear. However, a smaller threshold value in (4.2.11) implies that $\bar{d}$ is accepted as a descent direction even when this descent is quite weak. This may be desirable for choosing $\bar{d}$, over $-\mathfrak{C}_k^{-1}\nabla\Phi_k$, as the former is simpler to compute. For larger values of ξ, while termination is potentially easier to achieve, the descent condition is less easily satisfied, as the threshold is higher. This effect has been observed in Problem 14 only, as discussed in Section 5.4. The overall observation in Section 5.4 is that the actual solution is unaffected by the failure of $\bar{d}$ satisfying the descent condition for the higher threshold value of ξ. It is possible to choose a higher threshold value for the descent test in (4.2.11) than the solution accuracy. In the present examples, the same consistent value for both has been adopted as both are related to similar quantities.

5.2 Calculation of Minimum-norm Subgradient

The effect of the calculation of $\nabla\Phi_k$ is demonstrated by the convex-convex problems, Test Problems 8–16. For these problems, we can evaluate the multiple maximizers at any iteration and explicitly construct the minimum-norm subgradient. These problems demonstrate that when the satisfaction of $\langle\nabla_x f(x_k, \bar{y}), \bar{d}\rangle \geq -\xi$ occurred at the very last iteration, there is no improvement in the value of the norm of d, and the algorithm does not benefit from the calculation of $\nabla\Phi_k$. However, where this condition is satisfied in the last few iterations, as in Problem 15, the computed $\nabla\Phi_k$ results in an improvement in both the norm of d_k and Ψ_k. This result suggests that if a stringent termination criterion needs to be implemented, then QN1 which computes $\nabla\Phi_k$ improves the satisfaction of the termination criterion.

We note that several of the problems 8–16 have infinite numbers of maximizers at the solution. For example, For example, in Problem 8, the value $x_1 = -4$ ensures that the problem does not depend on y_1 and hence any $y_1, -2 \leq y_1 \leq 2$, is a solution. Similarly, in Problem 9, $x_2 = 0$ makes the problem independent of y_2 and hence the problem has infinite maximizers $y_* = [5, y_2], -5 \leq y_2 \leq 5$. The latter observation also applies to Problem 10. In these situations, by Caratheodory's Theorem (Theorem 1.1.1) at most $(m + 1)$ gradient vectors, arising from the maximizers, can be used to characterize the subgradient. (In practice, however, the results reported are based on the use of only the appropriate vertices or the upper and lower bounds for the corresponding y.)

5.3 Superlinear Convergence

In general, superlinear convergence rates are observed for QN1 and QN2 with the test problems, supporting the use of the anticipatory BFGS Hessian approximation. A unit stepsize is reached at early iterations, and the number of iterations and/or time taken to convergence by QN1 and QN2 are significantly less than the number it took Kiw.

5.4 Termination Criterion and Accuracy of the Solution

The test problems are solved using two levels of stringency of the termination criterion. In general, QN2 performs similarly to QN1 when ξ is in the order of 10^{-07} or 10^{-08}, but fails in Test Problems 9 and 15, for ξ in the order 10^{-09} or 10^{-10}. The full quasi-Newton algorithm, QN1, did not fail. Kiw failed to converge in a number of test problems even for the less stringent termination criterion.

In Problem 14, the computation of $\nabla\Phi_k$ is triggered in iteration 32 as the direction generated does not satisfy the descent condition for $\xi = 10^{-8}$. However, this descent condition is satisfied for $\xi = 10^{-10}$ and $\nabla\Phi_k$ is not computed. This is the only example observed where $\nabla\Phi_k$ is computed for the lower but not the higher accuracy as the descent condition $\langle \nabla_x f(x_k, \bar{y}), \bar{d} \rangle < -\xi$ is satisfied for the latter.

The results suggest that QN2 is a good approximation to the full algorithm QN1 which incorporates the evaluation of $\nabla\Phi_k$, with possible deterioration of performance if high accuracy is demanded with small values of ξ.

Table 1A Performance of QN1, QN2 and Kiw

	Problem	QN1	QN2	Kiw
1A	$k_\alpha \mid \alpha_k = 1; \forall k \geq k_\alpha$	4	4	$\alpha = 1$ not achieved
	$k_g \mid \langle \nabla_x f(x_{k_g}, \bar{y}), \bar{d} \rangle \geq -\xi$	7	–	–
	Number of iterations	7	7	22
	Time (seconds)	3.766	3.415	14.621
1B	$k_\alpha \mid \alpha_k = 1; \forall k \geq k_\alpha$	$\alpha = 1$ not achieved	4	$\alpha = 1$ not achieved
	$k_g \mid \langle \nabla_x f(x_{k_g}, \bar{y}), \bar{d} \rangle \geq -\xi$	7, 8	–	–
	Number of iterations	8	8	27
	Time (seconds)	5.267	3.806	19.478
2A	$k_\alpha \mid \alpha_k = 1; \forall k \geq k_\alpha$	3	3	$\alpha = 1$ not achieved
	$k_g \mid \langle \nabla_x f(x_{k_g}, \bar{y}), \bar{d} \rangle \geq -\xi$	17, 18	–	–
	Number of iterations	18	19	549
	Time (seconds)	8.853	7.932	333.790

Table 1A (continued)

	Problem	QN1	QN2	Kiw
2B	$k_\alpha \mid \alpha_k = 1; \forall k \geq k_\alpha$	3	3	$\alpha = 1$ not achieved
	$k_g \mid \langle \nabla_x f(x_{k_g}, \bar{y}), \bar{d} \rangle \geq -\xi$	24	–	–
	Number of iterations	24	24	2689
	Time (seconds)	11.987	10.004	1697.661
3A	$k_\alpha \mid \alpha_k = 1; \forall k \geq k_\alpha$	3	3	$\alpha = 1$ not achieved
	$k_g \mid \langle \nabla_x f(x_{k_g}, \bar{y}), \bar{d} \rangle \geq -\xi$	15	–	–
	Number of iterations	15	15	47
	Time (seconds)	7.270	6.168	47.398
3B	$k_\alpha \mid \alpha_k = 1; \forall k \geq k_\alpha$	3	3	$\alpha = 1$ not achieved
	$k_g \mid \langle \nabla_x f(x_{k_g}, \bar{y}), \bar{d} \rangle \geq -\xi$	16	–	–
	Number of iterations	16	16	85
	Time (seconds)	7.711	6.729	95.117
4A	$k_\alpha \mid \alpha_k = 1; \forall k \geq k_\alpha$	2	2	$\alpha = 1$ not achieved
	$k_g \mid \langle \nabla_x f(x_{k_g}, \bar{y}), \bar{d} \rangle \geq -\xi$	6	–	–
	Number of iterations	6	6	72
	Time (seconds)	3.555	3.115	33.248
4B	$k_\alpha \mid \alpha_k = 1; \forall k \geq k_\alpha$	$\alpha = 1$ not achieved	$\alpha = 1$ not achieved	$\alpha = 1$ not achieved
	$k_g \mid \langle \nabla_x f(x_{k_g}, \bar{y}), \bar{d} \rangle \geq -\xi$	11, 12, 13, 14	–	–
	Number of iterations	14	12	90
	Time (seconds)	10.805	16.154	44.263
5A	$k_\alpha \mid \alpha_k = 1; \forall k \geq k_\alpha$	5	5	$\alpha = 1$ not achieved
	$k_g \mid \langle \nabla_x f(x_{k_g}, \bar{y}), \bar{d} \rangle \geq -\xi$	7	–	
	Number of iterations	7	7	26
	Time (seconds)	4.357	3.946	11.096
5B	$k_\alpha \mid \alpha_k = 1; \forall k \geq k_\alpha$	5	5	$\alpha = 1$ not achieved
	$k_g \mid \langle \nabla_x f(x_{k_g}, \bar{y}), \bar{d} \rangle \geq -\xi$	8	–	
	Number of iterations	8	8	31
	Time (seconds)	4.877	4.186	14.742

Table 1B   Performance of QN1, QN2 and Kiw

	Problem	QN1	QN2	Kiw
6A	$k_\alpha \mid \alpha_k = 1; \forall k \geq k_\alpha$	7	7	$\alpha = 1$ not achieved
	$k_g \mid \langle \nabla_x f(x_{k_g}, \bar{y}), \bar{d} \rangle \geq -\xi$	15	–	–
	Number of iterations	15	15	19
	Time (seconds)	7.802	6.940	11.867
6B	$k_\alpha \mid \alpha_k = 1; \forall k \geq k_\alpha$	7	7	$\alpha = 1$ not achieved
	$k_g \mid \langle \nabla_x f(x_{k_g}, \bar{y}), \bar{d} \rangle \geq -\xi$	16	–	–
	Number of iterations	16	16	23
	Time (seconds)	8.882	7.200	14.911
7A	$k_\alpha \mid \alpha_k = 1; \forall k \geq k_\alpha$	10	10	$\alpha = 1$ not achieved
	$k_g \mid \langle \nabla_x f(x_{k_g}, \bar{y}), \bar{d} \rangle \geq -\xi$	16	–	–
	Number of iterations	16	16	65
	Time (seconds)	10.676	9.223	39.257
7B	$k_\alpha \mid \alpha_k = 1; \forall k \geq k_\alpha$	10	10	$\alpha = 1$ not achieved
	$k_g \mid \langle \nabla_x f(x_{k_g}, \bar{y}), \bar{d} \rangle \geq -\xi$	16	–	–
	Number of iterations	16	16	70
	Time (seconds)	10.515	9.194	42.290
8A	$k_\alpha \mid \alpha_k = 1; \forall k \geq k_\alpha$	16	14	2
	$k_g \mid \langle \nabla_x f(x_{k_g}, \bar{y}), \bar{d} \rangle \geq -\xi$	None	–	–
	Number of iterations	16	16	5
	Time (seconds)	8.052	6.149	2.524
8B	$k_\alpha \mid \alpha_k = 1; \forall k \geq k_\alpha$	$\alpha = 1$ not achieved	14	2
	$k_g \mid \langle \nabla_x f(x_{k_g}, \bar{y}), \bar{d} \rangle \geq -\xi$	18, 19	–	–
	Number of iterations	19	18	6
	Time (seconds)	10.856	7.351	2.724
9A	$k_\alpha \mid \alpha_k = 1; \forall k \geq k_\alpha$	22	22	$\alpha = 1$ not achieved
	$k_g \mid \langle \nabla_x f(x_{k_g}, \bar{y}), \bar{d} \rangle \geq -\xi$	23, 24, 25	–	–
	Number of iterations	25	25	684
	Time (seconds)	19.899	23.443	683.012
9B	$k_\alpha \mid \alpha_k = 1; \forall k \geq k_\alpha$	22	$\alpha = 1$ not achieved	$\alpha = 1$ not achieved
	$k_g \mid \langle \nabla_x f(x_{k_g}, \bar{y}), \bar{d} \rangle \geq -\xi$	23, 24, 25	–	–
	Number of iterations	25	36	777
	Time (seconds)	19.919	40.558	807.360

Table 1B (continued)

Problem		QN1	QN2	Kiw
			Failure: $\alpha_* = 0$ and $\Phi(x_*) = $ 165.7534	
10A	$k_\alpha \mid \alpha_k = 1; \forall k \geq k_\alpha$	24	24	$\alpha = 1$ not achieved
	$k_g \mid \langle \nabla_x f(x_{k_g}, \overline{y}), \overline{d} \rangle \geq -\xi$	None	–	–
	Number of iterations	26	26	7211
	Time (seconds)	25.366	21.972	14913.485
10B	$k_\alpha \mid \alpha_k = 1; \forall k \geq k_\alpha$	31	31	$\alpha = 1$ not achieved
	$k_g \mid \langle \nabla_x f(x_{k_g}, \overline{y}), \overline{d} \rangle \geq -\xi$	34	–	–
	Number of iterations	34	34	9284
	Time (seconds)	29.242	28.971	19580.756

Table 1C Performance of QN1, QN2 and Kiw

Problem		QN1	QN2	Kiw
11A	$k_\alpha \mid \alpha_k = 1; \forall k \geq k_\alpha$	8	8	$\alpha = 1$ not achieved
	$k_g \mid \langle \nabla_x f(x_{k_g}, \overline{y}), \overline{d} \rangle \geq -\xi$	19	–	–
	Number of iterations	19	19	113
	Time (seconds)	7.060	6.399	88.557
11B	$k_\alpha \mid \alpha_k = 1; \forall k \geq k_\alpha$	8	8	$\alpha = 1$ not achieved
	$k_g \mid \langle \nabla_x f(x_{k_g}, \overline{y}), \overline{d} \rangle \geq -\xi$	22	–	–
	Number of iterations	22	22	141
	Time (seconds)	9.835	7.571	109.378
12A	$k_\alpha \mid \alpha_k = 1; \forall k \geq k_\alpha$	7	7	$\alpha = 1$ not achieved
	$k_g \mid \langle \nabla_x f(x_{k_g}, \overline{y}), \overline{d} \rangle \geq -\xi$	14	–	–
	Number of iterations	14	14	52
	Time (seconds)	6.259	4.987	37.644
12B	$k_\alpha \mid \alpha_k = 1; \forall k \geq k_\alpha$	7	7	$\alpha = 1$ not achieved
	$k_g \mid \langle \nabla_x f(x_{k_g}, \overline{y}), \overline{d} \rangle \geq -\xi$	14	–	–
	Number of iterations	14	14	60

Table 1C (continued)

	Problem	QN1	QN2	Kiw
	Time (seconds)	6.259	4.987	45.676
13A	$k_\alpha \mid \alpha_k = 1; \forall k \ge k_\alpha$	4	4	$\alpha = 1$ not achieved
	$k_g \mid \langle \nabla_x f(x_{k_g}, \bar{y}), \bar{d} \rangle \ge -\xi$	7	–	–
	Number of iterations	7	7	42
	Time (seconds)	3.615	3.234	40.699
13B	$k_\alpha \mid \alpha_k = 1; \forall k \ge k_\alpha$	4	4	$\alpha = 1$ not achieved
	$k_g \mid \langle \nabla_x f(x_{k_g}, \bar{y}), \bar{d} \rangle \ge -\xi$	8	–	–
	Number of iterations	8	8	45
	Time (seconds)	3.926	3.375	41.630
14A	$k_\alpha \mid \alpha_k = 1; \forall k \ge k_\alpha$	$\alpha = 1$ not achieved	$\alpha = 1$ not achieved	$\alpha = 1$ not achieved
	$k_g \mid \langle \nabla_x f(x_{k_g}, \bar{y}), \bar{d} \rangle \ge -\xi$	32	–	–
	Number of iterations	32	32	6
	Time (seconds)	25.637	21.461	9.193 Failure: $\alpha_* = 0$ $\Phi(x_*) = 55.24477$
14B	$k_\alpha \mid \alpha_k = 1; \forall k \ge k_\alpha$	$\alpha = 1$ not achieved	$\alpha = 1$ not achieved	$\alpha = 1$ not achieved
	$k_g \mid \langle \nabla_x f(x_{k_g}, \bar{y}), \bar{d} \rangle \ge -\xi$	None	–	–
	Number of iterations	40	40	6
	Time (seconds)	32.257	27.210	9.193 Failure: $\alpha_* = 0$ $\Phi(x_*) = 55.24477$

Table 1D Performance of QN1, QN2 and Kiw

	Problem	QN1	QN2	Kiw
15A	$k_\alpha \mid \alpha_k = 1; \forall k \ge k_\alpha$	$\alpha = 1$ not achieved	$\alpha = 1$ not achieved	$\alpha = 1$ not achieved
	$k_g \mid \langle \nabla_x f(x_{k_g}, \bar{y}), \bar{d} \rangle \ge -\xi$	30	–	–
	Number of iterations	30	30	20
	Time (seconds)	28.591	25.887	35.421 Failure: $\alpha_* = 0$ $\Phi(x_*) = 64.50730$
15B	$k_\alpha \mid \alpha_k = 1; \forall k \ge k_\alpha$	$\alpha = 1$ not achieved	$\alpha = 1$ not achieved	$\alpha = 1$ not achieved

Table 1D (continued)

Problem	QN1	QN2	Kiw
$k_g \mid \langle \nabla_x f(x_{k_g}, \overline{y}), \overline{d} \rangle \geq -\xi$	34, 35	–	–
Number of iterations	35	35	20
Time (seconds)	39.938	41.620	35.421
		Failure: $\alpha_* = 0$	Failure: $\alpha_* = 0$
		$\Phi(x_*) = 62.70578$	$\Phi(x_*) = 64.50730$
16A $\quad k_\alpha \mid \alpha_k = 1; \forall k \geq k_\alpha$	$\alpha = 1$ not achieved	$\alpha = 1$ not achieved	$\alpha = 1$ not achieved
$k_g \mid \langle \nabla_x f(x_{k_g}, \overline{y}), \overline{d} \rangle \geq -\xi$	None	–	–
Number of iterations	53	53	11
Time (seconds)	56.631	50.272	17.546
			Failure: $\alpha_* = 0$
			$\Phi(x_*) = 12.79971$
16B $\quad k_\alpha \mid \alpha_k = 1; \forall k \geq k_\alpha$	$\alpha = 1$ not achieved	$\alpha = 1$ not achieved	$\alpha = 1$ not achieved
$k_g \mid \langle \nabla_x f(x_{k_g}, \overline{y}), \overline{d} \rangle \geq -\xi$	None	–	–
Number of iterations	60	60	11
Time (seconds)	61.909	56.221	17.546
			Failure: $\alpha_* = 0$
			$\Phi(x_*) = 12.79971$
17A $\quad k_\alpha \mid \alpha_k = 1; \forall k \geq k_\alpha$	$\alpha = 1$ not achieved	$\alpha = 1$ not achieved	$\alpha = 1$ not achieved
$k_g \mid \langle \nabla_x f(x_{k_g}, \overline{y}), \overline{d} \rangle \geq -\xi$	21	–	–
Number of iterations	21	21	29
Time (seconds)	1.492	0.601	2.583
17B $\quad k_\alpha \mid \alpha_k = 1; \forall k \geq k_\alpha$	$\alpha = 1$ not achieved	$\alpha = 1$ not achieved	$\alpha = 1$ not achieved
$k_g \mid \langle \nabla_x f(x_{k_g}, \overline{y}), \overline{d} \rangle \geq -\xi$	23	–	–
Number of iterations	23	23	35
Time (seconds)	1.512	0.611	3.014
18A $\quad k_\alpha \mid \alpha_k = 1; \forall k \geq k_\alpha$	$\alpha = 1$ not achieved	$\alpha = 1$ not achieved	$\alpha = 1$ not achieved
$k_g \mid \langle \nabla_x f(x_{k_g}, \overline{y}), \overline{d} \rangle \geq -\xi$	12	–	–
Number of iterations	12	12	36
Time (seconds)	1.412	0.551	5.187
18B $\quad k_\alpha \mid \alpha_k = 1; \forall k \geq k_\alpha$	$\alpha = 1$ not achieved	$\alpha = 1$ not achieved	$\alpha = 1$ not achieved
$k_g \mid \langle \nabla_x f(x_{k_g}, \overline{y}), \overline{d} \rangle \geq -\xi$	22	–	–
Number of iterations	22	22	46
Time (seconds)	1.712	0.681	6.830

Table 1E Performance of QN1, QN2 and Kiw

Problem	QN1	QN2	Kiw
19A $k_\alpha \mid \alpha_k = 1; \forall k \geq k_\alpha$	$\alpha = 1$ not achieved	$\alpha = 1$ not achieved	$\alpha = 1$ not achieved
$\quad k_g \mid \langle \nabla_x f(x_{k_g}, \overline{y}), \overline{d} \rangle \geq -\xi$	None	–	–
$\quad$ Number of iterations	25	25	17
$\quad$ Time (seconds)	0.711	0.661	3.665
19B $k_\alpha \mid \alpha_k = 1; \forall k \geq k_\alpha$	$\alpha = 1$ not achieved	$\alpha = 1$ not achieved	$\alpha = 1$ not achieved
$\quad k_g \mid \langle \nabla_x f(x_{k_g}, \overline{y}), \overline{d} \rangle \geq -\xi$	28	–	–
$\quad$ Number of iterations	28	28	23
$\quad$ Time (seconds)	1.572	0.711	4.336
		Failure: $\alpha_* = 0$ $\Phi(x_*) = -3.000000$	
20A $k_\alpha \mid \alpha_k = 1; \forall k \geq k_\alpha$	3	3	$\alpha = 1$ not achieved
$\quad k_g \mid \langle \nabla_x f(x_{k_g}, \overline{y}), \overline{d} \rangle \geq -\xi$	13	–	–
$\quad$ Number of iterations	13	13	140
$\quad$ Time (seconds)	0.541	0.531	4.487
20B $k_\alpha \mid \alpha_k = 1; \forall k \geq k_\alpha$	3	3	$\alpha = 1$ not achieved
$\quad k_g \mid \langle \nabla_x f(x_{k_g}, \overline{y}), \overline{d} \rangle \geq -\xi$	13	–	–
$\quad$ Number of iterations	14	14	144
$\quad$ Time (seconds)	0.551	0.571	4.796
			Failure: $\alpha_* = 0$ $\Phi(x_*) = 2.249999$
21A $k_\alpha \mid \alpha_k = 1; \forall k \geq k_\alpha$	20	20	$\alpha = 1$ not achieved
$\quad k_g \mid \langle \nabla_x f(x_{k_g}, \overline{y}), \overline{d} \rangle \geq -\xi$	22	–	–
$\quad$ Number of iterations	22	22	423
$\quad$ Time (seconds)	0.591	0.591	44.274
21B $k_\alpha \mid \alpha_k = 1; \forall k \geq k_\alpha$	20	20	$\alpha = 1$ not achieved
$\quad k_g \mid \langle \nabla_x f(x_{k_g}, \overline{y}), \overline{d} \rangle \geq -\xi$	22	–	–
$\quad$ Number of iterations	22	22	1817
$\quad$ Time (seconds)	0.591	0.591	197.534

References

Charalambous, C. and J.W. Bandler (1976). "Nonlinear Minimax Optimization as a Sequence of Least pth Optimization with Finite Values of p", *International Journal of System Science*, 7, 377–391.

Conn, A.R. (1979), "An Efficient Second Order Method to Solve the Constrained Minmax Problem", Report, Department of Combinatorics and Optimization, University of Waterloo.

Demyanov, V.F. and V.N. Malozemov (1974). *Introduction to Minimax*, John Wiley, New York.

Kiwiel, K.C. (1987). "A Direct Method of Linearization for Continuous Minimax Problems", *Journal of Optimization Theory and Applications*, 55, 271–287.

Panin, V.M. (1981). "Linearization Method for Continuous Min-Max Problems", *Kibernetika*, 2, 75–78.

Polak, E., D.Q. Mayne and J.E. Higgins (1991). "Superlinearly Convergent Algorithm for Min-Max Problems", *Journal of Optimization Theory and Applications*, 69, 407–439.

Rustem, B. and Q. Nguyen (1998). "An algorithm for the Inequality Constrained Discrete Minimax Problem", *SIAM Journal on Optimization*, 8, 265–283.

Chapter 6

Minimax as a robust strategy for discrete rival scenarios

The discrete minimax problem arises when the worst-case is to be determined over a discrete set. The latter is characterized by a discrete number of scenarios. Minimax is thus the best strategy in view of the worst-case scenario.

In the presence of a discrete set of rival decision models, forecasts or scenarios purporting to describe the same system, the optimal decision needs to take account of all possible representations. The minimax problem arises when statistical or economic analysis cannot rule out all but one of the rival possibilities. We then need to consider the optimal strategy corresponding to the worst case. Optimality is no longer determined by a single scenario but by all scenarios simultaneously.

In this chapter, we discuss the discrete minimax problem and the robust character of its solution. We consider nonlinear equality and inequality constraints and use an augmented Lagrangian formulation to characterize the problem. The solution algorithm is discussed in Chapter 7.

1 INTRODUCTION TO RIVAL MODELS AND FORECAST SCENARIOS

Forecasting with rival models is usually resolved by some form of forecast pooling (see, e.g., Fuhrer and Haltmaier, 1986; Granger and Newbold, 1977; Lawrence et al., 1986; Makridakis and Winkler, 1983). In policy optimization, a similar approach leads to the pooling of objective functions derived from the rival models (see Rustem, 1987, 1994). Chow (1979) initially formulated a robust policy approach for two rival economic models. This approach obtains the optimal policy based only on one model and evaluates its effect if the second model turns out to actually represent the system. A "payoff matrix" is constructed and the strategy chosen is the optimal strategy based on the model that inflicts the lesser damage when implemented on the rival model. There is a discrete choice set of policy strategies. Each member of the set is an optimal policy derived using only one of the models as the true representation of the economy.

In this chapter, we extend the intuitive concept considered by Chow to the determination of the worst-case rival scenario *simultaneously* with the minimization over x. Policy choice is no longer restricted to be discrete, dependent on any one model, as optimality is no longer based on one model only. The best decision is computed given that the worst case is determined using all the scenarios. An example of this approach, with two rival macroeconomic models, is discussed in Becker et al. (1986).

We introduce the minimax strategy by considering the pooling of rival objective functions. Let ϑ denote pooling weights, with

$$\vartheta \in \mathbb{E}_+^{m^{sce}} \equiv \left\{\vartheta \in \Re^{m^{sce}} \mid \vartheta \geq 0, \langle 1, \vartheta \rangle = 1\right\}$$

where $1 \in \Re^{m^{sce}}$ denotes the vector with every component unity. Consider the optimal decision problem formulated as the constrained optimization of the pooled objective functions subject to nonlinear constraints

$$\min_x \{\langle \vartheta, f(x) \rangle \mid g(x) = 0, \ h(x) \leq 0\} \tag{1.1}$$

where $\vartheta \in \mathbb{E}_+^{m^{sce}}$, $x \in \Re^n$, $f : \Re^n \to \Re^{m^{sce}}$, $g : \Re^n \to \Re^e$, $h : \Re^n \to \Re^i$ and f, g and h are twice continuously differentiable functions. In (1.1) each element of f, denoted by f^j, represents a rival objective function corresponding to the jth rival model or scenario. The restrictions $g(x) = 0$ and $h(x) \leq 0$ are the equality and inequality constraints imposed on the overall decision problem. In the above formulation, the vector ϑ is fixed and defines the pooling weights. Also, generally, the number of scenarios or models is much less than the number of decision variables, that is, $n \gg m^{sce}$. One reason for this is that policy optimization is essentially dynamic in nature and the total number of decision variables is the product of the decision variables for one time period and the number of time periods. Further motivation for (1.1) is discussed in Rustem (1987).

Among possible choices of pooling weights, the robust pooling corresponds to the strategy that is invariant to whichever rival scenario, or model, actually turns out to represent the system. The pooling that corresponds to the robust policy is given by the solution of the minimax problem

$$\min_x \max_\vartheta \left\{\langle f(x), \vartheta \rangle \mid g(x) = 0, \ h(x) \leq 0, \ \vartheta \in \mathbb{E}_+^{m^{sce}}\right\} \tag{1.2}$$

where the worst-case rival model scenario is computed *simultaneously* with the minimization over x. It is shown in Lemma 3.1 below that the solution of (1.2) has a robust character. Whichever rival model turns out to represent the actual system, the optimal (minimax) strategy ensures that the objective function value will not deteriorate: it will be at least as good as the minimax value. The solution of (1.2) yields a value of ϑ corresponding to the robust minimax strategy. If the insurance provided by the robust, and cautious, policy has too high a cost in policy terms, then the decision maker could base the optimal policy on (1.1) with ϑ chosen in the neighborhood of the minimax value such that the policy can be made "acceptable" and at the same time as robust as possible. It must be said that *this strategy is only robust with respect to the known rival representations of the system.* If a further, possibly unknown, model of the economy exists and is not included in the set of rival models in (1.2), clearly the computed strategy cannot be expected to be robust with respect to this additional model.

In Chapter 7, we discuss an algorithm for solving the equality and inequality constrained the minimax problem (1.2). In Section 2, the equivalent discrete minimax formulation of (1.2) and its reformulation as a nonlinear programming problem is given. The optimality condition of (1.2) is consequently derived. In Section 3, the robustness of the minimax formulation is described and in Section 4, an augmented Lagrangian formulation is introduced as a convexification procedure for the underlying problem.

2 THE DISCRETE MINIMAX PROBLEM

The pooling minimax formulation (1.2) can be reformulated as a discrete minimax problem. The advantages of the latter are that its interpretation as a worst-case design problem is emphasized in terms of the discrete scenarios and that the complexity of the algorithm for solving the discrete minimax problem is the same as that of a nonlinear programming algorithm. This is in contrast to the continuous minimax algorithms in Chapters 2, 4 and 5, each iteration of which may require nonlinear programming solutions.

Lemma 2.1   *The minimax problem (1.2) is equivalent to*

$$\min_{x} \max_{i \in \{1,2,\ldots,m^{sce}\}} \left\{ f^i(x) \mid g(x) = 0, \ h(x) \leq 0 \right\}. \qquad (2.1)$$

Proof. The result follows from the fact that the maximum of m^{sce} numbers is equal to the maximum of their convex combination (Medanic and Andjelic, 1971, 1972; Cohen, 1981).   □

It should also be noted that (2.1) can be solved by the nonlinear programming problem

$$\min_{x,v} \{v \mid f(x) \leq 1v, \ g(x) = 0, \ h(x) \leq 0\} \qquad (2.2)$$

where $v \in \Re^1$ is the scalar variable introduced to represent the maximum value.

Remark Lemma 2.1 is used to represent (1.2) with (2.1)–(2.2) whenever this becomes necessary.

Let the Lagrangian function associated with (1.2) be given by

$$L(x, \vartheta, \mu^e, \mu^i, \lambda, \eta) = \langle f(x), \vartheta \rangle + \langle g(x), \mu^e \rangle + \langle h(x), \mu^i \rangle + \langle \vartheta, \lambda \rangle + (\langle 1, \vartheta \rangle - 1)\eta$$

(2.3)

where $\mu^e \in \Re^e$,

$$\mu^i \in \Re^i_+ \equiv \left\{ \mu^i \in \Re^i \mid \mu^i \geq 0 \right\}$$

$$\lambda \in \Re^{m^{sce}}_+ \equiv \left\{ \lambda \in \Re^{m^{sce}} \mid \lambda \geq 0 \right\}$$

and $\eta \in \Re^1$ are the multipliers associated with $g(x) = 0$, $h(x) \leq 0$, $\vartheta \geq 0$ and $\langle 1, \vartheta \rangle = 1$, respectively.

We consider the first order necessary conditions of optimality of (1.2) and (2.2). For the former, we apply the optimality conditions for nonlinear programming problems (see CN 2.2), mindful of the fact that the objective is minimized with respect to x and maximized with respect to ϑ. Let (x_*, ϑ_*) solve problem (1.2). Then there are values $(\mu^e_*, \mu^i_*, \lambda_*, \eta_*)$ which satisfy the following first order necessary conditions:

$$\nabla f(x_*)\vartheta_* + \nabla g(x_*)\mu^e_* + \nabla h(x_*)\mu^i_* = 0$$

(2.4)

$$f(x_*) + \lambda_* + 1\eta_* = 0$$

(2.5)

$$h(x_*) \leq 0; \quad g(x_*) = 0, \quad \langle \mu^i_*, h(x_*) \rangle = 0, \quad \mu^i \in \Re^i_+$$

(2.6)

$$\langle \vartheta_*, \lambda_* \rangle = 0, \quad \lambda_* \in \Re^{m^{sce}}_+, \quad \vartheta_* \in \mathbb{E}^{m^{sce}}_+$$

(2.7)

As (2.2) is a nonlinear programming problem in $(x, v) \in \Re^{n+1}$, the optimality conditions are obtained by simple application of the first optimality conditions in CN 2.2. Let (x_*, ϑ_*) solve problem (2.2). Then, by applying CN 2.2, we conclude that there are values $(\vartheta_*, \overline{\mu}^e_*, \overline{\mu}^i_*)$ which satisfy

$$\nabla f(x_*)\overline{\vartheta}_* + \nabla g(x_*)\overline{\mu}^e_* + \nabla h(x_*)\overline{\mu}^i_* = 0$$

(2.8)

$$h(x_*) \leq 0, \quad g(x_*) = 0, \quad \langle \overline{\mu}^i_*, h(x_*) \rangle = 0, \quad \overline{\mu}^i \in \Re^i_+$$

(2.9)

$$f(x_*) \leq 1v_*, \quad \overline{\vartheta}^j_* \left(f^j(x_*) - v_* \right) = 0, \quad j = 1, \dots, m^{sce}, \quad \overline{\vartheta}_* \in \mathbb{E}^{m^{sce}}_+$$ (2.10)

where $\overline{\vartheta}$ is the multiplier corresponding to the constraint $f(x) \leq 1v$. The value v_* is the maximum among the elements of vector f and corresponds to $-\eta_*$ in (2.5). The vector λ_* is the slack

$$\lambda_* = 1v_* - f(x_*).$$

By complimentarity, the shadow price $\overline{\vartheta}_*$ and the slack satisfy $\langle \overline{\vartheta}_*, \lambda_* \rangle = 0$. Hence, $(\overline{\vartheta}_*, \overline{\mu}_*^e, \overline{\mu}_*^i)$ is equivalent to $(\vartheta_*, \mu_*^e, \mu_*^i)$ and thus the optimality conditions for the two problems are equivalent.

3 THE ROBUST CHARACTER OF THE DISCRETE MINIMAX STRATEGY

The minimax strategy is essentially the best decision determined simultaneously with the worst-case discrete (rival) scenario. Its character is thus based on the assertion that the worst case can and will occur. The question that arises is the performance of the minimax strategy if the worst case does not happen. The answer to that is that if the worst case does not happen, the performance of the system, measured by the objective value, will be superior. In other words, the decision maker will be better off. This is discussed in detail in Lemma 3.1.

3.1 Naive Minimax

A related strategy is naive minimax. To explain what we mean by that term consider three scenarios represented by three objective functions: f^1, f^2, f^3. Assuming that scenario i will occur, the optimal strategy is given by

$$\min \left\{ f^i(x) \mid g(x) = 0, \ h(x) \le 0 \right\}$$

and this can be evaluated for $i = 1, 2, 3$. We denote the solution of each problem as x_*^i. Having optimized each problem, it is then possible to evaluate the effect of adopting x_*^i if another scenario $j \ne i$ occurs. The cross-evaluations are given by

$$f^i(x_*^j), \quad i \ne j.$$

Adopting x_*^i, we can evaluate the worst damage if any scenario, other than i, is realized by considering

$$\max_{j \ne i} f^i(x_*^j) - f^i(x_*^i)$$

for $i = 1, 2, 3$. Evaluating the strategy that minimizes the worst-case damage corresponds to

$$\min_{i=1,2,3} \max_{j \ne i} f^i(x_*^j) - f^i(x_*^i).$$

Adopting the value x_*^i, $i = 1, 2, 3$, which results in the least damage, or deterioration, if having implemented decision x_*^i, scenario $j \ne i$ is realized, yields naive minimax. In this case, x_*^i is the optimal solution corresponding to the ith scenario and the naive minimax solution adopted is one of $i = 1, 2, 3$.

The *naive* strategy is therefore to choose the x_*^i which would be least damaging if some other scenario, $j \neq i$, occurs. Thus, *optimality in naive minimax is defined in terms of a single scenario.* By contrast, optimality for (1.2), or (2.1)–(2.2), is defined in terms of all the scenarios, as indicated in the conditions (2.4)–(2.7). Hence, in (1.2), or (2.1)–(2.2), the determination of x_* and the worst case is simultaneous and not sequential as in naive minimax. Furthermore, in (2.2), inequality $f(x) \leq 1v$, in which v corresponds to the maximum value being minimized, indicates that the minimax solution is noninferior to *naive* minimax

$$f^i(x_*^j) - f^i(x_*^i) \geq f^i(x_*^j) - v_*, \quad i = 1, 2, 3, \quad i \neq j$$

as x_*^i, $i = 1, 2, 3$, is also a feasible point of (2.2).

3.2 Robustness of the Minimax Strategy

The robustness of minimax arises from the fact that it is simply the best strategy determined simultaneously with, and corresponding to, the worst case. As optimality is defined in view of all the scenarios, the performance level of the minimax strategy is guaranteed and will improve if any scenario, other than the worst case, is realized. These characteristics are formalized in the following result.

Lemma 3.1 (Guaranteed Performance, Noninferiority, Robustness of Minimax) *Let*

(i) *there exist a minimax solution to (1.2), denoted by (x_*, ϑ_*), with associated multipliers $(\mu_*^e, \mu_*^i, \lambda_*, \eta_*)$;*

(ii) $f(x), g(x), h(x) \in \mathbb{C}^1$ *at x_*; and,*

(iii) *strict complementarity holds for $\vartheta \geq 0$ at the solution (i.e., $\vartheta_* = 0 \Rightarrow \lambda_* > 0$ and $\vartheta_* > 0 \Rightarrow \lambda_* = 0$).*

Then, for $i, j, \ell \in \{1, 2, ..., m^{sce}\}$, we have

(a) $f^i(x_*) = f^j(x_*), \forall i, j \ (i \neq j)$ *iff* $\vartheta_*^i, \vartheta_*^j \in (0, 1)$;

(b) $f^i(x_*) = f^j(x_*) > f^\ell(x_*), \forall i, j, \ell (\ell \neq i, j)$ *iff* $\vartheta_*^\ell = 0$ *and* $\vartheta_*^i, \vartheta_*^j \in (0, 1)$;

(c) $f^i(x_*) > f^j(x_*), \forall j, \ (j \neq i)$ *iff* $\vartheta_*^i = 1$;

(d) $f^i(x_*) < f^j(x_*), \forall j (j \neq i)$ *iff* $\vartheta_*^i = 0$.

Proof. Necessity in case (a) can be shown by considering (2.7), which, for $\vartheta_*^i, \vartheta_*^j \in (0, 1)$, yields

$$\vartheta_*^i \lambda_*^i = \vartheta_*^j \lambda_*^j = 0$$

and thence $\lambda_*^i = \lambda_*^j = 0$. Using (2.5) we have $f^i(x_*) = f^j(x_*)$. Sufficiency is established with $f^i(x_*) = f^j(x_*)$ and noting from (2.5)

$$\langle \vartheta_*, f(x_*) - \lambda_* + 1\eta_* \rangle = 0.$$

From (2.7), we have $\langle 1, \vartheta_* \rangle = 1$ and $\langle \vartheta_*, \lambda_* \rangle = 0$, and

$$\eta_* = -\langle f(x_*), \vartheta_* \rangle.$$

Premultiplying the equality (2.5) by 1 and using this equality yields

$$0 = \langle 1, f(x_*) \rangle - \langle 1, \lambda_* \rangle + \langle 1, 1 \rangle \eta_* = -\langle 1, \lambda_* \rangle.$$

By (2.7), $\lambda_* = 0$ and strict complementarity implies that $\vartheta_*^i, \vartheta_*^j \in (0, 1), \forall i, j$.

Case (b) can be shown by considering (2.7) for $\vartheta_*^i, \vartheta_*^j \in (0, 1)$, $\vartheta_*^\ell = 0$. We have

$$\vartheta_*^i \lambda_*^i = \vartheta_*^j \lambda_*^j = \vartheta_*^\ell \lambda_*^\ell = 0$$

thence $\lambda_*^i = \lambda_*^j = 0$ and, by strict complementarity, $\lambda_*^\ell > 0$. From (2.5) we have

$$0 = f^m(x_*) + \lambda_*^m + 1\eta_*, \quad m = i, j \tag{3.1}$$

$$0 = f^\ell(x_*) + \lambda_*^\ell + 1\eta_* \tag{3.2}$$

and combining these yields

$$f^\ell(x_*) - f^m(x_*) = -\lambda_*^\ell < 0, \quad m = i, j.$$

For sufficiency, let $f^i(x_*) = f^j(x_*) > f^\ell(x_*)$. Combining (3.1), (3.2) and using (2.7), we have

$$\vartheta_*^\ell \left(f^\ell(x_*) - f^m(x_*) \right) = \vartheta_*^\ell \left(\lambda_*^m - \lambda_*^\ell \right) = \vartheta_*^\ell \lambda_*^m \geq 0.$$

Since $f^\ell(x_*) - f^m(x_*) < 0$, we have $\vartheta_*^\ell = 0$. With $\vartheta_*^\ell = 0$, $\forall \ell, f^\ell(x_*) < f^m(x_*)$, we can use (a) for those i, j for which $f^i(x_*) = f^j(x_*)$ to establish $\lambda_*^i = \lambda_*^j = 0$. By strict complementarity this implies that $\vartheta_*^i, \vartheta_*^j \in (0, 1)$.

Case (c) can be established noting that for $\vartheta_*^i = 1$, we have $\lambda_*^i = 0$, $\vartheta_*^j = 0$, $\forall j \neq i$ and, by strict complementarity, $\lambda_*^j > 0$. From (2.5) and (2.7) we obtain

$$f^j(x_*) - f^i(x_*) \leq \lambda_*^i - \lambda_*^j = -\lambda_*^j < 0.$$

Conversely, $f^i(x_*) > f^j(x_*)$ implies

$$\vartheta_*^j\left(f^i(x_*) - f^i(x_*)\right) = \vartheta_*^j \lambda_*^i \geq 0$$

and thus $\vartheta_*^j = 0$, $\forall j \neq i$. Case (d) is the converse of (c). $\square$

If the min max over three functions $f^1(x), f^2(x)$ and $f^3(x)$ is being computed, then, at the solution, $f^1 = f^2 > f^3$ iff $\vartheta_1, \vartheta_2 \in (0, 1]$ and $\vartheta_3 = 0$ or $f^1 > f^2(x) \geq f^3$ iff $\vartheta_1 = 1$ and $\vartheta_2 = \vartheta_3 = 0$. Suppose that f^1, f^2, f^3 correspond to three rival forecast scenarios, one of which is the description of the actual system in the future. With $f^1 = f^2 > f^3$, at the minimax solution, the decision maker need not care, as far as the objective function values are concerned, if the actual state turns out to be f^1 or f^2. If it is f^3, then the decision maker is better off. The value of the multiplier vector ϑ reflects this. Lemma 3.1 states this in greater generality. Since ϑ is chosen to maximize the Lagrangian, the solution can be seen as a robust optimum in the sense of a worst-case design problem.

This result illustrates the way in which ϑ_* is related to $f(x_*)$. When some of the elements of ϑ_* are such that $\vartheta_*^i \in (0, 1)$ for some $i \in M \subseteq \{1, 2, ..., m^{sce}\}$, it is shown that the $f^i(x_*)$, $i \in M$, have the same value. *In this case, the optimal policy x_* yields the same objective function value whichever forecast scenario happens to be realized.* Thus, x_* is a *robust policy*. The investor is ensured that implementing x_* will yield an objective function value that is at least as good as the minimax optimum. This noninferiority of x_* may, on the other hand, amount to a cautious approach with a high cost. The investor can, in such circumstances, use ϑ_* as a guide and seek in its neighborhood a slightly less cautious scheme that is politically more acceptable. Choosing $\vartheta = \overline{\vartheta}$ from a reasonably close neighborhood of ϑ_*, the optimal strategy is based on (1.1). In (1.1), $\overline{\vartheta}$ is fixed and represents a pooling of all the scenarios. In general, $\overline{\vartheta} \in \mathfrak{R}_+^{m^{sce}}$ can be chosen arbitrarily to reflect the views and expectations of the investor. However, as all expected value optimization needs to be justified in view of the worst-case scenario, a choice in the neighborhood of ϑ_* would be desirable.

3.3 An Example

In order to motivate the minimax formulation, consider the multiple objective problem (1.1) where f is given by

$$f(x) \equiv \begin{bmatrix} f^1(x) \\ f^2(x) \\ f^3(x) \end{bmatrix} = \begin{bmatrix} (ax - b)^2 \\ (cx - d)^2 \\ \frac{1}{2}(cx^* - d)^2 - g \end{bmatrix}$$

where $a, b, c, d, \neq 0$ and $g > 0$. $f^1(x), f^2(x), f^3(x)$ are minimized by $x_*^1 = b/a$,

$x_*^2 = x_*^3 = d/c$, respectively, and

$$0 = f^1(x_*^1) = f^2(x_*^2) > f^3(x_*^3) = -g.$$

We note that, $f^3(x) < f^2(x)$, for all x. Thus, by Lemma (3.1), the solution of (1.2) is such that $\vartheta_*^3 = 0$.

Furthermore, since

$$\sum_{i=1}^{3} \vartheta^i = 1, \vartheta^1 = 1 - \vartheta^2$$

we can reduce the three-dimensional problem in $(\vartheta^1, \vartheta^2, \vartheta^3)$ a single dimension, ϑ. Hence, (1.1) can be written as

$$\min_x \left\{ \vartheta f^1(x) + (1 - \vartheta) f^2(x) \right\}. \tag{3.3}$$

Given a specified value of ϑ, $x(\vartheta)$ that minimizes (3.3) is given by the optimality condition

$$\frac{\partial \left(\vartheta f^1(x) + (1 - \vartheta) f^2(x) \right)}{\partial x} = 2\vartheta a(ax - b) + 2(1 - \vartheta)c(cx - d) = 0 \tag{3.4}$$

where $\vartheta \in [0, 1]$, and

$$x(\vartheta) = \frac{\vartheta ab + (1 - \vartheta)cd}{\vartheta a^2 + (1 - \vartheta)c^2}. \tag{3.5}$$

In addition, (1.2) indicates the conditions for determining ϑ. Assuming, for simplicity, that $\vartheta \in (0, 1)$ we only need to consider

$$\frac{\partial \left(\vartheta f^1(x) + (1 - \vartheta) f^2(x) \right)}{\partial \vartheta} = 0. \tag{3.6}$$

The other conditions relate to the constraints on ϑ involving a lagrangian and:

$$\vartheta \geq 0, \quad 1 - \vartheta \geq 0, \quad \mu^1 \vartheta = \mu^2 (1 - \vartheta) = 0, \quad \mu^1 \geq 0, \quad \mu^2 \geq 0 \tag{3.7}$$

and, by appropriate choice of the problem parameters, these may be satisfied. Hence, they are ignored in this example. Nevertheless, (3.4)–(3.7) are deduced from the optimality conditions discussed in CN 2.2, and (2.4)–(2.10) for the general discrete minimax problem (1.2). The significance of (3.6) is that, when $\vartheta \in (0, 1)$, the minimax solution aims to make both objectives equal in value. Hence the solution implied by (3.6) is

$$x_* = \frac{b \pm d}{a \pm c}. \tag{3.8}$$

The corresponding value of ϑ in (3.5) is given by the choice

$$\vartheta = \frac{\pm c}{a \pm c}.$$ (3.9)

In other words, for ϑ given by (3.9), (3.5) reduces to (3.8) with the minimax objective function values

$$f^1(x_*) = f^2(x_*) = \frac{(ad - bc)^2}{(a \pm c)^2}.$$ (3.10)

The third objective function, on the other hand, does not correspond to the maximum since

$$f^3(x_*) = \frac{1}{2}\frac{(ad - bc)^2}{(a \pm c)^2} - g = \frac{1}{2}f^1(x_*) - g = \frac{1}{2}f^2(x_*) - g < f^2(x_*).$$

In Figure 3.1, the values of $f^1(x(\vartheta))$, $f^2(x(\vartheta))$ and $\vartheta f^1(x(\vartheta)) + (1 - \vartheta)f^2(x(\vartheta))$ are plotted for a given set of the parameters a, b, c, d used in Table 3.2. The satisfaction of (3.6) and (3.10) are illustrated at the minimax point.

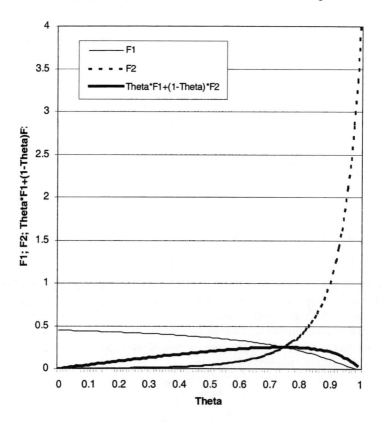

Figure 3.1 Variation of optimal values of F1 and F2 with ϑ.

Table 3.1 The payoff matrix for each strategy and scenario

	Scenario 1 (i.e., $(ax - b)^2$) realized	Scenario 2 (i.e., $(cx - d)^2$) realized	Scenario 3 (i.e., $\frac{1}{2}(cx - d)^2 - g$) realized
Optimal strategy applied based on scenario 1 (i.e., b/a)	0	$\dfrac{(bc - ad)^2}{a^2}$	$\dfrac{1}{2}\dfrac{(bc - ad)^2}{a^2} - g$
Optimal strategy applied: based on scenario 2 (i.e., d/c)	$\dfrac{(ad - bc)^2}{c^2}$	0	$-g$
Optimal strategy applied: based on scenario 3 (i.e., d/c)	$\dfrac{(ad - bc)^2}{c^2}$	0	$-g$
Minimax strategy applied: based on scenarios 1, 2, 3 (i.e., $(b \pm d)/(a \pm c)$)	$\dfrac{(ad - bc)^2}{(a \pm c)^2}$	$\dfrac{(ad - bc)^2}{(a \pm c)^2}$	$\dfrac{1}{2}\dfrac{(ad - bc)^2}{(a \pm c)^2} - g$

Table 3.2 The payoff matrix for $a = 1$, $b = 2$, $c = 3$, $d = 4$, $g = 5$

	Scenario 1 (i.e., $(x - 2)^2$) realized	Scenario 2 (i.e., $(3x - 4)^2$) realized	Scenario 3 (i.e., $\frac{1}{2}(3x - 4)^2 - 5$) realized
Scenario 1 optimal strategy applied (i.e., $x_* = 2$)	0	4	-3
Scenario 2 optimal strategy applied (i.e., $x_* = \frac{4}{3}$)	$\frac{4}{9}$	0	-5
Scenario 3 optimal strategy applied (i.e., $x_* = \frac{4}{3}$)	$\frac{4}{9}$	0	-5
Minimax strategy applied: based on scenarios 1, 2, 3 ($\vartheta_* = \frac{3}{4}$, $x_* = \frac{3}{2}$)	$\frac{1}{4}$	$\frac{1}{4}$	$-4\frac{7}{8}$

In order to compare these results with naive minimax, we construct Table 3.1. The significance of minimax may be indicated by assigning values to a, b, c, d, g. Table 3.2 illustrates such an example.

From the figures in Table 3.2, it can be verified that, scenario 3 is consistently more optimistic than scenario 2. Thus, scenario 3 cannot represent the worst case and column and row 3 and can be eliminated from consideration. An individual scenario (i.e., 1, 2 or 3) does yield the best result (i.e., 0 is attained), assuming that it is actually realized. Performance deteriorates to 4 or 4/9 if another scenario (i.e., scenario 2 or scenario 3) is realized. The minimax strategy ensures that performance is the same (i.e., 1/4) for both scenarios. This is a small deterioration from the optima of 0 for both individual scenarios but the strategy also provides protection against deterioration from 0 to 4, or $\frac{4}{9}$. In addition, the performance of minimax is assured to improve if any scenario other than the worst case is realized. This is illustrated in the performance of minimax strategy for scenario 3 (i.e., $-4\frac{7}{8}$).

Finally, *naive* minimax would choose the optimal strategy corresponding to scenario 2 (in contrast to the optimal strategy corresponding to scenario 1) as it gives rise to lesser damage if scenario 1 is realized.

4 AUGMENTED LAGRANGIANS AND CONVEXIFICATION OF DISCRETE MINIMAX

In Section 2, we formulated the Lagrangian associated with (1.2) as $L(x, \vartheta, \mu^e, \mu^i, \lambda, \eta)$, given by (2.3). The characterization of the solution of a constrained minimax problem such as (1.2) as a saddle point of the Lagrangian function is heavily dependent on the convexity properties of the underlying problem (see Demyanov and Malomezov, 1974; Arrow et al., 1973; Rockafeller, 1973). Motivated by the discussions in Arrow and Solow (1958), it has been shown that, in the case of the nonlinear programming problem (i.e., just the min case of the minimax problem below), convexity assumptions can be relaxed via a modified Lagrangian approach (see Arrow et al., 1973; Rockafeller, 1973). Thus, we adopt an augmented Lagrangian approach, which is essentially a convexification procedure, to solve the minimax problem. In this section, we invoke these results to characterize the solution of the minimax problem as a saddle point. The saddle-point formulation and the convexification are subsequently used by the algorithm discussed in Chapter 7.

The characterization of the minimax solution of (1.2) as a saddle point requires the relaxation of convexity assumptions (see Demyanov and Malomezov, 1974; Cohen, 1981). In order to achieve this characterization, we modify (2.3) by augmenting it with a penalty function. Hence, we define the augmented Lagrangian for (1.2) by

$$L^a(x, \vartheta, \mu^e, \mu^i, \lambda, \eta, c, \pi) = L(x, \vartheta, \mu^e, \mu^i, \lambda, \eta)$$

$$+ \frac{1}{2c}\|(ch(x) + \pi)_+\|_2^2 + \frac{c}{2}\|g(x)\|_2^2 \qquad (4.1)$$

where $0 \leq c \in \mathfrak{R}_+^1, 0 \leq \pi \in \mathfrak{R}_+^i$ and $(\cdot)_+$ is such that its ith element is given by

$$(\cdot)_+^i = \max\{(\cdot)^i, 0\}.$$

The augmented Lagrangian ensures that the local convexity assumed to establish the optimality conditions in Section 2 can, to some extent, be extended beyond this locality. This convexification is based on results established in Arrow et al. (1973); Rockafeller (1973). We establish the following result due to Finsler (1937) (also, Arrow et al., 1973; Bellman, 1995) while taking account of the difference in the penalty function

$$\frac{1}{2c}\|(ch(x) + \pi)_+\|_2^2$$

used in (4.1).

Lemma 4.1 (Finsler's Lemma) *Let Q be a real $n \times n$ matrix and let J^T be a real $n \times m$ matrix. Suppose $\langle x, Qx \rangle > 0$ for all $x \neq 0$ such that $J^T x = 0$. Then, for all $x \neq 0$ and for c sufficiently large, we have*

$$\langle x, (Q + cJJ^T)x \rangle > 0.$$

That is, if Q is positive definite on the null space of J^T, then for c sufficiently large $Q + cJJ^T$ is positive definite on the whole space.

Proof. Let

$$\mathbb{K} \equiv \{x \in \mathfrak{R}^n \mid \langle x, x \rangle = 1\}$$

and $\langle J^T x, J^T x \rangle \geq 0$, for all $x \in \mathbb{K}$. Then, we shall show that the following two statements are equivalent in $\mathbb{K}$:

(a) $\langle x, Qx \rangle > 0$ whenever $\langle J^T x, J^T x \rangle = 0$;

(b) for all c sufficiently large, $\langle x, Qx \rangle + c\langle J^T x, J^T x \rangle > 0$, for all x.

By assumption, (a) is satisfied. Hence, from (b), for all c sufficiently large,

$$\langle x, (Q + cJJ^T)x \rangle > 0$$

for $\langle x, x \rangle = 1$, and hence whenever $x \neq 0$.

To establish the equivalence of (a) and (b), we note that (b) trivially implies (a). We consider the converse. Let $\mathbb{K}'$ be a subset of $\mathbb{K}$ on which $\langle x, Qx \rangle \leq 0$. $\mathbb{K}'$ is compact. Since

$$\langle J^{\mathrm{T}}x, J^{\mathrm{T}}x \rangle \geq 0$$

on $\mathbb{K}$, we have

$$\langle x, Qx \rangle + c\langle J^{\mathrm{T}}x, J^{\mathrm{T}}x \rangle \geq \langle x, Qx \rangle > 0, \quad \forall x \in \mathbb{K}\backslash\mathbb{K}' \text{ and } \forall c \geq 0.$$

If $\mathbb{K}' = \emptyset$, the proof would be finished. If $\mathbb{K}' \neq \emptyset$, let

$$\xi_Q = \min\{\langle x, Qx \rangle \mid x \in \mathbb{K}'\}$$

$$\xi_J = \min\left\{\langle J^{\mathrm{T}}x, J^{\mathrm{T}}x \rangle \mid x \in \mathbb{K}'\right\}$$

and $\xi_j = \langle J^{\mathrm{T}}x', J^{\mathrm{T}}x' \rangle$, for some $x' \in \mathbb{K}'$. We have $\langle J^{\mathrm{T}}x, J^{\mathrm{T}}x \rangle \geq 0$, $\forall x \in \mathbb{K}'$. But if

$$\langle J^{\mathrm{T}}x', J^{\mathrm{T}}x' \rangle = 0$$

then, by (a) $\langle x', Qx' \rangle > 0$, contrary to the definition of $\mathbb{K}'$. Hence,

$$\xi_J = \langle J^{\mathrm{T}}x', J^{\mathrm{T}}x' \rangle > 0$$

and clearly for all c sufficiently large, we have

$$\xi_Q + c\xi_J > 0.$$

Then, $\forall x \in \mathbb{K}'$, by definition of a minimum, the inequality

$$\langle x, Qx \rangle + c\langle J^{\mathrm{T}}x, J^{\mathrm{T}}x \rangle \geq \xi_Q + c\xi_J > 0$$

follows for all c sufficiently large. We have thus established the result for $x \in \mathbb{K}\backslash\mathbb{K}'$ and $x \in \mathbb{K}'$. □

Let the jth element of the vector π be defined by

$$\pi^j(x) = \begin{cases} -ch^j(x) & \text{if } h^j(x) \leq 0 \\ \kappa & \text{if } h^j(x) > 0 \end{cases}$$

where $\kappa \geq 0$ is a constant and $c \geq 0$. We note that the minimax algorithm in Chapter 7 adjusts π iteratively, such that at convergence to the solution x_*, we have $\pi_* = -ch(x_*)$.

Theorem 4.1 *Suppose*

(i) π is determined as above, with $\pi_* = -ch(x_*)$;

(ii) x_*, ϑ_*, μ_*^e, μ_*^i, λ_*, η_* satisfy the second order sufficiency conditions in CN 2.2 (with $[\nabla g(x_*) \vdots \nabla \bar{h}(x_*)]$ linearly independent) for x_*, ϑ_* to be a strict local solution to (1.2);

(iii) *strict complementarity holds for $h(x) \leq 0$ (i.e., $h^i(x_*) > 0 \rightarrow (\mu_*^i)^i = 0$ and $h^i(x_*) = 0 \rightarrow (\mu_*^i)^i > 0$);*

(iv) *c is sufficiently large;*

(v) *the function $L^a(x_*, \vartheta_*, \mu_*^e, \mu_*^i, \lambda_*, \eta_*, c, \pi_*)$ has an unconstrained strict local minimum at x_*; and*

(vi) *for all i and j, the expressions*

$$\frac{\partial^2 \left(\frac{1}{2c}\|(ch(x) + \pi_*)_+\|_2^2\right)}{\partial(h^i)^2} \quad \text{and} \quad \frac{\partial^2 \left(\frac{c}{2}\|g(x)\|_2^2\right)}{\partial(g^j)^2} \cdot$$

are continuous at the points $(h^i(x_*), c)$ and $(g^j(x_*), c)$, respectively, where $h^i(x_*) \leq 0$ and $g^j(x_*) = 0$.

Then:

(a) *for all (μ^e, μ^i) sufficiently close to (μ_*^e, μ_*^i), $L^a(x, \vartheta, \mu^e, \mu^i, \lambda, \eta, c, \pi)$ is strictly convex in x close to x_*, and concave in μ^e, μ^i, λ, η and ϑ;*

(b) *the function $L^a(x_*, \vartheta_*, \mu_*^e, \mu_*^i, \lambda_*, \eta_*, c, \pi)$ has an unconstrained local minimum with respect to x and maximum with respect to ϑ.*

Remark The equivalent result for the minimax formulation (2.2) follows directly from Arrow et al. (1973); Rockafeller (1973).

Proof. For any $c > 0$, we have, by the first order necessary conditions of optimality of (1.1) (see CN 2.2),

$$\nabla L_x^a(x_*, \vartheta_*, \mu_*^e, \mu_*^i, \lambda_*, \eta_*, c, \pi_*) = \nabla f(x_*)\vartheta_* + \nabla g(x_*)[cg(x_*) + \mu_*^e]$$

$$+ \nabla h(x_*)\left[(ch(x_*) + \pi_*)_+ + \mu_*^i\right]$$

$$= \nabla f(x_*)\vartheta_* + \nabla g(x_*)\mu_*^e + \nabla h(x_*)\mu_*^i$$

$$= \nabla L(x_*, \vartheta_*, \mu_*^e, \mu_*^i)$$

$$= 0.$$

The Hessian of L^a is given by

$$\nabla_x^2 L^a(x_*, \vartheta_*, \mu_*^e, \mu_*^i, \lambda_*, \eta_*, c, \pi_*) = \sum_{j=1}^{m^{sce}} \nabla^2 f^j(x_*)\vartheta_*^j$$

$$+c\left\{\sum_{\iota=1}^{e}\nabla^2 g^\iota(x_*)g^\iota(x_*) + \nabla g(x_*)\nabla g^{\mathrm{T}}(x_*)\right\}$$

$$+\sum_{i=1}^{i}\nabla^2 h^i(x_*)(ch^i(x_*) + \pi_*^i)_+ + c\nabla h(x_*)\nabla h^{\mathrm{T}}(x_*)$$

$$+\sum_{\iota=1}^{e}\nabla^2 g^\iota(x_*)(\mu_*^e)^\iota + \sum_{i=1}^{i}\nabla^2 h^i(x_*)(\mu_*^i)^i$$

$$= \nabla_x^2 L(x_*, \vartheta_*, \mu_*^e, \mu_*^i)$$

$$+c\left\{\nabla g(x_*)\nabla g^{\mathrm{T}}(x_*) + \nabla h(x_*)\nabla h^{\mathrm{T}}(x_*)\right\}. \qquad (4.2)$$

Defining the index set

$$I_0 = \left\{i \mid h^i(x_*) = 0, (\mu^i)^i > 0\right\}$$

and the Jacobian matrix of active constraints

$$J = \left[\nabla g^1(x_*) \vdots \nabla g^2(x_*) \vdots ... \vdots \nabla g^e(x_*) \vdots \nabla h^i(x_*), ..., \ i \in I_0 \right]$$

by CN 2.2 we have

$$\langle v, \nabla_x^2 L(x_*, \vartheta_*, \mu_*^e, \mu_*^i) v\rangle > 0, \quad \forall v \neq 0, \ J^{\mathrm{T}}v = 0.$$

By Lemma 4.1, $\nabla_x^2 L(x_*, \vartheta_*, \mu^e_*, \mu_*^i) + \bar{v}JJ^{\mathrm{T}}$ is positive definite for some scalar $\bar{v}$ sufficiently large. Now, if c is sufficiently large, we have $c > \bar{v}$ and the matrix

$$\nabla^2 L^a(x_*, \vartheta_*, \mu_*^e, \mu_*^i, \lambda_*, \eta_*, c, \pi_*) = \nabla^2 L(x_*, \vartheta_*, \mu_*^e, \mu_*^i) + \bar{v}JJ^{\mathrm{T}}$$

$$+\left[c\left[\nabla g(x_*)\nabla g^{\mathrm{T}}(x_*) + \nabla h(x_*)\nabla h^{\mathrm{T}}(x_*)\right] - \bar{v}JJ^{\mathrm{T}}\right]$$

is positive definite since

$$\langle v, \left[\nabla_x^2 L(x_*, \vartheta_*, \mu_*^e, \mu_*^i) + \bar{v}JJ^{\mathrm{T}} + \left[c\left[\nabla g(x_*)\nabla g^{\mathrm{T}}(x_*) + \nabla h(x_*)\nabla h^{\mathrm{T}}(x_*)\right]\bar{v}JJ^{\mathrm{T}}\right]\right]v\rangle$$

$$\geq \langle v, \left[\nabla_x^2 L(x_*, \vartheta_*, \mu_*^e, \mu_*^i) + \bar{v}JJ^{\mathrm{T}}\right]v\rangle$$

$$> 0.$$

The inequality constraints, not in I_0, add a nonnegative term to the right-hand side of the above inequality. By continuity, $\nabla_x^2 L^a(x, \vartheta, \mu^e, \mu^i, \lambda_*, \eta_*, c, \pi)$ is positive definite for $(x, \vartheta, \mu^e, \mu^i, \lambda, \eta)$ sufficiently close to $(x_*, \vartheta_*, \mu_*^e, \mu_*^i, \lambda_*, \eta_*)$.

The proof for x, ϑ, μ, η follows from Demyanov and Malomezov (1974, Theorem 5.1).

We now state the saddle point property of $L^a(\cdot)$ in the neighborhood of the minimax solution. This is largely a consequence of Theorem 4.1.

Theorem 4.2 *Let the conditions of Theorem 4.1 be satisfied. Then, we have the saddle point*

$$L^a(x_* \vartheta, \mu^e, \mu^i, \lambda, \eta, c, \pi_*) \leq L^a(x_*, \vartheta_*, \mu_*^e, \mu_*^i, \lambda_*, \eta_*, c, \pi_*)$$

$$\leq L^a(x, \vartheta_*, \mu_*^e, \mu_*^i, \lambda_*, \eta_*, c, \pi_*)$$

for every x in some neighborhood of x_* and for every ϑ, μ^e, μ^i, λ, η.

Remark The equivalent result for the minimax formulation (2.2) follows directly from Arrow et al. (1973) and Rockafeller (1973).

Proof. The proof follows from Theorem 4.1 above for the nonlinear programming part involving x, μ^e, μ^i and Demyanov and Malomezov (1974, Theorem 5.2) for x, ϑ, λ, η.

The left inequality is a consequence of Theorem 4.1. Also, the way π is determined ensures

$$\pi_* = -ch(x_*).$$

The right inequality follows from the fact that $g(x_*) = 0$ and $h(x_*) \leq 0$ and any deviation of the multipliers at their optimal value will result in a value of $L^a(x_*, \mu^e, \mu^i, \lambda, \eta, c, \pi_*)$ that is either the same as, or inferior to (i.e., less than), $L^a(x_*, \mu_*^e, \mu_*^i, \lambda, \eta, c, \pi_*)$. $\square$

References

Arrow, K.J., F.J. Gould and S.M. Howe (1973). "General Saddle Point Result for Constrained Optimization", *Mathematical Programming*, 5, 225–234.

Arrow, K.J. and R.M. Solow (1958). "Gradient Methods for Constrained Maxima, with Weakened Assumptions", in: K. Arrow, L. Hurwicz and H. Uzawa (editors), *Studies in Linear and Nonlinear Programming*, Stanford University Press, Stanford, CA, 166–176.

Becker, R.G., B. Dwolatzky, E. Karakitsos and B. Rustem (1986). "The Simultaneous Use of Rival Models in Policy Optimization", *The Economic Journal*, 96, 425–448.

Bellman, R. (1995). *Introduction to Matrix Analysis*, SIAM, Philadelphia.

Chow, G.C. (1979). "Effective Use of Econometric Models in Macroeconomic Policy Formulation", in: S. Holly, B. Rustem and M. Zarrop (editors), *Optimal Control for Econometric Models*, Macmillan, London, 31–39.

Cohen, G. (1981). "An Algorithm for Convex Constrained Minimax Optimization Based on Duality", *Applied Mathematics & Optimization*, 7, 347-372.

Demyanov, V.F. and V.N. Malomezov (1974). *Introduction to Minimax*, John Wiley, New York.

Demyanov, V.F. and A.B. Pevnyi (1972). "Some Estimates in Minmax Problems", *Kibernetika*, 1, 107–112.

Finsler, P. (1937). "Über das Vorkommen definiter und semidefiniter Formen in schären quadratischen Formen", *Commentarii Mathematicii Helveticii*, 9, 188-192.

Fuhrer, J. and J. Haltmaier (1986). "Minimum Variance Pooling of Forecasts at Different Levels of Aggregation", Special Studies Paper 208, Federal Reserve Board, Washington, DC.

Granger, C. and P. Newbold (1977). *Forecasting Economic Time Series*, Academic Press, New York.

Lawrence, M.J., R.H. Edmunson and M.J. O'Connor (1986). "The Accuracy of Combining Judgemental and Statistical Forecasts", *Management Science*, 32, 1521–1532.

Makridakis, S. and R. Winkler (1983). "Averages of Forecasts: Some Empirical Results" *Management Science*, 29, 987–996.

Medanic, J. and M. Andjelic (1971). "On a Class of Differential Games without Saddle-point Solutions", *Journal of Optimization Theory and Applications*, 8, 413–430.

Medanic, J. and M. Andjelic (1972). "Minmax Solution of the Multiple Target Problem", *IEEE Transactions on Automatic Control*, 17, 597–604.

Rockafeller, R.T. (1973). "A Dual Approach to Solving Nonlinear Programming Problems by Unconstrained Minimization", *Mathematical Programming*, 5, 354–373.

Rustem, B. (1987). "Methods for the Simultaneous Use of Multiple Models in Optimal Policy Design", in: C. Carraro and D. Sartore (editors), *Developments in Control Theory for Economic Analysis*, Martinus Nijhoff Kluwer, Dordrecht, 157–186.

Rustem, B. (1994). "Robust Min-max Decisions with Rival Models", in: D. Belsley (editor), *Computational Techniques for Econometrics and Economic Analysis*, Kluwer, Dordrecht, 109–136.

Chapter 7

Discrete minimax algorithm for equality and inequality constrained models

In this chapter, we consider a sequential quadratic programming algorithm for the discrete minimax problem. As described in Chapter 6, the constraints are used to formulate an augmented Lagrangian. The algorithm involves a sequential quadratic programming subproblem, an adaptive penalty parameter selection rule to regulate the emphasis on constraint satisfaction, an Armijo-type stepsize strategy, convergent to unit steps, that ensures progress towards optimality and feasibility of the constraints.

The algorithm is formulated for general nonlinear constrained problems in which the objective and the constraints are twice continuously differentiable. In the case of linear constraints, the algorithm simplifies considerably. The global convergence of the general algorithm is shown, along with the convergence of the stepsize to unity. It is also shown that the penalty parameter does not grow indefinitely. The local Q-superlinear convergence of the algorithm is demonstrated. The algorithm for linear constraints is also discussed and the convergence properties of this special case are deduced.

1 INTRODUCTION

In this chapter, we consider an algorithm for solving the discrete minimax problem

$$\min_{x} \max_{i \in \{1,2,\dots,m^{sce}\}} \left\{ f^i(x) \mid g(x) = 0, \ h(x) \leq 0 \right\} \qquad (1.1)$$

introduced in Chapter 6. As we have already discussed, this formulation applies to decision models with a discrete set of scenarios. Throughout this chapter, the discussion relies on the equivalence of the above discrete minimax problem with a continuous minimax and a nonlinear programming problem. Thus, using the terminology introduced in Chapter 6, it is established in Lemma 6.2.1 that (1.1) is equivalent to

$$\min_{x} \max_{\vartheta} \left\{ \langle f(x), \vartheta \rangle \mid g(x) = 0, \ h(x) \leq 0, \ \vartheta \in \mathbb{E}_{+}^{m^{sce}} \right\} \qquad (1.2)$$

where

$$\vartheta \in \mathbb{E}_+^{m^{sce}} \equiv \left\{ \vartheta \in \mathfrak{R}^{m^{sce}} \mid \vartheta \geq 0, \ \langle 1, \vartheta \rangle = 1 \right\}. \tag{1.3}$$

Finally, it is shown in Section 6.2 that (1.1) is also equivalent to the nonlinear programming problem

$$\min_{x,v} \{ v \mid f(x) \leq 1v; g(x) = 0, \ h(x) \leq 0 \}. \tag{1.4}$$

These equivalences are explicitly noted whenever they are invoked in the discussion below.

Formulation (1.4) indicates that a specialized algorithm for (1.1) or (1.2) or (1.4) would be worth considering if it is an improvement on the application of existing nonlinear programming algorithms (e.g., Rustem, 1998) to (1.4). Otherwise, the discrete minimax problem (1.1) can be solved using a nonlinear programming algorithm applied to (1.4).

Algorithms for solving mostly the unconstrained discrete minimax problem have been considered by a number of authors, including Charalambous and Conn (1978), Coleman (1978), Conn (1979), Conn and Li (1992), Demyanov and Malozemov (1974), Demyanov and Pevnyi (1972), Dutta and Vidyasagar (1977), Hald and Madsen (1981), Han (1978, 1981), Murray and Overton (1980), Polak et al. (1988), and Womersley and Fletcher (1986).

In the constrained case, discussed in some of the above studies, convergence results to unit step lengths, global convergence and local convergence rates have not been established (e.g., Coleman, 1978; Dutta and Vidyagasar, 1977).

In this chapter, we discuss an algorithm for the general equality and inequality constrained discrete minimax problem. The equality constrained case is based on the algorithm in Rustem (1992) and the inequality constrained case is based on Rustem and Nguyen (1998). In the combined equality and inequality constrained algorithm below, the dual approach in (1.2), formulated originally by Medanic and Andjelic (1971, 1972) and Cohen (1981), is initially adopted. Subsequently, both (1.1) and (1.2) are utilized to formulate the algorithm. The approach uses the augmented Lagrangian formulation (6.4.1) to directly solve the equality and inequality constrained case. The algorithm involves a sequential quadratic programming subproblem, an adaptive penalty parameter selection rule and a stepsize strategy, convergent to unit steps, that ensures progress towards optimality and feasibility for the inequality constraints. The method used for handling the constraints is based on the nonlinear programming algorithm discussed in Rustem (1993). Numerical results are discussed in Rustem (1994b) and Rustem and Nguyen (1998).

The stepsize strategy used by the algorithm aims at measuring progress towards feasibility and the minimax solution. The algorithm defines the direction of progress as a quasi-Newton step obtained from a quadratic subproblem. The penalty parameter of the augmented Lagrangian function is determined adaptively. The growth in the penalty parameter is required only to ensure a descent property. It is shown that this penalty parameter does not grow indefinitely. Convergence of the overall algorithm, convergence to unit steplengths and Q-superlinear convergence rates are discussed.

In the unconstrained case, the algorithm discussed in this chapter is similar to that of Han (1978, 1981). One difference is the stepsize strategy, (3.9a) below. Both in the unconstrained and the constrained cases, the conditions for the attainment of unit stepsizes are established below. These are related to the accuracy of the projection of the Hessian approximation used by the quadratic subproblem. Various characterizations of the Q-superlinear convergence of the algorithm are discussed in Section 6. The attainment of unit stepsizes are illustrated with numerical examples in Rustem (1994a) and Rustem and Nguyen (1998). These results highlight the properties of the algorithm related to the penalty parameter and the unit stepsize achievement.

2 BASIC CONCEPTS

Let $\mathfrak{C}(\cdot)$ denote the Hessian of L^a, with respect to x, evaluated at $(\cdot)$ (see (6.4.2)).

We also denote by $\mathcal{H}(\cdot)$, $\mathcal{G}(\cdot)$ the matrices

$$\mathcal{H}(x) = [\nabla_x h^1(x), ..., \nabla_x h^i(x)]$$

$$\mathcal{G}(x) = [\nabla_x g^1(x), ..., \nabla_x g^e(x)].$$

Sometimes, $\mathcal{H}(x)$, $\mathcal{G}(x)$ evaluated at x_k will be denoted by $\mathcal{H}_k$, $\mathcal{G}_k$ and $h(x_k)$, $g(x_k)$ will be denoted by h_k, g_k, respectively. Thus, local linearizations of $h(x)$ and $g(x)$, at x_k, can be written as

$$h(x) \stackrel{\sim}{=} h_k + \mathcal{H}_k^{\mathrm{T}}[x - x_k]$$

$$g(x) \stackrel{\sim}{=} g_k + \mathcal{G}_k^{\mathrm{T}}[x - x_k].$$

Let the linearized active inequality constraints

$$\left\{ i \in \{1, ..., i\} \mid h_k^i + \langle \nabla h^i(x_k), x - x_k \rangle = 0 \right\} \tag{2.1}$$

be given by

$$\overline{h}_k + \overline{\mathcal{H}}_k^{\mathrm{T}}[x - x_k] = 0. \tag{2.2}$$

Assumption 2.1   *The columns of*

$$\overline{\mathcal{G}}_k \equiv \left[\mathcal{G}_k \vdots \overline{\mathcal{H}}_k \right] \tag{2.3a}$$

are assumed to be linearly independent.

We therefore write the active constraints as

$$\overline{G}_k^T[x - x_k] + \overline{g}_k = 0 \qquad (2.3b)$$

where

$$\overline{g}_k = \begin{bmatrix} g_k \\ \overline{h}_k \end{bmatrix}. \qquad (2.3c)$$

Assumption (2.1) is used to simplify the quadratic subproblem used in the algorithm in Section 3 for solving (1.3). It can be relaxed by increasing the complexity of the quadratic subproblem.

The augmented Lagrangian (6.4.1) does not possess a continuous Hessian. In order to overcome the difficulties arising from this, we shall, at appropriate junctures, consider the effect of replacing the Hessian of (4.2.2) by that of

$$\overline{L}^a(x, \vartheta, \mu^e, \mu^i, \lambda, \eta, c, \pi) = L(x, \vartheta, \mu^e, \mu^i, \lambda, \eta) + \frac{1}{2c}\left\|\left(ch(x)+\pi\right)\right\|_2^2 + \frac{c}{2}\|g(x)\|_2^2$$

$$(2.4)$$

In contrast to (2.2), the Hessian of $\overline{L}^a$ exists, provided f, g and h are suitably differentiable. The role of π in replacing the Hessian of (1.2) by that of (2.4) is crucial and resembles the smearing procedure adopted by Polak and Tits (1981). The difference is in the choice of π and c. It is shown below that this choice allows the treatment of the active constraints locally as equality constraints.

In nonlinear programming algorithms, the penalty parameter c in the augmented Lagrangian (6.4.1) is either taken as a constant, or increased by a prefixed rate, or is adapted as the algorithm progresses. Of the latter type of strategy, Biggs (1974), Mayne and Polak (1982), Polak and Mayne (1981) and Polak and Tits (1981) are specific examples. We also adopt an adaptive strategy below which departs from other works mainly in the relationship of c and the descent property of the direction of search, discussed in Lemmas 3.2 and 3.4 below. In particular, c is only adjusted to ensure that the direction of search is a descent direction for the penalty function that governs the stepsize strategy (3.9) below. This approach, discussed originally in Rustem (1992, 1994b), is an extension of a strategy for nonlinear programming in Rustem (1986, 1993).

In Section 3, we introduce the discrete minimax algorithm and establish its basic properties. In Section 4, we establish its global convergence.

3 THE DISCRETE MINIMAX ALGORITHM

3.1 Inequality Constraints

We first define the way in which the vector π is determined to ensure the existence of the Hessian of the augmented Lagrangian at specific points of interest. Let the ith element of the vectors h and π be respectively denoted by h^i and π^i. Let

$$\pi^j_{k+\frac{1}{2}} = \begin{cases} -c_{k+1}h^i_k & \text{if } h^i_k \leq 0 \\ \pi^j_k & \text{if } h^i_k > 0 \end{cases} \tag{3.1a}$$

and

$$\pi^j_{k+1} = \begin{cases} -c_{k+1}h^i_{k+1} & \text{if } h^i_{k+1} \leq 0 \\ \pi^j_{k+1/2} & \text{if } h^i_{k+1} > 0 \end{cases} \tag{3.1b}$$

where π_0 is set in Step 0. c_{k+1} is set in Step 3 below, with $c_0 \in (0, \infty]$ a given constant. The reason for distinguishing between $\pi_{k+1}, \pi_{k+1/2}$ is to ensure (3.13b) below. It is shown below that c_k is not altered at the later stages of the algorithm and $c_k = c_*$. For such a constant c_*, it can be verified from (3.1) that $\pi_{k+\frac{1}{2}} = \pi_k$ since

$$\pi^j_k = \pi^j_{k+\frac{1}{2}} = \begin{cases} -c_*h^i_k & \text{if } h^i_k \leq 0 \\ \pi^j_k & \text{if } h^i_k > 0. \end{cases}$$

3.2 Quadratic Programming Subproblem

To construct the algorithm, consider the objective function

$$\mathcal{V}(x, \vartheta) = \langle \vartheta, f(x) \rangle \tag{3.2}$$

and its linear approximations, with respect to x, at a point x_k,

$$\mathcal{V}_k(x, \vartheta) = \langle \vartheta, f(x_k) + \nabla f(x_k)^{\mathrm{T}}(x - x_k) \rangle \tag{3.3a}$$

where

$$\nabla f(x) = \left[\nabla f^1(x), ..., \nabla f^{m^{sce}}(x) \right].$$

We shall sometimes denote $f(x)$ and $\nabla f(x)$, evaluated at x_k, by f_k and ∇f_k, respectively. Thus, for $d = x - x_k$ (3.3a) can be written as

$$\mathcal{V}_k(x_k + d, \vartheta) = \langle \vartheta, f_k + \nabla f_k^{\mathrm{T}} d \rangle. \tag{3.3b}$$

The quadratic objective used to compute the direction of progress is given by

$$\mathcal{V}_k(x_k + d, \vartheta) + \langle d, [c_k G_k g_k + \mathcal{H}_k(c_k h_k + \pi_k)_+] \rangle + \frac{1}{2} \langle d, \hat{\mathbb{C}}_k d \rangle. \tag{3.4}$$

The matrix $\hat{\mathbb{C}}_k$ is a positive definite approximation to

$$\sum_{j=1}^{m^{sce}} \vartheta_k^j \nabla^2 f^j(x_k) + \sum_{\iota=1}^{e} (\mu_k^e)^{\iota} \nabla^2 g^{\iota}(x_k) + \sum_{i=1}^{i} (\mu_k^i)^{i} \nabla^2 h^i(x_k) + c \Big[\mathcal{H}_k \mathcal{H}_k^{\mathrm{T}} + G_k G_k^{\mathrm{T}} \Big].$$

(3.5)

We note that (3.5) is obtained by excluding from $\mathfrak{C}_k$, the Hessian of $\bar{L}^a$, the second derivatives due to the penalty term in the augmented Lagrangian (Tapia, 1986). The values ϑ_k, μ_k^e, μ_k^i are given by the solution of the quadratic subproblem of the previous iteration. The direction of progress at each iteration of the algorithm is determined by the quadratic subproblem

$$\min_d \max_{\vartheta} \Big\{ \mathcal{V}_k(x_k + d, \vartheta) + \langle d, [c_k G_k g_k + \mathcal{H}_k(c_k h_k + \pi_k)_+] \rangle + \frac{1}{2} \langle d, \hat{\mathfrak{C}}_k d \rangle \Big|$$

$$G_k^{\mathrm{T}} d + g_k = 0, \ \mathcal{H}_k^{\mathrm{T}} d + h_k \leq 0, \ \vartheta \in \mathbb{E}_+^{m^{sce}} \Big\}.$$

(3.6a)

An equivalent but simpler subproblem is given by the quadratic program

$$\min_{d,v} \Big\{ v + \langle d, [c_k G_k g_k + \mathcal{H}_k(c_k h_k + \pi_k)_+] \rangle + \frac{1}{2} \langle d, \hat{\mathfrak{C}}_k d \rangle \big| G^{\mathrm{T}}_k d + g_k = 0,$$

$$\mathcal{H}_k^{\mathrm{T}} d + h_k \leq 0, \ \nabla f_k^{\mathrm{T}} d + f_k \leq 1v \Big\}.$$

(3.6b)

Also, (3.6b) involves fewer variables. It is shown below that the multipliers associated with

$$\nabla f_k^{\mathrm{T}} d + f_k \leq 1v$$

are the values ϑ that solve (3.6a) since the solution of either subproblem satisfies common optimality properties.

Assumption 3.1   *The quadratic subproblems have a finite solution.*

The assumption implies that the gradients of all equality constraints and the inequality constraints satisfied as equalities at the solution of (3.6) are linearly independent. This may be relaxed by increasing the complexity of the quadratic subproblem.

3.3 Stepsize Strategy

Let the value of d, ϑ, v solving (3.6) be denoted by d_k, ϑ_{k+1}, v_{k+1}. The stepsize to be taken along d_k is determined using the equivalent minimax formulation (2.1). Consider the max-function for the discrete minimax problem

$$\Phi(x) = \max_{j \in \{1,2,\dots,m^{sce}\}} \left\{ f^j(x) \right\} \qquad (3.7)$$

and

$$\Phi_k(x) = \max_{j \in \{1,2,\dots,m^{sce}\}} \left\{ f^j(x_k) + \langle \nabla f^j(x_k), x - x_k \rangle \right\}$$

Let $\Phi_k(x_k + d_k)$ be given by

$$\Phi_k(x_k + d_k) = \max_{j \in \{1,2,\dots,m^{sce}\}} \left\{ f^j(x_k) + \langle \nabla f^j(x_k), d_k \rangle \right\}. \qquad (3.8)$$

We define the merit function to be used in the stepsize strategy by

$$\ell(x, c, \pi) = \Phi(x) + \frac{c}{2} \|g(x)\|_2^2 + \frac{1}{2c} \left\| \left(ch(x) + \pi \right)_+ \right\|_2^2.$$

The stepsize strategy determines the smallest nonnegative integer t_k such that $\alpha_k = \bar{\alpha}_k^t, \bar{\alpha} \in (0,1)$, yields

$$x_{k+1} = x_k + \alpha_k d_k$$

which satisfies the inequality

$$\ell(x_{k+1}, c_{k+1}, \pi_{k+1}) - \ell(x_k, c_{k+1}, \pi_{k+1/2}) \le \rho \alpha_k \Psi(d_k, c_{k+1}, \pi_{k+1/2}) \qquad (3.9a)$$

where $\rho \in (0,1)$ is a given scalar and $\Psi(d_k, c_{k+1}, \pi_{k+1/2})$

$$\Psi(d_k, c_{k+1}, \pi_{k+1/2}) = \Phi_k(x_k + d_k) - \Phi(x_k) \qquad (3.9b)$$

$$+ \left\langle d_k, \left[c_{k+1} G_k g_k + \mathcal{H}_k \left(c_{k+1} h_k + \pi_{k+\frac{1}{2}} \right)_+ \right] \right\rangle + \frac{1}{2} \langle d_k, \hat{\mathfrak{C}}_k d_k \rangle.$$

The stepsize α_k determined by (3.9) ensures that x_{k+1} simultaneously improves the objective function and the feasibility with respect to the constraints. The penalty term used to measure feasibility is a quadratic consistent with the augmented Lagrangian (6.4.1). It is shown in Theorem (4.1) below that (3.9) can always be fulfilled by the algorithm.

The determination of the penalty parameter c is an important aspect of the algorithm. This is specified in the following algorithm description.

3.4 The Algorithm

Step 0: Given $x_0, c_0 \in [0, \infty)$, and small positive numbers $\delta, \rho, \epsilon, \bar{\alpha}$ such that $\delta \in (0, \infty), \rho \in (0,1), \epsilon \in (0, \frac{1}{2}], \bar{\alpha} \in (0,1), \hat{\mathfrak{C}}_0$, set $k = 0$.

Step 1: Compute $\nabla f_k, G_k, \mathcal{H}_k$. Solve the quadratic subproblem (3.6b) (adopting subproblem (3.6a) also yields an equivalent algorithm) to obtain $d_k, \vartheta_{k+1}, v_{k+1}, \mu_{k+1}^e, \mu_{k+1}^i$ (in (3.6a) we compute ϑ_{k+1} and in (3.6b) we compute v_{k+1}).

Step 2: Test for optimality: If optimality is achieved, stop. Else go to Step 3.

Step 3: If

$$\Phi_k(x_k + d_k) - \Phi(x_k) - c_k\left[\|g_k\|_2^2 + \|(h_k)_+\|_2^2\right] - \langle(h_k)_+, \pi_k\rangle + \left(\epsilon + \frac{1}{2}\right)\langle d_k, \hat{\mathbb{C}}_k d_k\rangle$$

$$\leq 0, \tag{3.10}$$

then $c_{k+1} = c_k$. Else set

$$c_{k+1} = \max\left\{\frac{\Phi_k(x_k + d_k) - \Phi(x_k) + \left(\epsilon + \frac{1}{2}\right)\langle d_k, \hat{\mathbb{C}}_k d_k\rangle - \langle(h_k)_+, \pi_k\rangle}{\|g_k\|_2^2 + \|(h_k)_+\|_2^2}, c_k + \delta\right\}.$$
$$\tag{3.11}$$

Step 4: Find the smallest nonnegative integer t_k such that $\alpha_k = \bar{\alpha}^{t_k}$ with

$$x_{k+1} = x_k + \alpha_k d_k$$

such that the inequality (3.9) is satisfied.

Step 5: Update $\hat{\mathbb{C}}_k$ to compute $\hat{\mathbb{C}}_{k+1}$, set $k = k + 1$ and go to Step 1.

In Step 3, the penalty parameter c_{k+1} is adjusted to ensure that progress towards feasibility is maintained. In particular, c_{k+1} is chosen to make d_k a descent direction for the function $\ell(x, c, \pi)$.

The matrix $\hat{\mathbb{C}}_k$ is approximated using Powell's (1978) modification to a quasi-Newton formula due to Broyden (1969, 1970), Fletcher (1970), Goldfarb (1970) and Shanno (1970) for approximating the Hessian of a Lagrangian. This modified BFGS (Broyden–Fletcher–Goldfarb–Shanno) formula is given by

$$\hat{\mathbb{C}}_{k+1} = \hat{\mathbb{C}}_k - \frac{\hat{\mathbb{C}}_k \delta_x \delta_x^{\mathrm{T}} \hat{\mathbb{C}}_k}{\langle \delta_x, \hat{\mathbb{C}}_k \delta_x\rangle} + \frac{\nu\nu^{\mathrm{T}}}{\langle \delta_x, \nu\rangle}$$

where

$$\delta_x = x_{k+1} - x_k$$

$$\gamma = \nabla_x \bar{L}^a(x_{k+1}, \vartheta_{k+1}, \mu_{k+1}^e, \mu_{k+1}^i, \lambda_{k+1}, \eta_{k+1}, c_{k+1}, \pi_{k+1}$$

$$-\nabla_x \bar{L}^a(x_k, \vartheta_{k+1}, \mu_{k+1}^e, \mu_{k+1}^i, \lambda_{k+1}, \eta_{k+1}, c_{k+1}, \pi_{k+1/2})$$

$$\nu = \theta\gamma + (1 - \theta)\hat{\mathbb{C}}_k \delta_x$$

$$\theta = \begin{cases} 1 & \text{if } \langle \delta_x, \gamma \rangle \geq 0.2 \langle \delta_x, \hat{\mathbb{C}}_k \delta_x \rangle \\ \dfrac{0.8 \langle \delta_x, \hat{\mathbb{C}}_k \delta_x \rangle}{\langle \delta_x, (\hat{\mathbb{C}}_k \delta_x - \gamma) \rangle} & \text{if } \langle \delta_x, \gamma \rangle < 0.2 \langle \delta_x, \hat{\mathbb{C}}_k \delta_x \rangle. \end{cases}$$

3.5 Basic Properties

The following two lemmas establish the basic properties of the penalty term.

Lemma 3.1 *Let $\bar{\pi} \geq 0$ be a constant, $c_{k+1} \geq 0$, and let*

$$\pi_{k+1}^j(x) = \begin{cases} -c_{k+1} h^i(x) & \text{if } h^i(x) \leq 0 \\ \bar{\pi} & \text{if } h^i(x) > 0. \end{cases}$$

Then, we have

$$\left(c_{k+1} h^i(x) + \pi_{k+1}^i(x) \right)_+ = \left(c_{k+1} h^i(x) + \pi_{k+1}^i(x) \right) \geq 0 \qquad (3.12)$$

and

$$(h_k)_+^T \pi_{k+\frac{1}{2}} = (h_k)_+^T \pi_k \qquad (3.13a)$$

and

$$\left\| (c_{k+1} h_{k+1} + \pi_{k+1})_+ \right\|_2^2 - \left\| (c_{k+1} h_k + \pi_{k+1/2})_+ \right\|_2^2$$

$$\leq \left\| \left(c_{k+1} h_{k+1} + \pi_{k+\frac{1}{2}} \right) \right\|_2^2 - \left\| \left(c_{k+1} h_k + \pi_{k+\frac{1}{2}} \right) \right\|_2^2. \qquad (3.13b)$$

Proof. When $h^j(x) \leq 0$, $\pi_{k+1}^j(x) = -c_{k+1} h^j(x) \geq 0$ and when $h^j(x) > 0$, $\pi_{k+1}^j(x) = \bar{\pi} \geq 0$.

Thus, we have

$$h^j(x) \leq 0 \Rightarrow \left(c_{k+1} h^j(x) + \pi_{k+1}^j(x) \right)_+ = \left(c_{k+1} h^j(x) + \pi_{k+1}^j(x) \right) = 0$$

$$h^j(x) > 0 \Rightarrow \left(c_{k+1} h^j(x) + \pi_{k+1}^j(x) \right)_+ = \left(c_{k+1} h^j(x) + \pi_{k+1}^j(x) \right) \geq 0$$

from which (3.12) follows directly.

Equality (3.13a) follows from (3.1) since $\pi_{k+1/2}^j \neq \pi_k^j$ if $h_k^j \leq 0$, in which case $(h_k^j)_+ = 0$.

To establish inequality (3.13b), consider (3.1a,b) which yields

$$0 = \left(c_{k+1} h_{k+1}^j + \pi_{k+1}^j \right)_+ \leq \left(c_{k+1} h_{k+1}^j + \pi_{k+1/2}^j \right)_+, \quad \text{for } h_{k+1}^j \leq 0.$$

Also, we have $\pi_{k+1}^j = \pi_{k+1/2}^j$,

$$\left(c_{k+1} h_{k+1}^j + \pi_{k+1}^j \right)_+ = \left(c_{k+1} h_{k+1}^j + \pi_{k+1/2}^j \right)_+, \quad \text{for } h_{k+1}^j > 0.$$

This yields the inequality

$$\left(c_{k+1}h_{k+1}^j + \pi_{k+1}^j\right)_+ \le \left(c_{k+1}h_{k+1}^j + \pi_{k+1/2}^j\right)_+.$$

Furthermore, we have $(\cdot)_+^2 \le (\cdot)^2$. Finally, using (3.1), we obtain the equality

$$\left(c_{k+1}h_k^j + \pi_{k+1/2}^j\right)_+ = \left(c_{k+1}h_k^j + \pi_{k+1/2}^j\right).$$

Hence we have the desired result. □

Inequality (3.13b) ensures that the descent discussion in Lemma 3.3 and Theorem 4.1 can always be related to the stepsize strategy in Step 4, when the penalty parameter is not necessarily a constant. Starting with a penalty term, which is not differentiable, we ensure equivalence with $\|(c_{k+1}h_k + \pi_{k+1/2})\|_2^2$ at points generated by the algorithm to invoke the twice differentiability of the latter in the convergence results of the next section.

We summarize the optimality conditions of the quadratic subproblem (3.6). Subsequently, we establish the descent property of d_k and that c_k determined by (3.11) is not increased indefinitely.

Lemma 3.2 *The Lagrangian function for the quadratic subproblem (3.6a) is given by*

$$L_k(x, \vartheta, \mu^e, \mu^i, \lambda, \eta, c_k, \pi_k) = \mathcal{V}_k(x_k + d, \vartheta) + \frac{1}{2}\langle d, \hat{\mathbb{C}}_k d\rangle$$

$$+\langle d, [c_k \mathcal{G}_k g_k + \mathcal{H}_k(c_k h_k + \pi_k)_+]\rangle$$

$$+\left\langle \mu^e, \mathcal{G}_k^T d + g_k\right\rangle + \left\langle \mu^i, \mathcal{H}_k^T d + h_k\right\rangle$$

$$-\langle \vartheta, \lambda\rangle + (\langle 1, \vartheta\rangle - 1)\eta \qquad (3.14a)$$

The first order necessary conditions of optimality for problem (3.6a) are

$$\nabla f_k \vartheta_{k+1} + \hat{\mathbb{C}}_k d_k + \mathcal{G}_k[c_k g_k + \mu_{k+1}^e] + \mathcal{H}_k\left[(c_k h_k + \pi_k)_+ + \mu_{k+1}^i\right] = 0 \quad (3.14b)$$

$$f_k + \nabla f_k^T d_k - \lambda_{k+1} + I\eta_{k+1} = 0 \qquad (3.14c)$$

$$\vartheta_{k+1} \in \mathbb{E}_+^{m^{sce}}, \quad \lambda_{k+1} \ge 0, \quad \vartheta_{k+1}^j \lambda_{k+1}^j = 0, \quad \forall j \in \{1, ..., m^{sce}\} \quad (3.14d)$$

$$\mathcal{G}_k^T d + g_k = 0, \quad \mathcal{H}_k^T d_k + h_k \le 0$$

$$\left\langle \mu_{k+1}^i, \mathcal{H}_k^T d_k + h_k\right\rangle = 0, \quad \mu_{k+1}^i \ge 0. \qquad (3.14e)$$

and for the quadratic programming subproblem (3.6b) the necessary conditions are given by (3.14b), (3.14e) and

$$f_k + \nabla f_k^{\mathrm{T}} d_k \le l v_{k+1}$$

$$\vartheta^j_{k+1} \left(f_k^j + \langle \nabla f_k^j, d_k \rangle - v_{k+1} \right) = 0, \quad j = 1, \ldots, m^{sce}, \quad \vartheta_{k+1} \in \mathbb{E}_+^{m^{sce}}. \quad (3.14\mathrm{f})$$

Lemma 3.3 (The Descent Property of d_k) *Let*

$$\begin{bmatrix} g_k \\ (h_k)_+ \end{bmatrix} \ne 0$$

(the case when

$$\begin{bmatrix} g_k \\ (h_k)_+ \end{bmatrix} = 0$$

is discussed in Lemma 3.5) and let $\hat{\mathfrak{C}}_k$ be positive semi-definite. The direction d_k satisfying the linearized inequality constraints (3.14e) of the quadratic subproblem (3.6) and c_{k+1} chosen as in Step 3 of the above algorithm ensure the inequality

$$\Psi(d_k, c_{k+1}, \pi_{k+1/2}) \le -\epsilon \langle d_k, \hat{\mathfrak{C}}_k d_k \rangle. \quad (3.15)$$

Proof. Using (3.12), (3.14a), (3.14e) and noting that (3.12) equally applies to π_k and $\pi_{k+1/2}$, we can write

$$\langle d_k, \mathcal{H}_k(c_{k+1} h_k + \pi_{k+1/2})_+ \rangle \le -\langle h_k, (c_{k+1} h_k + \pi_{k+1/2})_+ \rangle$$

$$= -c_k \|(h_k)_+\|_2^2 - \langle (h_k)_+, \pi_k \rangle$$

and c_{k+1} chosen in Step 3 ensures that (3.9b) is bounded by

$$\Psi(d_k, c_{k+1}, \pi_{k+1/2}) \le \Phi_k(x_k + d_k) - \Phi(x_k)$$

$$-c_{k+1} \left[\|g_k\|_2^2 + \|(h_k)_+\|_2^2 \right] - \langle (h_k)_+, \pi_k \rangle + \frac{1}{2} \langle d_k, \hat{\mathfrak{C}}_k d_k \rangle \le -\epsilon \langle d_k, \hat{\mathfrak{C}}_k d_k \rangle. \quad (3.16)$$

Thereby (3.15) is established. $\square$

Lemma 3.4 *Let d_k and ϑ_{k+1} be the solutions of the quadratic subproblem (3.6), then*

$$\Phi_k(x_k + d_k) = \max_{\vartheta \in \mathbb{E}_+^{m^{sce}}} \mathcal{V}_k(x_k + d_k, \vartheta) = \mathcal{V}_k(x_k + d_k, \vartheta_{k+1}).$$

Proof. The first equality follows, from the fact that $\Phi_k(x_k + d_k)$ is the maximum of m^{sce} numbers (evaluated at $x_k + d_k$) and this is equal to the maximum of the convex combination of these numbers.

It can be verified using (3.14a), or its equivalent for subproblem (3.6b), that for d_k, μ_{k+1}^e, μ_{k+1}^i, λ_{k+1}, η_{k+1}, ν_{k+1} solving (3.6) we have

$$\vartheta_{k+1} = \arg \max_{\vartheta \in \mathbb{E}_+^{m^{sce}}} L_k(x_k + d_k, \vartheta, \mu_{k+1}^e, \mu_{k+1}^i, \lambda_{k+1}, \eta_{k+1}, c_k, \pi_k)$$

$$= \arg \max_{\vartheta \in \mathbb{E}_+^{m^{sce}}} \mathcal{V}_k(x_k + d_k, \vartheta). \tag{3.17}$$

The implication of (3.17) is that since at d_k, μ_{k+1}^e, μ_{k+1}^i, λ_{k+1}, η_{k+1} the maximization of L_k is independent of terms not included in $\mathcal{V}_k$, ϑ_{k+1} maximizes both functions. This establishes the second equality. The same result also follows for subproblem (3.6b), since ϑ is the multiplier of

$$\nabla f^{\mathrm{T}}_k d + f_k \le \mathbf{1} v. \quad \square$$

It is shown in Lemma (3.5) below that (3.16) is nonpositive and the descent property of Lemma (3.3) also holds when

$$\begin{bmatrix} g_k \\ (h_k)_+ \end{bmatrix} = 0.$$

Lemma 3.5 (The Finiteness of the Penalty Parameter) *Let $f, g, h \in \mathbb{C}^1$, $\hat{\mathbb{C}}_k$ be positive semi-definite and $(d_k, \vartheta_{k+1}, \mu_{k+1}^e, \mu_{k+1}^i, \lambda_{k+1}, \eta_{k+1})$, computed by (3.6a), and $(d_k, \nu_{k+1}, \vartheta_{k+1}, \mu_{k+1}^e, \mu_{k+1}^i)$, computed by (3.6b), be bounded. Then*

(i) If the nonlinear constraints of (1.2) are satisfied, that is, $g(x_k) = 0$; $h(x_k) \le 0$, then the descent property (3.15) of Lemma 3.3 is satisfied for any choice of $c_k \in [0, \infty)$.

(ii) Let $c_k \in [0, \infty)$ and $g(x_k) = 0$; $h(x_k) \le 0$, for some k. Then, Step 3 of the algorithm chooses $c_{k+1} = c_k$. For this choice of c_{k+1}, the direction d_k ensures the descent property (3.15).

(iii) $\forall k$, $\exists c_{k+1} \in [0, \infty)$ satisfying Step 3 of the algorithm.

(iv) Let the sequence $\{x_k\}$ generated by the algorithm be bounded. Then, c_k is increased finitely often and there exists an integer $k_* \ge 0$ and exists $c_* \in [0, \infty)$ such that the algorithm chooses $c_k = c_*, \forall k \ge k_*$.

Remark The penalty c_k is the value used in the quadratic subproblems when computing d_k, ϑ_{k+1}, μ_{k+1}^e, μ_{k+1}^i, ν_{k+1}.

Proof. To show (i) consider (3.14b,e) and (3.1) for $g_k = 0$; $h_k \leq 0$,

$$0 = \left\langle d_k, \left[\nabla f_k \vartheta_{k+1} + \hat{\mathbb{C}}_k d_k + \mathcal{G}_k[c_k g_k + \mu_{k+1}^e] + \mathcal{H}_k\left[(c_k h_k + \pi_k)_+ + \mu_{k+1}^i\right]\right]\right\rangle$$

$$= \langle d_k, \nabla f_k \vartheta_{k+1}\rangle + \langle d_k, \hat{\mathbb{C}}_k d_k\rangle - \langle h_k, \mu_{k+1}^i\rangle$$

and hence we have

$$\langle d_k, \nabla f_k \vartheta_{k+1}\rangle = -\langle d_k, \hat{\mathbb{C}}_k d_k\rangle. \tag{3.18}$$

Consider (3.16) for $g_k = 0$; $h_k \leq 0$

$$\Psi(d_k, c_{k+1}, \pi_{k+1/2}) + \epsilon\langle d_k, \hat{\mathbb{C}}_k d_k\rangle$$

$$= \Phi_k(x_k + d_k) - \Phi(x_k) + \left(\frac{1}{2} + \epsilon\right)\langle d_k, \hat{\mathbb{C}}_k d_k\rangle$$

$$- c_{k+1}\langle g_k, g_k\rangle + \langle d, \mathcal{H}_k(c_{k+1}h_k + \pi_{k+1/2})_+\rangle$$

$$= \max_{\vartheta \in \mathbb{E}_+^{m^{sce}}} \langle\vartheta, f_k + \nabla f_k^{\mathrm{T}} d_k\rangle - \max_{\vartheta \in \mathbb{E}_+^{m^{sce}}} \langle\vartheta, f_k\rangle + \left(\frac{1}{2} + \epsilon\right)\langle d_k, \hat{\mathbb{C}}_k d_k\rangle$$

$$= \langle\vartheta_{k+1}, f_k + \nabla f_k^{\mathrm{T}} d_k\rangle - \max_{\vartheta \in \mathbb{E}_+^{m^{sce}}} \langle\vartheta, f_k\rangle + \left(\epsilon + \frac{1}{2}\right)\langle d_k, \hat{\mathbb{C}}_k d_k\rangle$$

$$\leq \left(\epsilon - \frac{1}{2}\right)\langle d_k, \hat{\mathbb{C}}_k d_k\rangle \leq 0. \tag{3.19}$$

Equality (3.19) follows from the definition of $\Phi_k(x_k + d_k)$ and that $\Phi(x_k)$ is the maximum of the convex combination of m^{sce} numbers $f^1(x_k), \dots, f^{m^{sce}}(x_k)$. The subsequent equality follows from Lemma 3.4. The last two inequalities follow from (3.18) and the fact that ϑ_{k+1} does not necessarily optimize $\max_{\vartheta \in \mathbb{E}_+^{m^{sce}}} \langle\vartheta, f_k\rangle$. Thus, the descent condition (3.15) is satisfied for $g_k = 0$, $h_k \leq 0$.

To show (ii), consider the condition (3.10) which is satisfied for all $c_k \in [0, \infty)$ since descent is ensured for $g_k = 0$, $h_k \leq 0$. Thus, the algorithm chooses $c_{k+1} = c_k$. For this choice, it can simply be verified that Lemma 3.3 and (i) above hold. (iii) follows simply from (i) and (ii) since Step 3 of the algorithm generates a finite c_{k+1}. To establish (iv), we note that in Step 3, $c_{k+1} \in [0, \infty)$ such that the descent condition of Lemma 3.3 and (i) above hold. If c_k is increased infinitely often then, by Step 3, we have $\{c_{k+1}\} \rightarrow \infty$. As there exists a $c_{k+1} \in [0, \infty)$ that the algorithm can choose at every iteration, then there exists a $c_*[0, \infty)$ such that the test of Step 3 is always satisfied for $c_k = c_*, \forall k \geq k_*$. k_* is the iteration at which the algorithm reaches c_*. $\square$

The algorithm increases the penalty parameter until c_* is attained. An increase of the penalty parameter is not required after this since the condition in Step 3 is satisfied with $c_k = c_*$.

4 CONVERGENCE OF THE ALGORITHM

We consider in this section the global convergence of the method. The equality constrained case is discussed in Rustem (1992). The inequality constrained case is discussed in Rustem and Nguyen (1998). Inequality constraints require a substantially different approach to establish these results. Convergence of the stepsize sequence $\{\alpha_k\}$ to unity and the superlinear rates of convergence are consequences of results in Rustem (1992, 1993).

We first establish that the stepsize strategy (3.9) can always be satisfied. If the satisfaction of (3.9) was assumed, then the final inequality in the proof, given by (4.8) below, could be replaced by (3.9). Lemmas 3.3 and 3.5 (i) would provide the monotonicity of the merit function $\ell(x, c, \pi)$ in Theorem 4.1.

Assumption 4.1 $f, g, h \in \mathbb{C}^2(\mathbb{R}^n)$.

Theorem 4.1   *Let*

  (i)   *Assumption (4.1) be satisfied;*

 (ii)   *the approximate Hessian be such that, for each k,*

$$m\|y\|_2^2 \le y^{\mathrm{T}} \hat{\mathbb{C}}_k y \le M\|y\|_2^2, \quad M \ge m > 0, \quad 0 \neq y \in \mathbb{R}^n \qquad (4.1)$$

(iii)   *for each k, there exists a bounded optimal point* $(d_k, \vartheta_{k+1}, \mu_{k+1}^e, \mu_{k+1}^i,$
        $\lambda_{k+1}, \eta_{k+1}) \in \mathbb{R}^n \times \mathbb{R}^{m^{sce}} \times \mathbb{R}^e \times \mathbb{R}^i \times \mathbb{R}^{m^{sce}} \times \mathbb{R}^1$ *of the quadratic*
        subproblem (3.6a); similarly, there exists a bounded Kuhn–Tucker
        point $(d_k, \nu_{k+1}, \vartheta_{k+1}, \mu_{k+1}^e, \mu_{k+1}^i) \in \mathbb{R}^n \times \mathbb{R}^1 \times \mathbb{R}^{m^{sce}} \times \mathbb{R}^e \times \mathbb{R}^i$ *for*
        (3.6b);

(iv)   *there exist an integer* k_* *and a scalar* $c_* \ge 0$ *such that (3.10) is satisfied,*
        for all $k \ge k_*$ *with* $c_k = c_*$.

Then, the stepsize computed in Step 4 is such that $\alpha_k \in (0, 1]$ *and the sequence* $\{x_k, \vartheta_k\}$, *computed by the algorithm, satisfies the stepsize strategy (3.9) and generates a corresponding sequence* $\{\ell(x_k, c_*, \pi_k)\}$ *which is monotonically decreasing for* $k \ge k_*$.

Remarks   Along with assumptions (i) and (ii), boundedness in (iii) is ensured when the active constraint normals at the solution of the quadratic subproblem are linearly independent. Another condition implied by (iii) is that the feasible set of the quadratic subproblem is nonempty.

By Lemmas 3.3 and 3.5, c_k chosen by Step 3 is finite. Also, there exists $c_* \in [0, \infty)$, for which (3.10) is satisfied with $c_k = c_*, \forall k \geq k_*$. The integer k_* is the iteration at which c_* is attained. The result can be extended to show that there exists $\alpha_k \in (0, 1]$ satisfying the stepsize strategy *before* the algorithm achieves the constant $c_k = c_*$. This can be done by considering the expansion of $1/(2c_{k+1})\|(c_{k+1}h(x) + \pi_{k+\frac{1}{2}})\|_2^2$ instead of $1/(2c_*)\|(c_*h(x) + \pi_k)\|_2^2$ below. The assumption of $c_k = c_*$ is required to establish the overall monotonicity.

Proof. For $c_k = c_*$, (3.1) yields $\pi_{k+1/2} = \pi_k$. Using this in (3.13b) and considering the second order expansion of $[c_*/2\|g(x)\|_2^2 + 1/2c_*\|(c_*h(x) + \pi_k)\|_2^2]$, we have

$$\left[\frac{c_*}{2}\|g_{k+1}\|_2^2 + \frac{1}{2c_*}\left\|\left(c_*h_{k+1} + \pi_{k+1}\right)_+\right\|_2^2\right] - \left[\frac{c_*}{2}\|g_k\|_2^2 + \frac{1}{2c_*}\left\|\left(c_*h_k + \pi_k\right)_+\right\|_2^2\right]$$

$$\leq \left[\frac{c_*}{2}\|g_{k+1}\|_2^2 + \frac{1}{2c_*}\left\|\left(c_*h_{k+1} + \pi_k\right)\right\|_2^2\right] - \left[\frac{c_*}{2}\|g_k\|_2^2 + \frac{1}{2c_*}\left\|\left(c_*h_k + \pi_k\right)\right\|_2^2\right]$$

$$\leq \alpha_k\left\langle d_k, \left[c_*\,G_k g_k + \mathcal{H}_k\left(c_*h_k + \pi_k\right) + \frac{\alpha_k}{2}\,\hat{\mathbb{C}}_k d_k\right]\right\rangle + \alpha_k^2 \nu_k \| d_k \|_2^2 \quad (4.2a)$$

for $x(t) = x_k + t\alpha_k d_k$ and

$$\mathcal{A}_1(x(t), c_*) \equiv c_*\left[G(x(t))G(x(t))^{\mathrm{T}} + \mathcal{H}(x(t))\mathcal{H}(x(t))^{\mathrm{T}}\right] \quad (4.2b)$$

$$\mathcal{A}_2(x(t), c_*) \equiv c_* \sum_{\iota=1}^{e} \nabla^2 g^\iota(x(t))g^\iota(x(t)) + \sum_{i=1}^{i} \nabla^2 h^i(x(t))\left[c_* h^i(x(t)) + \pi_k^i\right]$$

$$(4.2c)$$

$$\nu_k = \int_0^1 (1-t)\left\|\mathcal{A}_1(x(t), c_*) + \mathcal{A}_2(x(t), c_*) - \hat{\mathbb{C}}_k\right\| dt. \quad (4.2d)$$

By definition of $\Phi(x)$ we have,

$$\Phi(x_{k+1}) = \max_{\vartheta \in \mathbb{E}_+^{m^{sce}}} \left\{\langle \vartheta, f(x_k + \alpha_k d_k)\rangle\right\} \quad (4.3a)$$

$$\leq \max_{\vartheta \in \mathbb{E}_+^{m^{sce}}} \left\{\langle \vartheta, f(x_k) + \alpha_k \nabla f_k^{\mathrm{T}} d_k\rangle\right\}$$

$$+\alpha_k^2 \| d_k \|_2^2 \int_0^1 (1-t)\left\|\max_{\vartheta \in \mathbb{E}_+^{m^{sce}}}\left\{\sum_{j=1}^{m^{sce}} \vartheta^j \nabla^2 f^j(x(t))\right\}\right\| dt. \quad (4.3b)$$

Using (3.3) we can write

$$\max_{\vartheta \in \mathbb{E}_+^{m^{sce}}} \left\{ \langle \vartheta, f(x_k) + \alpha_k \nabla f_k^T d_k \rangle \right\} = \max_{\vartheta \in \mathbb{E}_+^{m^{sce}}} \mathcal{V}_k(x_k + \alpha_k d_k, \vartheta). \quad (4.4)$$

Since $\max_{\vartheta \in \mathbb{E}_+^{m^{sce}}} \mathcal{V}_k(x_k + d, \vartheta)$ is convex with respect to d, we have for $\alpha_k \in [0, 1]$,

$$\max_{\vartheta \in \mathbb{E}_+^{m^{sce}}} \mathcal{V}_k(x_k + \alpha_k d_k, \vartheta) \le \max_{\vartheta \in \mathbb{E}_+^{m^{sce}}} \mathcal{V}_k(x_k, \vartheta)$$

$$+ \alpha_k \left\{ \max_{\vartheta \in \mathbb{E}_+^{m^{sce}}} \mathcal{V}_k(x_k + d_k, \vartheta) - \max_{\vartheta \in \mathbb{E}_+^{m^{sce}}} \mathcal{V}_k(x_k, \vartheta) \right\}. \quad (4.5)$$

Furthermore, we have

$$\Phi(x_k) = \max_{\vartheta \in \mathbb{E}_+^{m^{sce}}} \mathcal{V}_k(x_k, \vartheta) \quad (4.6b)$$

$$\Phi_k(x_k) = \max_{\vartheta \in \mathbb{E}_+^{m^{sce}}} \mathcal{V}_k(x_k + d, \vartheta). \quad (4.6b)$$

Using (4.6), (4.5), (4.4), Lemmas 3.3, 3.5 (i) and combining (4.2) and (4.3b) yields

$$\ell(x_{k+1}, c_*, \pi_{k+1}) \le \ell(x_k, c_*, \pi_k) + \alpha_k \Psi(d_k, c_*) \left\{ 1 - \frac{\alpha_k}{\epsilon m} \xi_k \right\} \quad (4.7)$$

$$\xi_k = \int_0^1 (1 - t) \left\| \max_{\vartheta \in \mathbb{E}_+^{m^{sce}}} \left\{ \sum_{j=1}^{m^{sce}} \vartheta^j \nabla^2 f^j(x(t)) \right\} + \mathcal{A}_1(x(t), c_*) + \mathcal{A}_2(x(t), c_*) - \hat{\mathbb{C}}_k \right\| dt.$$

The scalar $\rho \in (0, 1)$ in the stepsize strategy (3.9) determines α_k such that

$$\rho \le 1 - \frac{\alpha_k}{\epsilon m} \xi_k \le 1.$$

By Lemma 3.3 and Lemma 3.5 (i), the descent property $\Psi(d_k, c_*, \pi_k) \le 0$ holds. Thus, there exists a $\alpha_k \in (0, 1]$ to ensure (4.7), and thence (3.9). Suppose α^0 is the largest $\alpha \in [0, 1]$ satisfying inequalities (4.7) and (3.9). It follows that all $\alpha \le \alpha^0$ also satisfy these conditions and that the strategy in Step 4 selects $\alpha_k \in [\bar{a}\alpha^0, \alpha^0]$. Since $\Psi(d_k, c_*, \pi_k) \le 0$, the required monotonic decrease follows. $\square$

Lemma 4.1 *Let the assumptions of Theorem (4.1) be satisfied, and let, for some k_0 and $k \ge k_0$, the set*

$$\mathcal{F} = \left\{ x \in \mathbb{R}^n \mid \ell(x, c_*, \pi_k(x)) \le \ell\left(x_{k_0}, c_*, \pi_{k_0}(x_{k_0})\right) \right\}$$

be bounded. We then have

$$\lim_{k \to \infty} \Psi(d_k, c_*, \pi_k) = 0. \tag{4.8}$$

Proof. For $c_k = c_*$, we have, for a constant $\hat{\pi} \geq 0$,

$$\pi_k^i(x) = \begin{cases} -c_* h^i(x) & \text{if } h^i(x) \leq 0 \\ \hat{\pi} & \text{if } h^i(x) > 0. \end{cases}$$

Given $\rho \in (0, 1)$, by (4.7), the choice

$$\alpha^0 = \min\left\{1, \frac{(1-\rho)\epsilon m}{\xi_k}\right\}$$

always satisfies stepsize strategy (3.9). Clearly, $\alpha_k = \overline{\alpha}^{t_k}$ chosen in Step 4 is in the range $\alpha_k \in [\overline{\alpha}\alpha^0, \alpha^0]$ and thereby also satisfies (3.9). As f and h are twice continuously differentiable and $\mathcal{F}$ is compact, there is a scalar $\overline{M} < \infty$ such that $\xi_k \leq \overline{M}$. Thus, as $m > 0$, we have $\alpha_k \geq \overline{\alpha} > 0, \forall k$, for (3.9). The boundedness of $\ell(x, c_*, \pi_k(x))$ on $\mathcal{F}$ and $\Psi(d_k, c_*, \pi_k) \leq 0$ imply

$$0 \leq \rho \sum_k \alpha_k |\Psi(d_k, c_*, \pi_k)| \leq \sum_k \ell(x_k, c_*, \pi_k) - \ell(x_{k+1}, c_*, \pi_{k+1}) < \infty$$

which yields (4.8). $\square$

Lemma (4.2) *Let the assumptions of Theorem 4.1 be satisfied. Inequality (3.15), Lemmas 3.5 (i) and 4.1 imply $\lim_{k \to \infty} \|d_k\| = 0$.*

Proof. Using (4.8) with (3.15) yields the required result. $\square$

Theorem 4.2 *Let the first order necessary conditions of problem (2.1) be satisfied by $x_*, \vartheta_*, \mu_*^e, \mu_*^i, \nu_*$. Let the assumptions of Theorem 4.1 be satisfied. Then:*

(i) *the algorithm either terminates at $x_*, \vartheta_*, \mu_*^e, \mu_*^i, \nu_*$ or it generates an infinite sequence $\{x_k\}$ in which there exists a subsequence $\{x_k\}$ with $k_* \leq k \in K \subset \{0, 1, ...\}$ such that $\{\| d_k \|\} \to 0$ and thus every accumulation point x_* of the infinite sequence $\{x_k\}$ satisfies the necessary condition of optimality of the minimax problem;*

(ii) *if, furthermore, strict complementarity holds at the solution of (3.6b), for large k, ϑ_k, μ_k^i predict the active inequality constraints at x_*.*

Proof. By Lemmas 3.5, 4.1 and 4.2, there exists a subsequence $\{x_k\}$ such that $\{\| d_k \|\} \to 0$. Let there exist $x_* \in \mathbb{R}^n, \vartheta_* \in \mathbb{E}_+^{m^{sce}}, \mu^e \in \mathbb{R}^e, \mu_*^i \in \mathbb{R}_+^i, \nu_* \in \mathbb{R}^1$ such that $\{x_k\} \to x_*, \{\vartheta_k\} \to \vartheta_*, \{\mu_k^e\} \to \mu_*^e, \{\mu_k^i\} \to \mu_*^i, \{\nu_k\} \to \nu_*,$

$k_* \leq k \subset K$. The existence of such points is ensured since, by Theorem 4.1, the algorithm decreases $\ell(x_k, c_*, \pi_k)$ sufficiently at each iteration, thereby ensuring $x_k \in \mathcal{F}$, for compact $\mathcal{F}$.

In order to show that $x_*, \vartheta_*, \mu_*^e, \mu_*^i, \nu_*$ satisfy the first order necessary conditions of optimality, we consider the optimality conditions (3.14) of the subproblem (3.6). Applying Lemma 4.2, for $k \to \infty$, $k_* \leq k \in K$, we derive the first order conditions of (2.1) which define $x_*, \vartheta_*, \mu_*^e, \mu_*^i, \nu_*$. Thus, (3.14b) reduces to

$$0 = \nabla f(x_*)\vartheta_* + c_* \mathcal{G}_*[g(x_*)+\mu_*^e] + \mathcal{H}_*\Big[\big(c_* h(x_*) + \pi_*\big)_+ + \mu_*^i\Big]$$

$$= \nabla f(x_*)\vartheta_* + \mathcal{G}_*\mu_*^e + \mathcal{H}_*\mu_*^i$$

since for $h(x_*) \leq 0$, (3.1) yields $\pi_* = -c_* h(x_*)$ and

$$g(x_*) = 0, \quad h(x_*) \leq 0, \quad \langle \mu_*^i, \, h(x_*) \rangle = 0, \quad \mu_*^i \geq 0$$

$$f(x_*) \leq 1\nu_*, \quad \vartheta_*^j(f^j(x_*) - \nu_*) = 0, \quad j = 1, ..., m^{sce}, \quad \vartheta_* \in \mathbb{E}_+^{m^{sce}}.$$

To show (ii), we note that, in view of the last two conditions, for sufficiently large k, with strict complementarity holding, none of the inactive constraints, that is,

$$h^i_k + d_k^{\mathrm{T}}\nabla h_k^i < 0, \quad (\mu_{k+1}^i)^i = 0$$

$$f^j_k + d_k^{\mathrm{T}}\nabla f_k^j < \nu_{k+1}, \quad \vartheta_{k+1}^j = 0$$

are predicted to be active at x_*, $h_*^i < 0$, $(\mu_*^i)^i = 0$, $f_*^j < \nu_*$, $\vartheta_*^j = 0$. $\square$

The convergence of the stepsizes to unity and the Q-superlinear convergence rate of the algorithm can be studied by invoking Theorem 4.2 (ii) and considering the inequality constraints active at the solution. In Section 5, convergence of the stepsize α_k to unity is related to the accuracy of the Hessian approximation used in the quadratic subproblem. In Section 6, it is shown that the one-sided-projected-Hessian condition ensures the Q-superlinear convergence of $\{x_k\}$ and $\{x_k, \vartheta_k, \nu_k\}$ and the two-step Q-superlinear convergence of $\{x_k, \vartheta_k, \mu_k^e, \mu_k^i, \nu_k\}$. An extended one-sided-projected-Hessian necessary and sufficient condition is also established for the Q-superlinear convergence of $\{x_k\}$ and $\{x_k, \nu_k\}$.

5 ACHIEVEMENT OF UNIT STEPSIZES

We consider first the attainment of unit stepsizes. The conditions that ensure unit stepsizes turn out to be the closeness of the algorithm to the solution as well as the accuracy of the Hessian approximation used near this solution. The condition that ensures a superlinear convergence rate is shown in the next section to be the increasing accuracy of the Hessian approximation.

We shall first establish Lemma 5.1 below which ensures, for a set of specially constructed multipliers,

$$\begin{bmatrix} \tilde{\mu}_{k+1}^e(\tilde{c}_{k+1}) \\ \tilde{\mu}_{k+1}^i(\tilde{c}_{k+1}) \end{bmatrix}$$

a specially constructed $\tilde{c}_{k+1}$, and $\bar{g}_k$ given by (2.3), that the inequality

$$\left\langle \bar{g}_k, \begin{bmatrix} \tilde{\mu}_{k+1}^e(\tilde{c}_{k+1}) \\ \tilde{\mu}_{k+1}^i(\tilde{c}_{k+1}) \end{bmatrix} \right\rangle \geq 0$$

is satisfied. Also, as the solution is approached, it is shown that

$$\left\| \begin{bmatrix} \tilde{\mu}_{k+1}^e(\tilde{c}_{k+1}) \\ \tilde{\mu}_{k+1}^i(\tilde{c}_{k+1}) \end{bmatrix} - \begin{bmatrix} \tilde{\mu}_{k+1}^e \\ \tilde{\mu}_{k+1}^i \end{bmatrix} \right\| \to 0$$

$$[(\tilde{c}_{k+1}) - c_*] \to 0.$$

The values $\tilde{\mu}^e$, $\tilde{\mu}^i$ and $\tilde{c}$ are constructed to show that unit stepsizes are attained such that $\alpha_k = 1, \forall k \geq k_\alpha \geq k_*$: they are *not* used in any computational procedure in the algorithm. The integer k_α is the iteration at which $\alpha_k = 1$ is achieved and maintained thereafter.

Consider problem (3.6a,b) in terms of the constraints active at its solution

$$\min_d \max_\vartheta \left\{ \mathcal{V}_k(x_k + d, \vartheta) + \left\langle d, \left[c_k G_k g_k + \overline{\mathcal{H}}_k (c_k \bar{h}_k + \pi_k)_+ \right] \right\rangle \right.$$

$$\left. + \frac{1}{2} \langle d, \hat{\mathfrak{C}}_k d \rangle \, \Big| \, \overline{G}_k^T d + g_k = 0, \overline{\mathcal{H}}_k^T d + \bar{h}_k = 0, \vartheta \in \mathbb{E}_+^{m^{sce}} \right\}.$$

Let $\bar{\mu}_{k+1}^i$ denote the multipliers corresponding to the active inequality constraints (2.3) at the solution of (3.14). Following Assumption 2.1, that the columns of $\overline{G}_k$ defined in (2.3) are linearly independent, the assumption in Sections 3–4 that $\hat{\mathfrak{C}}_k$ is positive definite, considering only the active inequalities at the solution of (3.14), using (3.14b) and (3.14e) we have

$$d_k = -P_k[\hat{\mathfrak{C}}_k]\hat{\mathfrak{C}}_k^{-1}\nabla f_k \vartheta_{k+1} - \hat{\mathfrak{C}}_k^{-1}\overline{G}_k \left(\overline{G}_k^T \hat{\mathfrak{C}}_k^{-1} \overline{G}_k \right)^{-1} \begin{bmatrix} g_k \\ \bar{h}_k \end{bmatrix} \qquad (5.1)$$

$$P_k[\hat{\mathfrak{C}}_k] = I - \hat{\mathfrak{C}}_k^{-1}\overline{G}_k \left(\overline{G}_k^T \hat{\mathfrak{C}}_k^{-1} \overline{G}_k \right)^{-1} \overline{G}_k^T \qquad (5.2)$$

$$\left[\begin{array}{c} \mu_{k+1}^e \\ \mu_{k+1}^i \end{array}\right] = -\left[\overline{G}_k^T \hat{\mathfrak{C}}_k^{-1} \overline{G}_k\right]^{-1}\left[\overline{G}_k^T \hat{\mathfrak{C}}_k^{-1} \nabla f_k \vartheta_{k+1} - \overline{g}_k\right] - \left[\begin{array}{c} c_k g_k \\ (c_k \overline{h}_k + \overline{\pi}_k)_+ \end{array}\right].$$

(5.3)

Furthermore, for $\tilde{c}_{k+1} \in (-\infty, +\infty)$ let

$$\left[\begin{array}{c} \tilde{\mu}_{k+1}^e(\tilde{c}_{k+1}) \\ \tilde{\mu}_{k+1}^i(\tilde{c}_k + 1) \end{array}\right] = -\left[\overline{G}_k^T \hat{\mathfrak{C}}_k^{-1} \overline{G}_k\right]^{-1}\left[\overline{G}_k^T \hat{\mathfrak{C}}_k^{-1} \nabla f_k \vartheta_{k+1} - \overline{g}_k\right] - \tilde{c}_{k+1} \overline{g}_k. \quad (5.4)$$

In the discussion below, we sometimes refer to $\tilde{\mu}_{k+1}^e(\tilde{c}_{k+1}), \tilde{\mu}_{k+1}^i(\tilde{c}_{k+1})$, as $\tilde{\mu}_{k+1}^e, \tilde{\mu}_{k+1}^i$.

Lemma 5.1   *Let the assumptions of Theorem 4.2 be satisfied, $\{x_k\}$ be convergent and let the columns of $\overline{G}_k$ be linearly independent. Then*
  (i) for $\tilde{\mu}_{k+1}^e, \tilde{\mu}_{k+1}^i$ given by (5.4), there exists $\tilde{c}_{k+1} \in (-\infty, +\infty)$ such that

$$\left\langle \overline{g}_k, \left[\begin{array}{c} \tilde{\mu}_{k+1}^e(\tilde{c}_{k+1}) \\ \tilde{\mu}_{k+1}^i(\tilde{c}_k + 1) \end{array}\right]\right\rangle \geq 0$$

  and

(ii)

$$\left\|\left[\begin{array}{c} \tilde{\mu}_{k+1}^e(\tilde{c}_{k+1}) \\ \tilde{\mu}_{k+1}^i(\tilde{c}_k + 1) \end{array}\right] - \left[\begin{array}{c} \mu_{k+1}^e \\ \mu_{k+1}^i \end{array}\right]\right\| \to 0, \quad (\tilde{c}_{k+1} - c_*) \to 0 \quad (5.5)$$

Proof. To show (i), consider (5.4) and let

$$a_k = \left\langle \overline{g}_k, \left(\overline{G}_k^T \hat{\mathfrak{C}}_k^{-1} \overline{G}_k\right)^{-1}\left\{\overline{G}_k^T \hat{\mathfrak{C}}_k^{-1} \nabla f_k \vartheta_{k+1} - \overline{g}_k\right\}\right\rangle, \quad b_k = \|\overline{g}_k\|_2^2$$

$$\left\langle \overline{g}_k, \left[\begin{array}{c} \tilde{\mu}_{k+1}^e \\ \tilde{\mu}_{k+1}^i \end{array}\right]\right\rangle = a_k - \tilde{c}_{k+1} b_k, \quad \tilde{c}_{k+1} = \begin{cases} c_* & \text{if } a_k - c_* b_k \geq 0 \\ \frac{a_k}{b_k} & \text{if } a_k - c_* b_k < 0, \end{cases}$$

By definition, $b_k \geq 0$. Also, if $b_k = 0$ we have $a_k = 0$. Thus, this choice of $\tilde{c}$ ensures (i). A natural initial value is $\tilde{c}_0 = 0$. As $\{b_k\} \to 0$, it is clear from the above that $\{\tilde{c}_k\} \to c_*$. To show (ii), consider (5.3) and (5.4). According to Lemma 4.2 and Theorem 4.2, for $k \geq k_*$, we have

$$\left\|\left[\begin{array}{c} \mu_{k+1}^e \\ \mu_{k+1}^i \end{array}\right] - \left[\begin{array}{c} \tilde{\mu}_{k+1}^e(\tilde{c}_{k+1}) \\ \tilde{\mu}_{k+1}^i(\tilde{c}_k + 1) \end{array}\right]\right\| = \left\|\left[\begin{array}{c} c_k g_k \\ (c_k \overline{h}_k + \overline{\pi}_k)_+ \end{array}\right]\right\| - \left\|\left[\begin{array}{c} g_k \\ \overline{h}_k \end{array}\right]\right\| \to 0$$

since $c_*, \tilde{c}_{k+1}$ are finite, $g_k \to 0$, $\overline{h}_k \to 0$ and $\overline{\pi}_k = -c_* \overline{h}_k \to 0$. $\square$

Theorem 5.1 *Let the assumptions of Lemma 5.1 be satisfied and let $\mathfrak{C}_*$ denote the value of the Hessian (3.5) at the solution. There is a small number $\sigma > 0$ such that if $\phi_k \leq \sigma$, for all large k, with*

$$\phi_k = \frac{\left\| P_k^{\mathrm{T}}\left[\mathfrak{C}_* - \hat{\mathfrak{C}}_k\right] P_k d_k \right\|}{\|d_k\|} + \frac{\left\| \bar{g}_k^{\mathrm{T}} (G_k^{\dagger})^{\mathrm{T}}\left[\mathfrak{C}_* - \hat{\mathfrak{C}}_k\right]\left\{ G_k^{\dagger}\bar{g}_k - 2P_k d_k \right\} \right\|}{\|d_k\|^2}$$

and $P_k = P_k[\hat{\mathfrak{C}}_k]$, given by (5.2), $G_k^{\dagger} = \hat{\mathfrak{C}}_k^{-1}\overline{G}_k(\overline{G}_k^{\mathrm{T}}\hat{\mathfrak{C}}_k^{-1}\overline{G}_k)^{-1}$ or $P_k = P_k[I]$, $G_k^{\dagger} = \overline{G}_k(\overline{G}_k^{\mathrm{T}}\overline{G}_k)^{-1}$. Then, stepsize strategy (3.9) is satisfied for $\alpha_k = 1$, $\forall k \geq k_\alpha \geq k_*$.

Proof. Let the matrix $G(x, \mu)$ be defined by

$$\mathcal{A}_3(x, \mu^e, \overline{\mu}^i) \equiv \left[\sum_{\iota=1}^{e} (\mu^e)^\iota \nabla^2 g^\iota(x) + \sum_{i=1}^{i} (\overline{\mu}^i)^i \nabla^2 \overline{h}^i(x) \right].$$

We show that there is a $k \geq k_\tau$ with x_{k+1} with $\tau_k = 1$ satisfies (3.10).

Consider the second order expansion of

$$\left\langle \bar{g}(x), \begin{bmatrix} \mu_{k+1}^e \\ \mu_{k+1}^i \end{bmatrix} \right\rangle$$

$$\left\langle \bar{g}_{k+1}, \begin{bmatrix} \mu_{k+1}^e \\ \mu_{k+1}^i \end{bmatrix} \right\rangle = \left\langle \bar{g}_k, \begin{bmatrix} \mu_{k+1}^e \\ \mu_{k+1}^i \end{bmatrix} \right\rangle + \left\langle d_k, \overline{G}_k \begin{bmatrix} \mu_{k+1}^e \\ \mu_{k+1}^i \end{bmatrix} \right\rangle$$

$$+ \frac{1}{2}\left\langle d_k, \mathcal{A}_3(x_*, \mu_*^e, \overline{\mu}_*^i) d_k \right\rangle + v_k^1$$

where

$$v_k^1 = \int_0^1 (1-t)\left\langle d_k, \left[\mathcal{A}(x_k + td_k, \mu_{k+1}^e, \overline{\mu}_{k+1}^i) - \mathcal{A}_3(x_*, \mu_*^e, \overline{\mu}_*^i)\right] d_k \right\rangle dt$$

where $\mathcal{A}_3(x_*, \mu_*^e, \overline{\mu}_*^i)$ is the constraint second derivatives at x_*, to be subsumed in $\mathfrak{C}_*$. Using the fact that d_k satisfies the linearized constraints, $\overline{G}_k^{\mathrm{T}} d_k + \bar{g}_k = 0$, and using Lemma 5.1 we have

$$\left\langle \bar{g}_{k+1}, \begin{bmatrix} \mu_{k+1}^e \\ \mu_{k+1}^i \end{bmatrix} \right\rangle = \frac{1}{2}\left\langle d_k, \mathcal{A}_3(x_*, \mu_*^e, \overline{\mu}_*^i) d_k \right\rangle + v_k^1$$

$$\leq \left\langle \bar{g}_{k+1}, \begin{bmatrix} \tilde{\mu}_{k+2}^e(\tilde{c}_{k+2}) \\ \tilde{\mu}_{k+2}^i(\tilde{c}_k + 2) \end{bmatrix} \right\rangle + \frac{1}{2}\left\langle d_k, \mathcal{A}_3(x_*, \mu_*^e, \overline{\mu}_*^i) d_k \right\rangle + v_k^1.$$

Now consider the expression

$$\Gamma_k = \left\langle \overline{g}_{k+1}, \begin{bmatrix} \tilde{\mu}^e_{k+2}(\tilde{c}_{k+2}) \\ \tilde{\mu}^i_{k+2}(\tilde{c}_k + 2) \end{bmatrix} \right\rangle - \left\langle \overline{g}_{k+1}, \begin{bmatrix} \mu^e_{k+2} \\ \mu^i_{k+2} \end{bmatrix} \right\rangle$$

$$+ \left\langle \overline{g}_{k+1}, \begin{bmatrix} \mu^e_{k+2} \\ \mu^i_{k+2} \end{bmatrix} \right\rangle - \left\langle \overline{g}_{k+1}, \begin{bmatrix} \mu^e_{k+1} \\ \mu^i_{k+1} \end{bmatrix} \right\rangle. \tag{5.7}$$

Using the second order expansion of $\overline{g}(x)$ and (3.14,e) we can write

$$\|\overline{g}_{k+1}\| \leq \left\| \overline{g}_k + \overline{G}_k^T d_k \right\| + e_1 \|d_k\|_2^2 = e_1 \|d_k\|_2^2 \tag{5.8}$$

for some $e_1 \geq 0$. Furthermore, as

$$\left\{ \left\| \begin{bmatrix} \mu^e_k \\ \overline{\mu}^i_k \end{bmatrix} - \begin{bmatrix} \mu^e_{k+1} \\ \overline{\mu}^i_{k+1} \end{bmatrix} \right\| \right\} \to 0$$

and, by Lemma 5.1,

$$\left\{ \left\| \begin{bmatrix} \tilde{\mu}^e_{k+1}(\tilde{c}_{k+1}) \\ \tilde{\mu}^i_{k+1}(\tilde{c}_k + 1) \end{bmatrix} - \begin{bmatrix} \mu^e_{k+1} \\ \overline{\mu}^i_{k+1} \end{bmatrix} \right\| \right\} \to 0$$

$\|\Gamma_k\|$ can be expressed as

$$\|\Gamma_k\| \leq \beta_k \|d_k\|_2^2 \tag{5.9a}$$

$$\lim_{k \to \infty} \beta_k = \lim_{k \to \infty} e_1 \left[\left\| \begin{bmatrix} \tilde{\mu}^e_{k+2}(\tilde{c}_{k+2}) \\ \tilde{\mu}^i_{k+2}(\tilde{c}_k + 2) \end{bmatrix} - \begin{bmatrix} \mu^e_{k+2} \\ \overline{\mu}^i_{k+2} \end{bmatrix} \right\| + \left\| \begin{bmatrix} \mu^e_{k+2} \\ \overline{\mu}^i_{k+2} \end{bmatrix} - \begin{bmatrix} \mu^e_{k+1} \\ \overline{\mu}^i_{k+1} \end{bmatrix} \right\| \right]$$

$$= 0. \tag{5.9b}$$

Consider expression (4.3a). Let the value of ϑ corresponding to the solution of the right-hand side be given by ϑ^*_k thus, using Lemma 6.2.1,

$$\Phi(x_{k+1}) + c_* \langle g(x_{k+1}), g(x_{k+1}) \rangle = \{ \langle \vartheta^*_k, f(x_k + d_k) \rangle \} + c_* \langle \overline{g}_{k+1}, \overline{g}_{k+1} \rangle.$$

By Theorem 4.2, $\{x_k\} \to x_*$, then clearly $\{\vartheta^*_k\} \to \vartheta_*$. Thus, using (4.3) we can write

$$\Phi(x_{k+1}) + c_* \langle g(x_{k+1}), g(x_{k+1}) \rangle$$

$$\leq \left\{ \langle \vartheta^*_k, f(x_k) + \nabla f^T_k d_k \rangle \right\} + \frac{1}{2} \langle d_k, \left[\hat{\mathfrak{C}}_k - \hat{\mathfrak{C}}_k + \mathfrak{C}^g_* \right] d_k \rangle$$

$$+ \frac{c_*}{2} \left[\langle \overline{g}_k, \overline{g}_k \rangle + \langle \overline{g}_k, \overline{G}^T_k d_k \rangle \right] + \nu_k^2$$

$$\leq \max_{\vartheta \in \mathbb{E}_+^{m^{sce}}} \left\{ \left\langle \vartheta, f(x_k) + \nabla f_k^T d_k \right\rangle \right\} + \frac{1}{2} \left\langle d_k, \left[\hat{\mathfrak{C}}_k - \hat{\mathfrak{C}}_k + \mathfrak{C}_*^g \right] d_k \right\rangle$$

$$+ \frac{c_*}{2} \left[\langle \overline{g}_k, \overline{g}_k \rangle + \left\langle \overline{g}_k, \overline{G}_k^T d_k \right\rangle \right] + \nu_k^2 \tag{5.10}$$

where, with $x(t) = x_k + t d_k$,

$$\nu_k^2 = \int_0^1 (1-t) \left\langle d_k, \left\{ \left\{ \sum_{i=1}^{m^{sce}} \vartheta_k^{*i} \nabla^2 f^i(x(t)) \right\} \right. \right.$$

$$\left. \left. + c_* \overline{G}(x(t)) \overline{G}(x(t))^T + \mathcal{A}_2(x(t), c_*) - \mathfrak{C}_*^g \right\} d_k \right\rangle dt$$

where $\mathcal{A}_2$ is given by (4.2c) and $\mathfrak{C}_*^g$ is the the second derivative matrix evaluated at x_*, ϑ_*

$$\mathfrak{C}_*^g = \sum_{i=1}^{m^{sce}} \vartheta_*^i \nabla^2 f^i(x_*) + c \overline{G}_* \overline{G}_*^T.$$

Using (5.6), (5.7) and (5.10) we can rewrite (4.7) as

$$\Phi(x_{k+1}) + \frac{c_*}{2} \langle g(x_{k+1}), g(x_{k+1}) \rangle$$

$$\leq \Phi(x_k) + \frac{c_*}{2} \langle \overline{g}_k, \overline{g}_k \rangle + \left\{ \Phi_k(x_k + d_k) - \Phi(x_k) + c_* \left\langle \overline{g}_k, \overline{G}_k^T d_k \right\rangle \right.$$

$$\left. + \frac{1}{2} \left\langle d_k, \left[\hat{\mathfrak{C}}_k - \hat{\mathfrak{C}}_k + \mathfrak{C}_* \right] d_k \right\rangle + \Gamma_k + \nu_k^1 + \nu_k^2 \right\}. \tag{5.11}$$

We note that

$$\left\| \nu_k^1 + \nu_k^2 \right\| = \| d_k \|_2^2 \xi_k^*$$

$$\xi_k^* = \left\| \int_0^1 (1-t) \left\{ \left\{ \sum_{i=1}^{m^{sce}} \vartheta_k^{*i} \nabla^2 f^i(x(t)) \right\} \right. \right.$$

$$+ c_* \overline{G}(x(t)) \overline{G}(x(t))^T + \mathcal{A}_3 \left(x(t), \mu_{k+1}^e, \overline{\mu}_{k+1}^i \right)$$

$$\left. \left. + \mathcal{A}_2(x(t), c_*) - \mathfrak{C}_*^g - \mathcal{A}_3(x_*, \mu_*^e, \overline{\mu}_*^i) \right\} \right\| dt$$

and $\mathfrak{C}_* = \mathfrak{C}_*^g + \mathcal{A}_3(x_*, \mu_*^e, \overline{\mu}_*^i)$. Using (5.9), the descent property (3.14) in Lemma 3.2 and also Lemma 3.4 (i), we can write (5.11) as

$$\Phi(x_{k+1}) + \frac{c_*}{2}\langle g(x_{k+1}), g(x_{k+1})\rangle$$

$$\leq \Phi(x_k) + \frac{c_*}{2}\langle \overline{g}_k, \overline{g}_k\rangle + \left\{\Phi_k(x_k + d_k) - \Phi(x_k) + c_*\langle \overline{g}_k, \overline{G}_k^{\mathsf{T}}d_k\rangle\right.$$

$$\left. + \frac{1}{2}\langle d_k, \left[\hat{\mathbb{C}}_k - \hat{\mathbb{C}}_k + \mathbb{C}_*\right]d_k\rangle + \beta_k\|d_k\|_2^2 + \|d_k\|_2^2 \xi_k^*\right\}$$

$$\leq \Phi(x_k) + \frac{c_*}{2}\langle \overline{g}_k, \overline{g}_k\rangle + \Psi(d_k, c_*)\left[1 - \frac{1}{m\epsilon}\left[\frac{1}{2}\phi_k + \beta_k + \xi_k^*\right]\right] \quad (5.12)$$

where the last expression is obtained by invoking the equality $d_k = P_k - G_k^{\dagger}\overline{g}_k$. Since, by Theorem 4.2, $\{x_k\} \to x_*$, $\{\mu_k^e\} \to \mu_*^e$, $\{\mu_k^i\} \to \mu_*^i$, $\{\eta_k\} \to \eta_*$, $\{v_k\} \to v_*$, we have $\{\xi_k^*\} \to 0$ and $\{\beta_k\} \to 0$. The scalar

$$\rho \in (0,1)$$

in (3.9) is bounded by

$$\rho \leq 1 - \frac{1}{m\epsilon}\left[\frac{1}{2}\phi_k + \beta_k + \xi_k^*\right]. \quad (5.13)$$

If $\sigma > 0$ is such that $(1/m\epsilon)[(1/2)\sigma + \beta_k + \xi_k^*] \leq 1 - \rho$ (in view of $\xi_k \to 0$, $\beta_k \to 0$, this defines the number ρ) then (5.13) holds, and therefore, because $\Psi(d_k, c_*) \leq 0$, (3.9) is satisfied with $\alpha_k = 1$. $\quad\square$

6 SUPERLINEAR CONVERGENCE RATES OF THE ALGORITHM

The superlinear convergence of the algorithm is discussed only for the quadratic programming subproblem (3.6b). An interesting aspect of the subproblem and the convergence rate is that the Hessian of the quadratic objective function is singular. The results below discuss the superlinear convergence of $\{x_k\}$, $\{x_k, v_k\}$, $\{x_k, \vartheta_k, v_k\}$, $\{x_k, \vartheta_k, \mu_k^e, \overline{\mu}_k^i, v_k\}$ and relate these to the accuracy of the projected Hessian. The reason for discussing the convergence rate of $\{x_k\}$, as well as the rates of $\{x_k, v_k\}$, $\{x_k, \vartheta_k, v_k\}$, $\{x_k, \vartheta_k, \mu_k^e, \overline{\mu}_k^i, v_k\}$ is that, as mentioned by Boggs et al. (1982), the Q-superlinear convergence of the latter three sequences only imply the R-superlinear convergence of the former. The situation is further complicated by the fact that the superlinear convergence results for the latter two sequences are not necessarily the same as the results for the former two sequences when projected Hessians are being considered.

The effect of the inequality constraints through the use of the penalty term in the quadratic subproblem (3.6a) is ignored since it is shown in Rustem (1998, Chapter 8) in the context of nonlinear programming that this term does not affect the convergence rate estimates. Instead, we adopt the equality constrained framework established in Theorem 4.2 that all constraints are active at the solution and consider only the constraints $\bar{g}(x) = 0$ and the linear approximation $\bar{g}_k + \bar{G}_k^T d_k = 0$.

Assumptions 6.1

(i) *There exists $x_*, \vartheta_*, \mu_*^e, \bar{\mu}_*^i, \lambda_*, \eta_*$ satisfying the first order conditions of (1.2) or, equivalently, there exists $x_*, \vartheta_*, \mu_*^e, \bar{\mu}_*^i, \lambda_*, \nu_*$ satisfying the first order conditions of (1.4).*

(ii) *Strict complementarity holds for the inequality constraints of the quadratic programming subproblem (3.6b).*

(iii) $v^T \mathbb{C}_* v > 0, \forall v \in \mathbb{R}^n, \bar{G}_*^T v = 0.$

(iv) *The coefficient matrix*

$$\nabla \hat{\Gamma}_k = \begin{bmatrix} \hat{\mathbb{C}}_k & \nabla \bar{f}_k & \bar{G}_k & 0 \\ \nabla \bar{f}_k^T & 0 & 0 & \bar{1} \\ \bar{G}_k^T & 0 & 0 & 0 \\ 0 & \bar{1}^T & 0 & 0 \end{bmatrix} \qquad (6.1)$$

arising from the first order necessary conditions of the quadratic subproblem (3.6b), is nonsingular. The matrix $\nabla \bar{f}_k$ corresponds to the columns of ∇f_k of the inequality constraints in (3.6b) satisfied as equalities at the solution x_* of the minimax problem. Similarly, the vector $\bar{1}$ corresponds to the right hands of the inequality constraints in (3.6b) satisfied as equalities at x_*.

(v) *The columns of $\bar{G}_k, [\nabla f_k^T, \bar{1}]^T$ are linearly independent at the solution x_*.*

As a result of the convergence of Algorithm 3.1, established in Theorem 4.2, there is a stage, near the solution of the minimax problem, when the inequality constraints satisfied as equalities at the solution, are also satisfied as equalities for the quadratic program (3.5b). At this stage, with strict complementarity holding, the multipliers $\vartheta_k^i > 0$ if the constraint i is satisfied as an equality at the solution x_*. Also, $\langle \nabla f_k^i, d_k \rangle + f_k^i < \nu_{k+1}$ and $\vartheta_k^i = 0$ if the constraint is strictly satisfied at the solution. Thus, at this stage, the multipliers can predict the constraints active at the solution. These active constraints are written as $\langle \nabla \bar{f}_k^i, d_k \rangle + \bar{f}_k^i = \nu_{k+1}$. Since the other constraints do not affect the solution x_*, they can be ignored for the convergence rate analysis. Thus, the quadratic subproblem (3.6b) can be rewritten as

$$\min_{d,v} \left\{ v + c_* \langle \bar{G}_k \bar{g}_k, d \rangle + \frac{1}{2} \langle d, \hat{\mathbb{C}}_k d \rangle \,\middle|\, \bar{G}_k^T d + \bar{g}_k = 0, \nabla \bar{f}_k^T d + \bar{f}_k = \bar{1} v \right\}.$$

The first order optimality conditions for (6.2) can be written as

$$\nabla \hat{\Gamma}_k \begin{bmatrix} d_k \\ \overline{\vartheta}_{k+1} - \overline{\vartheta}_k \\ \begin{bmatrix} \mu^e_{k+1} \\ \overline{\mu}^i_{k+1} \end{bmatrix} - \begin{bmatrix} \mu^e_k \\ \overline{\mu}^i_k \end{bmatrix} \\ -\Delta v_k \end{bmatrix} = -\Gamma_k \qquad (6.3)$$

where

$$\Delta v_k = v_{k+1} - v_k$$

$$\Gamma_k = \begin{bmatrix} \nabla \overline{f}_k \overline{\vartheta}_k + \overline{G}_k \begin{bmatrix} \begin{bmatrix} \mu^e_k \\ \overline{\mu}^i_k \end{bmatrix} + c_* \overline{g}_k \end{bmatrix} \\ \overline{g}_k \\ \overline{f}_k - \overline{1} v_k \\ \langle \overline{1}, \overline{\vartheta}_k \rangle - 1 \end{bmatrix} \qquad (6.4)$$

and $\overline{\vartheta}_k$ relate to the strictly positive elements of ϑ_k which correspond to the inequality constraints satisfied as equalities.

The quadratic subproblem computes d_k such that $\overline{G}^{\mathsf{T}}_{k-1} d_{k-1} + \overline{g}_{k-1} = 0$. Thus, using the second order expansion of $\overline{g}(x)$ at x_k, we have for some $e_1 \geq 0$,

$$\| \overline{g}_k \| = e_1 \| d_{k-1} \|^2 . \qquad (6.5)$$

Since the inequality constraints are satisfied as equalities, we have

$$\overline{1} v_k = \nabla \overline{f}^{\mathsf{T}}_{k-1} d_{k-1} + \overline{f}_{k-1}. \qquad (6.6)$$

The second order expansion of $\overline{f}(x)$ about x_{k-1} yields, for some $e_2 \geq 0$,

$$\overline{f}_k - \overline{f}_{k-1} - \nabla \overline{f}^{\mathsf{T}}_{k-1} d_{k-1} = e_2 \| d_{k-1} \|^2 .$$

Thus, $\overline{f}_k - \overline{1} v_k$ in (6.4) can be written as

$$\overline{f}_k - \overline{1} v_k = e_2 \| d_{k-1} \|^2 . \qquad (6.7)$$

The first order expansion of the gradient, with respect to x, of the Lagrangian of either minimax formulation,

$$\nabla \overline{f}(x) \overline{\vartheta} + \overline{G}(x) \begin{bmatrix} \begin{bmatrix} \mu^e \\ \overline{\mu}^i \end{bmatrix} + c_* \overline{g}(x) \end{bmatrix}$$

yields

$$\nabla \overline{f}_k \overline{\vartheta}_k + \overline{G}_k \left[\begin{bmatrix} \mu_k^e \\ \overline{\mu}_k^i \end{bmatrix} + c_* \overline{g}_k \right] = \nabla \overline{f}_{k-1} \overline{\vartheta}_{k-1} + \overline{G}_{k-1} \left[\begin{bmatrix} \mu_{k-1}^e \\ \overline{\mu}_{k-1}^i \end{bmatrix} + c_* \overline{g}_{k-1} \right]$$

$$+ \hat{\mathfrak{C}}_{k-1} d_{k-1} + \Delta \mathfrak{C}(t) d_{k-1}$$

(6.8a)

$$\Delta \mathfrak{C}(t) = \int_0^1 \left\{ \mathfrak{C}(x_{k-1} + t d_{k-1}) - \mathfrak{C}_* + \mathfrak{C}_* - \hat{\mathfrak{C}}_{k-1} \right\} dt. \qquad (6.8b)$$

Clearly, as $\{x_k\} \to x_*$,

$$\left\{ \int_0^1 \{ \mathfrak{C}(x_{k-1} + t d_{k-1}) - \mathfrak{C}_* \} dt \right\} \to 0$$

and, as shown below, the condition for superlinear convergence ensures that the rest of (6.8b) also vanishes at the solution.

It should be noted from the first order optimality condition (3.14b), for either quadratic subproblem, at x_{k-1} is

$$\nabla \overline{f}_{k-1} \overline{\vartheta}_{k-1} + \overline{G}_{k-1} \left[\begin{bmatrix} \mu_{k-1}^e \\ \overline{\mu}_{k-1}^i \end{bmatrix} + c_* \overline{g}_{k-1} \right] = 0$$

and thus (6.8a) can be written as

$$\nabla \overline{f}_k \overline{\vartheta}_k + \overline{G}_k \left[\begin{bmatrix} \mu_k^e \\ \overline{\mu}_k^i \end{bmatrix} + c_* \overline{g}_k \right] = \Delta \mathfrak{C}(t) d_{k-1}. \qquad (6.9)$$

We can now state the superlinear convergence theorem. In this result conditions are discussed which are similar to the characterization of Boggs et al. (1982), Powell (1983) for nonlinear programming. The results below are established without invoking a prior linear convergence rate argument. As opposed to the necessary and sufficient result of Boggs et al. (1982), the result below only establishes the necessary part in the case of minimax problems. The analogous necessary and sufficient condition is discussed in Theorem 6.2. The following lemma is used in the latter part of Theorem 6.1.

Lemma 6.1 *Let Assumptions (2.1), (4.1) and (6.1) hold. Then, for some* $e_0 \in [0, \infty)$,

$$\|d_k\| \le e_0 \|x_* - x_k\|$$

and

$$\left\| \begin{matrix} d_k \\ \Delta v_k \end{matrix} \right\| \le e_0 \left\| \begin{bmatrix} x_* \\ v_* \end{bmatrix} - \begin{bmatrix} x_k \\ v_k \end{bmatrix} \right\|.$$

Proof. The first order optimality condition of the quadratic programming subproblem can be written as

$$
\left[\nabla\hat{\Gamma}_k\right]
\begin{bmatrix}
d_k \\
\overline{\vartheta}_{k+1} - \overline{\vartheta}_* \\
\begin{bmatrix} \mu^e_{k+1} \\ \overline{\mu}^i_{k+1} \end{bmatrix} - \begin{bmatrix} \mu^e_* \\ \overline{\mu}^i_* \end{bmatrix} + c_*\overline{g}_k \\
-(v_{k+1} - v_*)
\end{bmatrix}
= -
\begin{bmatrix}
\nabla\overline{f}_k\overline{\vartheta}_* + \overline{G}_k\begin{bmatrix} \mu^e_* \\ \overline{\mu}^i_* \end{bmatrix} \\
\overline{g}_k \\
\overline{f}_k - \overline{1}v_* \\
\langle \overline{1}, \overline{\vartheta}_*\rangle - 1
\end{bmatrix}
$$

Using the first order expansion of the right side of the above expression about $(x_*, \overline{\vartheta}_*, \mu^e_*, \overline{\mu}^i_*, v_*)$; by the nonsingularity of the coefficient matrix on the left and by the first order optimality condition of (2.1b), that is,

$$
\nabla\overline{f}(x_*)\overline{\vartheta}_* + \overline{G}_*\begin{bmatrix} \mu^e_* \\ \overline{\mu}^i_* \end{bmatrix} = 0, \quad \overline{g}(x_*) = 0, \quad \overline{f}(x_*) = \overline{1}v_*
$$

we have

$$
\|d_k\| \leq
\left\|
\begin{bmatrix}
d_k \\
\overline{\vartheta}_{k+1} - \overline{\vartheta}_* \\
\begin{bmatrix} \mu^e_{k+1} \\ \overline{\mu}^i_{k+1} \end{bmatrix} - \begin{bmatrix} \mu^e_* \\ \overline{\mu}^i_* \end{bmatrix} + c_*\overline{g}_k \\
-(v_{k+1} - v_*)
\end{bmatrix}
\right\|
\leq \left[\nabla\hat{\Gamma}_k\right]^{-1}\|x_* - x_k\|
$$

for some $\hat{e} \in [0, \infty)$. Hence, the first result follows from the above expression. To establish the second result, we use a similar argument with

$$
\left[\nabla\hat{\Gamma}_k\right]
\begin{bmatrix}
d_k \\
\overline{\vartheta}_{k+1} - \overline{\vartheta}_* \\
\begin{bmatrix} \mu^e_{k+1} \\ \overline{\mu}^i_{k+1} \end{bmatrix} - \begin{bmatrix} \mu^e_* \\ \overline{\mu}^i_* \end{bmatrix} + c_*\overline{g}_k \\
-\Delta v_k
\end{bmatrix}
= -
\begin{bmatrix}
\nabla\overline{f}_k\overline{\vartheta}_* + \overline{G}_k\begin{bmatrix} \mu^e_* \\ \overline{\mu}^i_* \end{bmatrix} \\
\overline{g}_k \\
\overline{f}_k - \overline{1}v_k \\
\langle \overline{1}, \overline{\vartheta}_k\rangle - 1
\end{bmatrix}
$$

where the right side is expanded to yield

$$
\left\|\begin{bmatrix} d_k \\ \Delta v_k \end{bmatrix}\right\| \leq
\left\|
\begin{bmatrix}
d_k \\
\overline{\vartheta}_{k+1} - \overline{\vartheta}_* \\
\begin{bmatrix} \mu^e_{k+1} \\ \overline{\mu}^i_{k+1} \end{bmatrix} - \begin{bmatrix} \mu^e_* \\ \overline{\mu}^i_* \end{bmatrix} + c_*\overline{g}_k \\
-\Delta v_k
\end{bmatrix}
\right\|
\leq \hat{e}\left[\nabla\hat{\Gamma}_k\right]^{-1}\left\|\begin{bmatrix} x_* \\ v_* \end{bmatrix} - \begin{bmatrix} x_k \\ v_k \end{bmatrix}\right\|. \quad \square
$$

Lemma 6.2

(i) Let $\{x_k\} \to x_*.\{x_k, v_k\} \to (x_*, v_*)$. Then, the sequence $\{x_k\}$ is Q-superlinearly convergent, that is,

$$\lim_{k \to \infty} \frac{\|x_* - x_{k+1}\|}{\|x_* - x_k\|} = 0$$

if and only if

$$\| d_k \| \le \kappa_k \| d_{k-1} \|, \quad \lim_{k \to \infty} \kappa_k = 0.$$

(ii) Let $\{x_k, v_k\} \to (x_*, v_*)$. Then the sequence $\{x_k, v_k\}$ is Q-superlinearly convergent, that is,

$$\lim_{k \to \infty} \frac{\left\| \begin{bmatrix} x_* \\ v_* \end{bmatrix} - \begin{bmatrix} x_{k+1} \\ v_{k+1} \end{bmatrix} \right\|}{\left\| \begin{bmatrix} x_* \\ v_* \end{bmatrix} - \begin{bmatrix} x_k \\ v_k \end{bmatrix} \right\|} = 0$$

if and only if

$$\left\| \begin{matrix} d_k \\ \Delta v_k \end{matrix} \right\| \le \kappa_k \left\| \begin{matrix} d_{k-1} \\ \Delta v_{k-1} \end{matrix} \right\|, \quad \lim_{k \to \infty} \kappa_k = 0.$$

Proof. We have

$$\| x_* - x_k \| \le \lim_{t \to \infty} \sum_{j=k}^{t-1} \| x_{j+1} - x_j \|$$

$$\le \kappa_k \| d_{k-1} \| (1 + \omega + \omega^2 + \omega^3 + \cdots)$$

$$\le \frac{\kappa_k}{1 - \omega} \{ \| x_k - x_* \| + \| x_* - x_{k-1} \| \}$$

for some $\omega \in [0, 1)$. As $\{\kappa_k\} \to 0, \omega$ is chosen such that $\kappa_k \le \omega < 1, \forall k \ge K_0$. K_0 is an integer and is such that $\kappa_k < 1, \forall k \ge K_0$. Rearranging the above expression yields

$$\frac{\| x_k - x_* \|}{\| x_k - x_* \| + \| x_{k-1} - x_* \|} \le \frac{\kappa_k}{1 - \omega}$$

which yields the required result.

Suppose that $\| x_* - x_k \| \le \kappa_k^0 \| x_* - x_{k-1} \|, \lim_{k \to \infty} \kappa_k^0 = 0$, with $\kappa_k^0 < 1$. This yields the inequality

$$\| x_* - x_k \| \le \kappa_k^0 \{ \| x_* - x_k \| + \| d_{k-1} \| \} \le \left(\frac{\kappa_k^0}{1 - \kappa_k^0} \right) \| d_{k-1} \|.$$

Lemma 2.1 yields the desired result for (i). We can establish (ii) following the same argument. $\square$

Theorem 6.1 *Let Assumptions 2.1, 4.1 and 6.1 be satisfied.*

(i) *The sequence $\{x_k\}$ satisfies $\| d_k \| \le \kappa_k \| d_{k-1} \|$ with $\lim_{k\to\infty} \kappa_k = 0$ and thence converges Q-superlinearly if, for $P_k = I - \overline{G}_k (\overline{G}_k^{\mathrm{T}} \overline{G}_k)^{-1} \overline{G}_k^{\mathrm{T}}$,*

$$\lim_{k\to\infty} \frac{\| P_k (\mathbb{C}_* - \hat{\mathbb{C}}_k) d_k \|}{\| d_k \|} = 0. \tag{6.10}$$

(ii) *Let $\Delta\vartheta_k = \vartheta_{k+1} - \vartheta_k$. The sequence $\{x_k, \vartheta_k, v_k\}$ satisfies*

$$\left\| \begin{array}{c} d_k \\ \Delta\vartheta_k \\ \Delta v_k \end{array} \right\| \le \kappa_k \| d_{k-1} \| \le \kappa_k \left\| \begin{array}{c} d_{k-1} \\ \Delta\vartheta_{k-1} \\ \Delta v_{k-1} \end{array} \right\| \tag{6.11}$$

and thence converges Q-superlinearly if (6.10) is satisfied.

(iii) *The sequence $\{x_k, \vartheta_k, \mu_k^e, \overline{\mu}_k^i, v_k\}$ converges at a two-step Q-superlinear rate if (6.10) holds.*

(iv) *The sequence $\{x_k, \vartheta_k, \mu_k^e, \overline{\mu}_k^i, v_k\}$ converges at a Q-superlinear rate if*

$$\lim_{k\to\infty} \frac{\| P_k (\mathbb{C}_* - \hat{\mathbb{C}}_k) d_k \|}{\| d_k \|} = 0 \quad and \quad \lim_{k\to\infty} \frac{\| (F_k^{\mathrm{T}} F_k)^{-1} F_k^{\mathrm{T}} (\mathbb{C}_* - \hat{\mathbb{C}}_k) d_k \|}{\| d_k \|} = 0 \tag{6.12}$$

where $F_k = \begin{bmatrix} \overline{G}_k & \nabla\overline{f}_k \end{bmatrix}$.

Proof. Using (6.9), expression (6.3) can be written as

$$\begin{bmatrix} \nabla\hat{\Gamma}_k \end{bmatrix} \begin{bmatrix} d_k \\ \Delta\vartheta_k \\ \Delta\begin{bmatrix} \mu_k^e \\ \overline{\mu}_k^i \end{bmatrix} + (\overline{G}_{k-1}^{\mathrm{T}} \overline{G}_{k-1})^{-1} \overline{G}_{k-1}^{\mathrm{T}} \Delta\mathbb{C}(t) d_{k-1} \\ -\Delta v_k \end{bmatrix} = - \begin{bmatrix} P_{k-1} \Delta\mathbb{C}(t) d_{k-1} \\ \overline{g}_k \\ \overline{f}_k - \overline{1} v_k \\ \langle \overline{1}, \vartheta_k \rangle - 1 \end{bmatrix}. \tag{6.13}$$

Since the matrix $\nabla\hat{\Gamma}_k$ is nonsingular, (6.13), (6.10), (6.5), (6.7) and $\{\| d_k \|\} \to 0$ yield

$$\| d_k \| \le \left\| \begin{array}{c} d_k \\ \Delta\vartheta_k \\ \Delta v_k \end{array} \right\| \le \kappa_k \| d_{k-1} \| \le \kappa_k \left\| \begin{array}{c} d_{k-1} \\ \Delta\vartheta_{k-1} \\ \Delta v_{k-1} \end{array} \right\|, \quad \text{with } \lim_{k\to\infty} \kappa_k = 0. \tag{6.14}$$

Thence, Lemma 6.2 establishes the Q-superlinear convergence of $\{x_k\}$, $\{x_k, \vartheta_k, v_k\}$. This establishes (i) and (ii).

In order to show (iii), we first show that

$$\left\| \begin{bmatrix} \vartheta_* - \vartheta_k \\ \begin{bmatrix} \mu_*^e \\ \overline{\mu}_*^i \end{bmatrix} - \begin{bmatrix} \mu_k^e \\ \overline{\mu}_k^i \end{bmatrix} \\ v_* - v_k \end{bmatrix} \right\| \leq \kappa_k^1 \| x_* - x_{k-2} \|, \quad \text{with } \lim_{k \to \infty} \kappa_k^1 = 0. \qquad (6.15)$$

From (6.3), (6.4), (6.5), (6.7) and (6.9) we have

$$\left\| \Delta \begin{bmatrix} \Delta \vartheta_k \\ \begin{bmatrix} \mu_k^e \\ \overline{\mu}_k^i \end{bmatrix} \\ \Delta v_k \end{bmatrix} \right\| \leq e_3 \| d_k \| + e_4 \| d_{k-1} \| + e_5 \| d_{k-1} \|^2 \qquad (6.16)$$

for some $e_3, e_4, e_5 \geq 0$. We can use $\| d_k \| \leq \kappa_k \| d_{k-1} \|$ in (6.16) to obtain

$$\left\| \Delta \begin{bmatrix} \Delta \vartheta_k \\ \begin{bmatrix} \mu_k^e \\ \overline{\mu}_k^i \end{bmatrix} \\ \Delta v_k \end{bmatrix} \right\| \leq \kappa_k^2 \| d_k \|, \quad \lim_{k \to \infty} \kappa_k^2 = 0. \qquad (6.17)$$

This leads to

$$\left\| \begin{bmatrix} \vartheta_* - \vartheta_k \\ \begin{bmatrix} \mu_*^e \\ \overline{\mu}_*^i \end{bmatrix} - \begin{bmatrix} \mu_k^e \\ \overline{\mu}_k^i \end{bmatrix} \\ v_* - v_k \end{bmatrix} \right\| \leq \lim_{t \to \infty} \sum_{j=k}^{t-1} \left\| \Delta \begin{bmatrix} \Delta \vartheta_j \\ \begin{bmatrix} \mu_j^e \\ \overline{\mu}_j^i \end{bmatrix} \\ \Delta v_j \end{bmatrix} \right\|$$

$$\leq \lim_{t \to \infty} \sum_{j=k}^{t-1} \kappa_j^2 \| d_{j-2} \|$$

$$\leq \kappa_k^2 \| d_{k-2} \| (1 + \omega + \omega^2 + \omega^3 + \cdots)$$

for some $\omega \in [0, 1)$. Combining this with Lemma 6.1 yields (6.15). Using (6.15) with the superlinear convergence of $\{x_k\}$, we have, for $\lim_{k \to \infty} \kappa_k^3 = 0$,

$$\left\| \begin{bmatrix} x_* - x_k \\ \vartheta_* - \vartheta_k \\ \begin{bmatrix} \mu_*^e \\ \overline{\mu}_*^i \end{bmatrix} - \begin{bmatrix} \mu_k^e \\ \overline{\mu}_k^i \end{bmatrix} \\ v_* - v_k \end{bmatrix} \right\| \leq \kappa_k \| x_* - x_{k-1} \| + \kappa_k^1 \| x_* - x_{k-2} \|$$

$$\leq \kappa_k^3 \left\| \begin{bmatrix} x_* - x_{k-2} \\ \vartheta_* - \vartheta_{k-2} \\ \begin{bmatrix} \mu_*^e \\ \overline{\mu}_*^i \end{bmatrix} - \begin{bmatrix} \mu_{k-2}^e \\ \overline{\mu}_{k-2}^i \end{bmatrix} \\ v_* - v_{k-2} \end{bmatrix} \right\|. \tag{6.18}$$

To establish (iv), we consider the first equation of (6.3) with (6.5), (6.7) and (6.9). We have

$$F_k \begin{bmatrix} \Delta \vartheta_k \\ \Delta \begin{bmatrix} \mu_k^e \\ \overline{\mu}_k^i \end{bmatrix} \end{bmatrix} = -\left[\hat{\mathbb{C}}_k d_k + \Delta\mathbb{C}(t)d_{k-1} \right]$$

$$\begin{bmatrix} \Delta \vartheta_k \\ \Delta \begin{bmatrix} \mu_k^e \\ \overline{\mu}_k^i \end{bmatrix} \end{bmatrix} = -(F_k^{\mathrm{T}}F_k)^{-1}F_k^{\mathrm{T}}\left[\hat{\mathbb{C}}_k d_k + \Delta\mathbb{C}(t)d_{k-1} \right].$$

Using $\| d_k \| \leq \kappa_k \| d_{k-1} \|$ with $\lim_{k\to\infty} \kappa_k = 0$ established above, we have

$$\left\| \begin{bmatrix} \Delta \vartheta_k \\ \Delta \begin{bmatrix} \mu_k^e \\ \overline{\mu}_k^i \end{bmatrix} \end{bmatrix} \right\| \leq \left[\kappa_k \| (F_k^{\mathrm{T}}F_k)^{-1}F_k^{\mathrm{T}}\hat{\mathbb{C}}_k \| + \frac{\| (F_k^{\mathrm{T}}F_k)^{-1}F_k^{\mathrm{T}}\Delta\mathbb{C}(t)d_{k-1} \|}{\| d_{k-1} \|} \right] \| d_{k-1} \|$$

which, with (6.11) and Lemma 6.2, yields the Q-superlinear convergence of $\{x_k, \vartheta_k, \mu_k^e, \overline{\mu}_k^i, v_k\}$. $\square$

We now consider an alternative Q-superlinear convergence theorem which establishes the *necessary and sufficient* condition for this rate. We first define the following matrices which are used below:

$$\hat{\mathbb{Q}}_k = \begin{bmatrix} \hat{\mathbb{C}}_k & 0^{n\times 1} \\ 0^{1\times n} & 0^{1\times 1} \end{bmatrix}, \quad \Delta\mathbb{Q}(t) = \begin{bmatrix} \Delta\mathbb{C}(t) & 0^{n\times 1} \\ 0^{1\times n} & 0^{1\times 1} \end{bmatrix}, \quad \mathfrak{F}_k = \begin{bmatrix} \nabla\overline{f}_k & \overline{G}_k \\ \overline{i}^{\mathrm{T}} & 0^{1\times e} \end{bmatrix}$$

and the corresponding projection operator is given by $\mathcal{P}_k = I - \mathfrak{F}_k(\mathfrak{F}_k^{\mathrm{T}}\mathfrak{F}_k)^{-1}\mathfrak{F}_k^{\mathrm{T}}$. The superscript on the 0s indicate the dimensions.

Theorem 6.2 *Under the same assumptions as Theorem 6.1, the sequences* $\{x_k\}$ *and* $\{x_k, v_k\}$ *satisfy* $\| d_k \| \leq \kappa_k \| d_{k-1} \|$ *and*

$$\left\| \begin{array}{c} d_k \\ \Delta v_k \end{array} \right\| \leq \kappa_k^4 \left\| \begin{array}{c} d_{k-1} \\ \Delta v_{k-1} \end{array} \right\|$$

respectively and thereby $\{x_k\}$ *converges Q-superlinearly if and only if*

$$\lim_{k \to \infty} \frac{\left\| \mathcal{P}_k \left[\begin{array}{c} \mathfrak{C}_* - \hat{\mathfrak{C}}_k \\ 0^{1 \times n} \end{array} \right] d_k \right\|}{\| d_k \|} = 0. \tag{6.19}$$

Proof. The optimality condition given by (6.3) can be rewritten using the above matrices and (6.5), (6.7), (6.9) as

$$\left[\begin{array}{cc} \hat{\mathfrak{Q}}_k & \mathfrak{F}_k \\ \mathfrak{F}_k^{\mathrm{T}} & 0 \end{array} \right] \left[\begin{array}{c} \Delta \vartheta_k \\ \Delta \left[\begin{array}{c} \mu_k^e \\ \bar{\mu}_k^i \end{array} \right] \end{array} \right] + (\mathfrak{F}_k^{\mathrm{T}} \mathfrak{F}_k)^{-1} \mathfrak{F}_k^{\mathrm{T}} \left[\begin{array}{c} \Delta \mathfrak{C}(t) \\ 0^{1 \times n} \end{array} \right] d_{k-1}$$

$$= - \left[\begin{array}{c} \mathcal{P}_k \left[\begin{array}{c} \Delta \mathfrak{C}(t) d_{k-1} \\ 0^{1 \times n} \end{array} \right] \\ \bar{f}_k - \bar{1} v_k \\ \bar{g}_k \end{array} \right]. \tag{6.20}$$

Using the same arguments as in Theorem 6.1 for (6.13) and that $\{\| \mathcal{P}_k - \mathcal{P}_{k-1} \|\} \to 0$, the Q-superlinear convergence of $\{x_k\}$ and $\{x_k, v_k\}$ can be demonstrated if (6.19) is satisfied.

Suppose, conversely that $\| d_k \| \leq \kappa_k \| d_{k-1} \|$ and

$$\left\| \begin{array}{c} d_k \\ \Delta v_k \end{array} \right\| \leq \kappa_k^4 \left\| \begin{array}{c} d_{k-1} \\ \Delta v_{k-1} \end{array} \right\|$$

respectively for the two sequences. Premultiplying the first equation of (6.20) and using the identity

$$\mathcal{P}_k \left[\begin{array}{c} d_k \\ \Delta v_k \end{array} \right] = \left[\begin{array}{c} d_k \\ \Delta v_k \end{array} \right] - \mathfrak{F}_k (\mathfrak{F}_k^{\mathrm{T}} \mathfrak{F}_k)^{-1} \mathfrak{F}_k^{\mathrm{T}} \left[\begin{array}{c} d_k \\ \Delta v_k \end{array} \right]$$

we have

$$\mathcal{P}_k\hat{\mathfrak{Q}}_k\mathcal{P}_k\begin{bmatrix} d_k \\ \Delta v_k \end{bmatrix} + \mathcal{P}_k\hat{\mathfrak{Q}}_k\mathfrak{F}_k(\mathfrak{F}_k^{\mathsf{T}}\mathfrak{F}_k)^{-1}\mathfrak{F}_k^{\mathsf{T}}\begin{bmatrix} d_k \\ \Delta v_k \end{bmatrix}$$

$$= -[\mathcal{P}_k - \mathcal{P}_{k-1}]\Delta\mathfrak{Q}(t)\begin{bmatrix} d_{k-1} \\ \Delta v_{k-1} \end{bmatrix} - \mathcal{P}_{k-1}\Delta\mathfrak{Q}(t)\begin{bmatrix} d_{k-1} \\ \Delta v_{k-1} \end{bmatrix}.$$

Assumption 6.1 (iv, v) ensures that $\left\|\mathcal{P}_k\hat{\mathfrak{Q}}_k\mathcal{P}_k\right\| \in (0, \infty)$. Also, as $\{\|\, d_k \,\|\} \to 0$, $\{\|\mathcal{P}_k - \mathcal{P}_{k-1}\|\} \to 0$.

Using the facts that

$$\|\, d_k \,\| \leq \left\|\begin{bmatrix} d_k \\ \Delta v_k \end{bmatrix}\right\| \quad \text{and} \quad \Delta\mathfrak{Q}(t)\begin{bmatrix} d_k \\ \Delta v_k \end{bmatrix} = \begin{bmatrix} \Delta\mathfrak{C}(t)d_{k-1} \\ 0 \end{bmatrix}$$

and dividing the above expression by $\|\, d_{k-1} \,\|$ and

$$\left\|\begin{bmatrix} d_{k-1} \\ \Delta v_{k-1} \end{bmatrix}\right\|$$

respectively yields (6.19) and the result follows from Lemma 6.2. $\quad\square$

7 THE ALGORITHM FOR ONLY LINEAR CONSTRAINTS

In the presence of linear constraints only, the algorithm can maintain feasibility at every iteration through the solution of the quadratic programming subproblem and convexity of the feasible region. Therefore, the penalty approach for constraints is not required. This simplifies the implementation of the algorithm and a number of the results in Sections 4–5 involving the vector

$$\begin{bmatrix} g(x) \\ \bar{h}(x) \end{bmatrix}$$

benefit from the fact that

$$\begin{bmatrix} g(x_k) \\ \bar{h}(x_k) \end{bmatrix} = 0$$

at every iteration.

Optimal portfolio problems in finance mostly involve linear equality or inequality constraints. This simplifies the algorithm considerably and and makes it relatively easy to implement. In this section, we describe the algorithm for linear constraints. Consider the classical mean-variance framework for given $\varrho \in [0, \infty]$

$$\min_x \left\{ f_\varrho(x) \,\middle|\, x \in \Omega \right\}$$

where $f_\varrho(x)$ is the quadratic objective function

$$f_\varrho(x) = -\langle \mathcal{E}(r), x \rangle + \varrho \langle x - \bar{x}, \mathfrak{C}(x - \bar{x}) \rangle,$$

$\mathcal{E}(r) \in \mathfrak{R}^n$ is the expected return vector of the set of investments, such as equities, being considered with

$$r = \mathcal{E}(r) + \epsilon$$

$\epsilon \sim \mathcal{N}(0, \mathfrak{C})$ is the random error, $\mathfrak{C} \in \mathfrak{R}^{n \times n}$ is the covariance matrix of the returns, $x \in \mathfrak{R}^n$ is the portfolio weights to be optimally determined, $\bar{x}$ denotes the benchmark weights which x should follow closely, and Ω is the feasible set of these weights, which includes the restrictions imposed by the investor. The error ϵ can be seen as either the error of the expected return from the historical mean or the error between the actual return and its forecast. Consequently, the covariance matrix used can be the historical covariance of the return vector or the covariance of the return forecast errors. The latter seems to be the risk measure more consistent with the return forecasts. As ϱ increases from zero, the optimal investment reflects the efficient risk-return trade-off indicated by this variation.

The main difficulty with the above mean variance optimization framework is the importance of the forecast return, $\mathcal{E}(r)$, and risk, $\mathfrak{C}$, estimates in the determination of the investment strategy. Although it seems natural that these data should be sufficiently precise for the optimal strategy to be useful, the inherent inaccuracy of these estimates is well known in finance.

In practice, therefore, the formulation of the classical optimal portfolio problem above is an oversimplification. Originating from rival economic and financial theories, there exist rival return forecasts purporting to represent the same financial system. As mentioned in Section 6.1, the problem of forecasting has been approached through forecast pooling by Fuhrer and Haltmaier (1986), Granger and Newbold (1977), Lawrence et al. (1986) and Makridakis and Winkler (1983). The extension of pooling to optimal policy design is often achieved using stochastic programming approaches by considering the probability of each scenario and evaluating approximate expected values (e.g., Kall and Wallace, 1994; Pardalos and Sandstrom, 1994). An alternative is discussed in Becker et al. (1986) where the robust pooling is computed using a minimax approach.

In the presence of rival forecasts, the investor may also wish to take account of *all existing rival scenarios in the design of optimal policy*. One strategy in such a situation is to adopt the worst-case design problem

$$\min_{x} \max_{i} \left\{ f_\varrho^i(x) \mid x \in \Omega; i = 1, \ldots, m^{sce} \right\}$$

where there are $i = 1, \ldots, m^{sce}$ scenarios, f_ϱ^i, denoting the objective for the ith scenario. Each scenario may be related to the forecast return $\mathcal{E}(r)$ or forecast variance $\mathfrak{C}$. This is an extension of the intuitive but suboptimal approach of naive minimax, discussed in Section 6.3.1 (Chow, 1979). The robustness of minimax stems from its basic property: it is the best decision determined simultaneously with the worst-case scenario, as discussed in Chapter 6. Minimax seeks the optimal strategy corresponding to the most adverse circumstance due to choice of scenario. All rival scenarios are assumed to be known. The minimax solution clearly does not provide general insurance against the eventuality that an unknown $(m^{sce} + 1)^{st}$ scenario might actually represent the system; it is just a robust strategy against known competing "scenarios".

To introduce the basic terminology, let $f(x) : \mathfrak{R}^n \to \mathfrak{R}^n$ be a twice continuously differentiable functions

$$f(x) = \begin{bmatrix} f_\varrho^1(x) \\ \vdots \\ f_\varrho^i(x) \\ \vdots \\ f_\varrho^{m^{sce}}(x) \end{bmatrix}.$$

Let the set $\mathbb{E}_+^n$ be given by

$$\mathbb{E}_+^n = \{x \in \mathfrak{R}^n | \langle x, I^x \rangle = 1, \ x \geq 0\}$$

and consider the feasible set

$$\Omega = \left\{x \in \mathbb{E}_+^n \ | \ J^{\mathrm{T}}x \leq J\right\}$$

where $J^{\mathrm{T}}x \leq J$ is a system of linear inequalities with J a matrix of n rows. Furthermore, we assume that $\Omega \neq \emptyset$.

Let $\mathfrak{C}(\cdot)$ denote the Hessian of the Lagrangian (6.2.3), with respect to x, evaluated at $(\cdot)$.

Consider the pooled objective function

$$\mathcal{V}(x, \vartheta) = \langle \vartheta, f(x) \rangle$$

and its linear approximation, with respect to x, at a point x_k,

$$\mathcal{V}_k(x, \vartheta) = \left\langle \vartheta, \left(f(x_k) + \nabla f(x_k)^{\mathrm{T}}(x - x_k) \right) \right\rangle \qquad (7.1)$$

where $\nabla f(x) \in \mathfrak{R}^n \times m^{sce}$ is the matrix

$$\nabla f(x) = \left[\nabla f_\varrho^1(x) \ \vdots \ ... \ \vdots \ \nabla f_\varrho^{m^{sce}}(x) \right].$$

We shall sometimes denote $f(x)$ and $\nabla f(x)$, evaluated at x_k, by f_k and ∇f_k, respectively. Thus, for $d = x - x_k$, (7.1) can be written as

$$\mathcal{V}_k(x_k + d, \vartheta) = \langle \vartheta, f_k + \nabla f_k^{\mathrm{T}} d \rangle.$$

The quadratic objective function used to compute the direction of progress is given by

$$\mathcal{V}_k(x, \vartheta) + \frac{1}{2} \langle x - x_k, \hat{\mathfrak{C}}_k(x - x_k) \rangle$$

where the matrix $\hat{\mathfrak{C}}_k$ is a symmetric positive semi-definite approximation to $\mathfrak{C}(x_k)$. The direction of progress at each iteration of the algorithm is determined by the quadratic subproblem

$$\min_x \max_\vartheta \left\{ \mathcal{V}_k(x, \vartheta) + \frac{1}{2} \langle x - x_k, \hat{\mathfrak{C}}_k(x - x_k) \rangle \,\middle|\, x \in \Omega, \ \vartheta \in \mathfrak{R}_+^{m^{sce}} \right\}. \quad (7.2)$$

Since the minimax subproblem is more complex, we also consider the quadratic programming subproblem

$$\min_{x, v \in \mathfrak{R}^{n+1}} \left\{ v + \frac{1}{2} \langle x - x_k, \hat{\mathfrak{C}}_k(x - x_k) \rangle \,\middle|\, x \in \Omega, \ \nabla f_k^{\mathrm{T}}(x - x_k) + f_k \leq 1^\lambda v \right\}. \quad (7.3)$$

The two subproblems are equivalent, but (7.3) involves fewer variables. It is shown in Rustem (1992) that the multipliers associated with $\nabla f_k^{\mathrm{T}}(x - x_k) + f_k \leq 1^\lambda v$ are the values ϑ and that the solution of either subproblem satisfies common optimality conditions.

Let the value of (x, ϑ, v) solving (7.2) and (7.3) be denoted by $(\hat{x}, \vartheta_{k+1}, v_{k+1})$. The stepsize along $(\hat{x} - x_k)$ is defined using the function

$$\Phi(x) = \max_{i \in \{1,2,\ldots,m^{sce}\}} \left\{ f_\varrho^i(x) \right\}$$

and

$$\Phi_k(x) = \max_{i \in \{1,2,\ldots,m^{sce}\}} \left\{ f_\varrho^i(x_k) + \langle \nabla f_\varrho^i(x_k), x - x_k \rangle \right\}.$$

The stepsize strategy determines α_k as the largest value of $\alpha = (\bar{\alpha})^j$, $\bar{\alpha} \in (0,1)$, $j = 0, 1, 2, \ldots$ such that x_{k+1} given by

$$x_{k+1} = x_k + \alpha_k(\hat{x} - x_k) \quad (7.4)$$

satisfies the inequality

$$\Phi(x_{k+1}) - \Phi(x_k) \leq \rho \alpha_k \Psi(\hat{x}) \quad (7.5a)$$

where $\rho \in (0,1)$ is a given scalar and

$$\Psi(\hat{x}) = \Phi_k(\hat{x}) - \Phi(x_k) + \frac{1}{2} \langle (\hat{x} - x_k), \hat{\mathfrak{C}}_k(\hat{x} - x_k) \rangle. \quad (7.5b)$$

The stepsize α_k determined by (7.5) basically ensures that x_{k+1} reduces the objective $\Phi(x)$ and, since Ω is convex, remains feasible.

The Algorithm

Step 0: Given $x_0 \in \Omega$, and small positive numbers $\rho, \overline{\alpha}$ such that $\rho \in (0,1)$, $\overline{\alpha} \in (0,1)$, the initial Hessian approximation, $\hat{\mathfrak{C}}_0$, set $k = 0$.

Step 1: Compute ∇f_k. Solve the quadratic subproblem (7.2) or (7.3) (choosing (7.2) or (7.3) defines a particular algorithm) to obtain $\hat{x}$, ϑ_{k+1}, and the associated multiplier vectors. In (7.3), we also compute ν_{k+1}.

Step 2: Test for optimality. If optimality is achieved, stop. Else, go to Step 3.

Step 3: Find the smallest nonnegative integer j_k such that $\alpha_k = \overline{\alpha}^{j_k}$, with x_{k+1} given by (7.4), such that the inequality (7.5) is satisfied.

Step 4: Update $\hat{\mathfrak{C}}_k$ to compute $\hat{\mathfrak{C}}_{k+1}$, set $k = k + 1$ and go to Step 1.

The main difference between the algorithm in this section and the general algorithm for nonlinear constraints is that in the case of linear constraints, feasibility is maintained throughout the algorithm. In this case, $\overline{G}_k^T d_k = \overline{g}_k = 0$, $P_k d_k = d_k$ and these can be used to simplify the results in Theorems 5.1, 6.1 and 6.2.

References

Becker, R.G., B. Dwolatzky, E. Karakitsos and B. Rustem (1986). "The Simultaneous Use of Rival Models in Policy Optimization", *The Economic Journal*, 96, 425–448.

Biggs, M.C.B. (1974). "The Development of a Class of Constrained Minimization Algorithms and their Application to the Problem of Power Scheduling", Ph.D. Thesis, University of London.

Boggs, P.T., J.W. Tolle and P. Wang (1982). "On the Local Convergence of Quasi-Newton Methods for Constrained Optimization", *SIAM Journal on Control and Optimization*, 20, 161–171.

Broyden C.G. (1969). "A New Method for Solving Nonlinear Simultaneous Equations", *Computer Journal*, 12, 95–100.

Broyden C.G. (1970). "The Convergence of a Class of Double-Rank Minimisation Algorithms 2. The New Algorithm", *Journal of the Institute of Mathematics and its Applications*, 6, 222–231.

Charalambous, C. and A.R. Conn (1978). "An efficient algorithm to solve the min-max problem directly", *SIAM Journal on Numerical Analysis*, 15, 162–187.

Chow, G.C. (1979). "Effective use of econometric models in macroeconomic policy formulation", in: S. Holly, B. Rustem and M. Zarrop (editors), *Optimal Control for Econometric Models*, Macmillan, London.

Cohen, G. (1981). "An Algorithm for Convex Constrained Minmax Optimization Based on Duality", *Applied Mathematics and Optimization*, 7, 347–372.

Coleman, T.F. (1978). "A note on 'New Algorithms for Constrained Minimax Optimization' ", *Mathematical Programming*, 15, 239–242.

Conn, A.R. (1979). "An Efficient Second Order Method to Solve the Constrained Minmax Problem", Report, Department of Combinatorics and Optimization, University of Waterloo Waterloo, Ontario.

Conn, A.R. and Y. Li (1992). "A Structure Exploiting Algorithm for Nonlinear Minimax Problems", *SIAM Journal on Optimization*, 2, 242–263.

Demyanov, V.F. and V.N. Malomezov (1974). *Introduction to Minmax*, John Wiley, New York.

Demyanov, V.F. and A.B. Pevnyi (1972). "Some Estimates in Minmax Problems" *Kibernetika*, 1, 107–112.

Dutta, S.R.K. and M. Vidyasagar (1977). "New Algorithms for Constrained Minmax Optimization", *Mathematical Programming*, 13, 140–155.

Fletcher, R. (1970). "A New Approach to Variable Metric Algorithms", *Computer Journal*, 13, 317–322.

Fuhrer, J. and J. Haltmaier (1986). "Minimum Variance Pooling of Forecasts at Different Levels of Aggregation", Special Studies Paper 208, Federal Reserve Board, Washington, DC.

Goldfarb D. (1970). "A Family of Variable Metric Algorithms Derived by Variational Means", *Mathematics of Computation*, 24, 23–26.

Granger, C. and P. Newbold (1977). *Forecasting Economic Time Series*, Academic Press, New York.

Hald, J.H. and K. Madsen (1981). "Combined LP and Quasi-Newton Methods for Minimax Optimization", *Mathematical Programming*, 20, 49–62.

Han, S.P. (1978). "Superlinear Convergence of a Minimax Method", Technical Report 78-336, Department of Computer Science, Cornell University, New York.

Han, S.P. (1981). "Variable Metric Methods for Minimizing a Class of Nondifferentiable Functions", *Mathematical Programming*, 20, 1–13.

Kall, P. and S.W. Wallace (1994). *Stochastic Programming*, Wiley, New York.

Lawrence, M.J., R.H. Edmunson and M.J. O'Connor (1986). "The Accuracy of Combining Judgemental and Statistical Forecasts", *Management Science*, 32, 1521–1532.

Makridakis, S. and R. Winkler (1983). "Averages of Forecasts: Some Empirical Results", *Management Science*, 29, 987–996.

Mayne, D.Q. and E. Polak (1982). "A Superlinearly Convergent Algorithm for Constrained Minimization Problems", *Mathemathics Programming Studies*, 16, 45–61.

Medanic, J. and M. Andjelic (1971). "On a Class of Differential Games without Saddle-point Solutions", *Journal of Optimization Theory and Applications*, 8, 413–430.

Medanic, J. and M. Andjelic (1972). "Minmax Solution of the Multiple Target Problem", *IEEE Transactions on Automatic Control*, AC-17, 597–604.

Murray, W. and M.L. Overton (1980). "A Projected Lagrangian Algorithm for Nonlinear Minmax Optimization", *SIAM Journal on Scientific and Statistical Computing*, 1, 345–370.

Pardalos, P.M. and M. Sandstrom (1994). "On the Use of Optimization Models for

Portfolio Selection: A Review of Some Computational Results", *Computational Economics*, 7, 227–244.

Polak, E. and D.Q. Mayne (1981). "A Robust Secant Method for Optimization Problems with Inequality Constraints", *Journal of Optimization Theory and Applications*, 33, 463–477.

Polak, E. and A.L. Tits (1981), "A Globally Convergent, Implementable Multiplier Method with Automatic Ppenalty Limitation", *Applied Mathematics and Optimization*, 6, 335–360.

Polak, E., D.Q. Mayne and J.E. Higgins (1988). "A Superlinearly Convergent Min-max Algorithm for Min-max Problems", Memorandum No: UCB/ERL M86/103, Department of Electrical Engineering, University of California, Berkeley, CA.

Powell, M.J.D. (1978). "A Fast Algorithm for Nonlinearly Constrained Optimization Problems", in: G.A. Watson (editor), *Numerical Analysis*, Lecture Notes in Mathematics, 630, Springer-Verlag, Berlin.

Powell, M.J.D. (1983) "Variable Metric Methods for Constrained Optimization", in: A. Bachem, M. Grotschel and B. Korte (editors), *Mathematical Programming: The State of The Art*, Springer-Verlag, Berlin.

Rustem, B. (1986). "Convergent Stepsizes for Constrained Optimization Algorithms", *Journal of Optimization Theory and Applications*, 49, 136–160.

Rustem, B. (1992). "A Constrained Min-max Algorithm for Rival Models of the Same Economic System", *Mathematical Programming*, 53, 279–295.

Rustem, B. (1993). "Equality and Inequality Constrained Optimization Algorithms with Convergent Stepsizes", *Journal of Optimization Theory and Applications*, 76, 429–453.

Rustem, B. (1994a). "Robust Min-max Decisions with Rival Models", in: D. Belsley (editor), *Computational Techniques for Econometrics and Economic Analysis*, Kluwer, Dordrecht.

Rustem, B. (1994b). "Convergent Stepsizes for Constrained Min-max Algorithms", *Advances in Dynamic Games and Applications*, 1, 168–194.

Rustem, B. (1998). *Algorithms for Nonlinear Programming and Multiple-Objective Decisions*, John Wiley, Chichester.

Rustem, B. and Q. Nguyen (1998). "An Algorithm for the Inequality Constrained Discrete Min-max Problem", *SIAM Journal on Optimization*, 8, 256–283.

Shanno, D.F. (1970). "Conditioning of Quasi-Newton Methods for Function Minimization", *Mathematics of Computation*, 24, 647–654.

Tapia, R.A. (1986). "On Secant Updates for use in General Constrained Optimization", Technical Report 84-3, Rice University.

Womersley, R.S. and R. Fletcher (1986). "An Algorithm for Composite Nonsmooth Optimization Problems", *Journal of Optimization Theory and Applications*, 48, 493–523.

Chapter 8

A continuous minimax strategy for options hedging

In this chapter, we consider how minimax can provide a robust hedging strategy for written call options. The contingent nature of the liability behind an option makes it important to address the management of this kind of liability. We formulate a minimax hedging strategy that minimizes the effect of a predefined worst-case scenario, mainly in terms of bounds on the underlying source of uncertainty, that is, the future price of the asset that underlies the option. We then identify variants, including multiperiod strategies, and discuss their performance relative to a standard strategy referred to as delta hedging. We also look into an alternative formulation of the minimax strategy using an evaluation of asset returns via the Capital Asset Pricing Model, or CAPM, and discuss its performance. We present the application of minimax to bond options and discuss the complexity involved in such an application. Finally, we include numerical results from an algorithmic point of view to demonstrate the performance of the minimax algorithm when applied to the hedging problem.

1 INTRODUCTION

We present an application of continuous minimax within the context of options hedging. An option is a contract that entitles the holder to buy or sell a specific number of shares of a given stock, at or within a certain period of time, for an agreed price[1]. The problem of hedging the risk of an option is mainly confined to the selling of the option where the seller incurs a liability contingent on the asset underlying the option. Because of the contingent nature of the liability, the seller of the option has to adjust her expectation of the magnitude of the liability, and in some cases the timing as well, and prepare her position such that she minimizes the potential negative impact to her of such a liability. Whereas the selling of an option is risky, with a potential loss of possibly unlimited magnitude, the buying of an option is mainly regarded as nonrisky in the sense that the buyer is acquiring an insurance policy for which the buyer pays a price to have the opportunity to exercise the option and benefit from it. The only risk to the buyer is the

---

[1] The price that a buyer of an option pays is called the premium.

potential loss of the amount she paid for the option if she decides not to exercise it.

The seller of the option can use a number of strategies to hedge the risk, depending on her appetite for risk taking. Clearly for a risk averse seller, finding a strategy that minimizes the potential loss in the value of her portfolio is important. Classical hedging strategies require the seller to be active in managing the options. We discuss a standard hedging strategy, called delta hedging, and use it as a reference strategy for comparing the strategies we develop in this chapter.

The minimax strategy optimizes the worst-case potential hedging error. We present the generic minimax formulation for hedging, hereafter in this chapter referred to as the *minimax hedging strategy* or *minimax*, and develop it into a number of specific strategies which we call *variants*. Just as with any other hedging strategy, minimax hedging is implemented over a time horizon, say, a 9-month period corresponding to the life of an option, and it may involve more than one rebalancing[2] date. At such a date, the hedge is adjusted to reflect the hedger's desired risk profile. At a rebalancing date, the minimax hedging strategy is computed using the algorithm in Chapter 4 to evaluate the worst-case potential hedging error and the corresponding solution.

In Sections 2 and 3, we give a general introduction to options and to an option pricing model. In Section 2, we describe call options on a stock from the point of view of the hedger who writes[3] one. In Section 3, we describe an option pricing model and a dynamic hedging strategy. In Section 4, we define the minimax hedging strategy as described in Howe et al. (1994). In Section 5, we present a simulation study designed to identify the properties of the variants of minimax and to ascertain whether minimax performs best for a set of options for which it is designed to perform best. In Section 6, we present an empirical illustration showing the performance of minimax when real data are used. In Sections 7 and 9, we give three extensions to the basic strategy given in Section 4; two of these extensions, presented in Section 7, are those of a multiperiod formulation as described in Howe et al. (1996), with the last extension, presented in Section 9, being motivated by the CAPM as described in Howe et al. (1998). We present simulation studies in Sections 8, 10 and 11. In Section 12, we discuss the application of minimax to bond options. Finally, Appendix B contains numerical results illustrating the performance of the continuous minimax algorithm.

---

[2] To rebalance is to change the asset allocation of a fund or a portfolio of assets; in the context of options hedging, the relevant portfolio includes an option and its underlying asset.

[3] When an option contract is sold, the seller of the contract is called the "writer" of the contract and the act of selling is referred to as "writing".

NOTATION

$B = B(y^S, t)$	call price
y_t^S	stock price at time t
$y_t^{S,\text{lower}}$	the lower bound on y_t^S
$y_t^{S,\text{upper}}$	the upper bound on y_t^S
X	exercise price
r	risk-free interest rate
t	current date
T	expiration date
$T - t$	time to maturity
σ	volatility
$\Theta(d)$	the cumulative normal distribution function
Δt	hedging interval
N	the contracted number of shares
x	number of shares to hold
K	transaction cost as percent of transaction volume

The subscripts:

0	time 0, the initiation date of the contract
t	time t, any time such that $0 < t < T$
T	time T, the expiration date
i	refers to stock i or option i

2 OPTIONS AND THE HEDGING PROBLEM

We apply minimax to the problem of hedging the risk of selling, or writing, a stock option, hereafter referred to as an option. An *option contract* is defined by the following: the premium B_0, that is, the price paid by the buyer; the *underlying stock S*; the contracted number of shares N; the date of the contract; the expiration date; the *exercise price X*, that is, the price at which the offer to buy or sell is to be made.

The two most widely traded options are called calls and puts: a *call* gives the holder the right to buy a specific number of shares; a *put* gives the holder the right to sell a specific number of shares. If the exercise of the option can take place only at the expiration date, it is called a European option. If the exercise can take place at any time on or before the exercise date, it is called an

American option. The hedging problem described below pertains to the writing or selling of European call options. For American call options, and in the case of the underlying stock not paying dividends, it is never optimal to exercise an American call before the expiration date[4]; our analysis equally applies.

Our analysis does not apply to put options. Although there is a certain symmetry between puts and calls, it is not perfect. The potential profit or loss, respectively, from buying or selling a call is unlimited whereas that from buying or selling a put is limited. In particular, when transaction costs are introduced, separate analyses are necessary.[5]

The hedging problem we consider is relevant to a writer of a European call option who receives the call premium but incurs a potential liability in case of exercise by the buyer at the expiration date. If the writer of the option does not own the contracted amount of stock, the potential liability is unlimited; this is shown by a graph of profit or loss against final stock price in Figure 2.1. Unlimited liability occurs when, at the time of exercise, the stock price is higher than the exercise price; a call option holder stands to gain the difference between these prices, which could potentially be very high, while the writer loses the difference, particularly in the situation where she does not have the stocks to deliver and would have to acquire the stocks from the open market at the prevailing high price.

The writer, hereafter also called the hedger, wishes to modify her exposure to risk: she would like to avoid a potentially large loss in case the final stock price is above the exercise price. She would hedge this risk by holding part or all of the contracted number of shares. If she chooses to hold all of the contracted number of shares at the time the contract was made, then she has implemented a covered write strategy. This is a static hedging strategy where a decision is made only at one point in time. If she chooses to hold part of the contracted number of shares and, in particular, adjust her holding based on the option's "delta"[6], then she is implementing a delta hedging strategy. This is a dynamic hedging approach where a decision, and a corresponding rebalancing, is made at several points in time.

In practice, hedgers can choose from a variety of strategies ranging from ad hoc to sophisticated ones based on option pricing theories. As hedging strategies with theoretical foundations including delta hedging are widely studied and generally acknowledged to be efficient, we assume these are used by the

---

[4] This result is discussed in Merton (1973).

[5] This is illustrated in Neuhaus (1990) where separate option pricing models were designed based on whether the call option is bought or sold.

[6] The "delta" of an option is the marginal change in the value of the option for a marginal change in the underlying stock's price. This is discussed in more detail in Section 3.

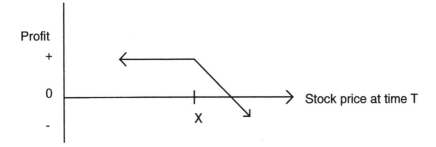

Figure 2.1 Profit or loss graph at the expiration date.

hedger. We address this in Section 4 where we develop the minimax hedging strategy: this strategy is based on the notion of a "minimax hedging error"[7].

3 THE BLACK AND SCHOLES OPTION PRICING MODEL AND DELTA HEDGING

In a dynamic strategy the hedger modifies her position in response to movements in the stock price. In this section, we present the Black and Scholes (1973) option pricing model (BS)[8]. The BS model will be used in delta hedging as well as in the minimax hedging strategy described in Section 4. In this section, we present a method of adjusting the BS option pricing model when transaction costs are included in the option valuation. Finally, we discuss delta hedging.

Black and Scholes (1973) derive a formula for the value of a European call option. The model states that at any time t, the hedger can set up a riskless portfolio, consisting of a position in the option and a position in the stock. In the hedging problem being considered in this chapter, the position in the option is a written call option and the position in the stock is the amount of stock purchased to cover the option. Since y_t^S and B_t are affected by the same underlying source of uncertainty[9], they are instantaneously perfectly correlated. The model also stipulates that the hedger does not need to introduce any cashflow into the portfolio in order to maintain the riskless nature of the portfolio. Therefore, this riskless portfolio can also be self-financing. The hedger can create a self-financing portfolio by financing any necessary

---

[7] The minimax hedging error is defined and discussed in section 4.

[8] Other option pricing models, such as the model developed by Cox and Ross (1976), may be used but this possibility is not explored here.

[9] For stock options, an example of a source of uncertainty that affects both the option and the underlying stock would be future earnings of the company.

purchases of shares of stock using sales of options. This can be expressed as follows: let V^H be the value of a "hedge" portfolio consisting of shares of stock held long[10] and N the contracted number of shares,

$$V^H = xy^S - NB. \qquad (3.1)$$

The dynamics of the stock price can be described by the following stochastic differential equation:

$$dy^S = \mu^{y^S} y^S dt + \sigma^{y^S} y^S dz \qquad (3.2)$$

where μ^{y^S} is the instantaneous rate of return on the stock; σ^{y^S} is the volatility of the rate of the return on the stock; dt is an increment of time; dz is an increment of Brownian motion (see CN 1).

Assuming $x = x(y^S, t)$ and $N = N(y^S, t)$, and using Ito's formula (see CN 2), the change in the value of the portfolio is

$$dV^H = dxy^S + xdy^S + dxdy^S - [dNB + NdB + dNdB] \qquad (3.3)$$

or

$$dV^H = xdy^S - NdB + [dx][y^S + dy^S] - [dN][B + dB]. \qquad (3.4)$$

If the hedge portfolio is self-financing, then the sum of the third and fourth terms on the right hand side of (3.4) equals zero, that is, $[dx][y^S + dy^S] - [dN][B + dB] = 0$ because all purchases or sales of assets are made at "new end of period prices". We now show that such a portfolio can also be riskless.

The dynamics of the option price can be described by a stochastic differential equation similar to (3.2) given by

$$dB = \mu^B B dt + \sigma^B B dz \qquad (3.5)$$

where μ^B is the instantaneous rate of return on the option; σ^B is the volatility of the rate of the return on the option.

A key assumption of Black and Scholes is that the option price is solely determined by the underlying stock price and time. Therefore, with the dependency of the option price on the stock price, (3.5) can be expressed as

$$dB = \left[\mu^{y^S} y \frac{\partial B}{\partial y} + \frac{\partial B}{\partial t} + \tfrac{1}{2} (\sigma^{y^S})^2 (y^S)^2 \frac{\partial^2 B}{\partial (y^S)^2} \right] dt + \sigma^{y^S} y^S \frac{\partial B}{\partial y^S} dz. \qquad (3.6)$$

Thus, (3.4) is written as

$$dV^H = x \left[\mu^{y^S} y^S \, dt + \sigma^{y^S} y \, dz \right] - N \left[\mu^B B \, dt + \sigma^B B \, dz \right]. \qquad (3.7)$$

In order to construct a riskless portfolio, the dz terms in (3.7) must cancel out. This can be achieved by choosing x as follows:

---

[10] The purchaser of shares of stock is said to be *long* in the stock.

$$x = N \frac{\partial B}{\partial y^S}. \tag{3.8}$$

Therefore, if we have N call options, we should choose $N \frac{\partial B}{\partial y^S}$ shares to immunize the risk of writing the calls. In other words, if we sell N call options, we should purchase $N \frac{\partial B}{\partial y^S}$ shares to make the portfolio riskless. However, the portfolio is only instantaneously riskless. In order to maintain a riskless portfolio, the writer has to rebalance continuously. $\partial B / \partial y^S$ is called the *delta* and the strategy for maintaining a riskless portfolio using it is called *delta hedging*. If the portfolio is riskless, then the instantaneous return is r. Substituting (3.6) and (3.8) into (3.7) leads to the Black and Scholes partial differential equation:

$$\frac{1}{2}(\sigma^{y^S})^2 (y^S)^2 \frac{\partial^2 B}{\partial (y^S)^2} + ry^S \frac{\partial B}{\partial y^S} - rB + \frac{\partial B}{\partial t} = 0. \tag{3.9}$$

The boundary conditions and initial conditions are

$$B(y^S, T) = \max(y^S - X, 0)$$

$$B(y^S, t) \le y^S \quad \forall t$$

$$B(0, t) = 0 \quad \forall t.$$

The solution to this system is the Black and Scholes formula.

The Black and Scholes Formula

$$B = y^S \Theta(d_1) - X e^{-r(T-t)} \Theta(d_2) \tag{3.10}$$

$$d_1 = \frac{\ln(y^S/x) + (r + (\sigma^2/2))(T - t)}{\sigma \sqrt{T - t}} \tag{3.11}$$

$$d_2 = \frac{\ln(y^S/X) + (r - (\sigma^2/2))(T - t)}{\sigma \sqrt{T - t}} = d_1 - \sigma \sqrt{T - t} \tag{3.12}$$

where $\Theta(d)$ is the cumulative normal distribution function (see CN 3), that is,

$$\Theta(d) = \int_{-\infty}^{d} \frac{1}{\sqrt{2\pi}} e^{-z^2/2} \, dz. \tag{3.13}$$

The above analysis also assumes *nonsatiation*, that is, investors prefer more wealth to less, and that the option is neither a dominant[11] nor a dominated

---

[11] From Merton (1973), Security A is dominant over Security B if, on some known date in the future, the return on A will exceed the return on B for some possible states of the world, and will be at least as large as on B, in all possible states of the world.

security. Explicit assumptions about equilibrium or about investors' preferences are not necessary. The fundamental requirement is the nonexistence of arbitrage opportunities or opportunities that would enable market participants to generate riskless profits. In this sense, the above result is *preference independent* because all assets are perfect substitutes for each other instantaneously and the strategy for maintaining a riskless portfolio is independent of the hedger's attitude to risk.

The BS option pricing model does not include transaction costs. Other models incorporate transaction costs that result in modified option prices compared to the prices from the BS model. Leland (1985) develops an option pricing model that includes transaction costs[12] and this is extended by Neuhaus (1990). In Leland's (1985) model, the hedging errors, including transaction costs, will almost surely approach zero as $\Delta t \rightarrow 0$. Leland (1985) and Neuhaus (1990) both incorporate transaction costs in their models by modifying the volatility. We present below the modification by Leland (1985). This is more closely related to discrete delta hedging than is the Neuhaus (1990) modification.

The revised volatility $\hat{\sigma}$ is given by

$$\hat{\sigma} = \sqrt{\sigma^2 \left[\frac{2}{\pi} \frac{K}{\sigma \sqrt{\Delta t}} \right]} \qquad (3.14)$$

where K^{13} is the roundtrip[14] transaction cost expressed as a proportion of trading volume.

We replace σ by $\hat{\sigma}$ in the BS option pricing model when we include transaction costs in the analysis. We define $K = 2\hat{K}$, where $\hat{K}$ is half the roundtrip transaction cost.

Delta, D, is the change in option price per unit change in stock price, that is

$$D = \frac{\partial B}{\partial y^S}. \qquad (3.15)$$

Using the BS option pricing model (3.10), delta is given analytically by

$$D = \Theta(d_1). \qquad (3.16)$$

As discussed above, Black and Scholes (1973) argue that the writer can realize a riskless portfolio by delta hedging; the writer computes the delta of an option

---

[12] Other option pricing models with transaction costs have been developed by other authors. We used Leland's (1985) model because it is applicable to discrete rebalancing and it fits in the framework of the minimax hedging strategy. See Boyle and Emanuel (1980), Gilster and Lee (1984), Panas (1993), Neuhaus (1990), and Davis and Norman (1990).

[13] In Leland (1985), transaction costs varied from $K = 0.0$ to $K = 0.04$.

[14] From Rosenberg (1993), a roundtrip trade is defined as any complete transaction made up of a buy followed by a sale of the same stock or vice versa.

and bases the number of shares to hold on this value. For N, the contracted number of shares, and x_t, the number of shares to hold at any time, under delta hedging, we have

$$x_t = D_t N. \tag{3.17}$$

The hedger must rebalance continuously to keep the portfolio riskless. This need arises from the fact that delta changes with time, and the hedge portfolio is only instantaneously riskless. Such a portfolio strategy is called a "delta-neutral" strategy. However, because she cannot rebalance continuously, she uses discrete delta hedging where she rebalances at discrete intervals of time. With discrete delta hedging, she incurs a hedging error which, for an interval of time, is the net position of a hedge portfolio brought about by changes in y_t^S. The hedging error (HE) for a portfolio of a written call option and stock held long, for the interval t to $t + 1$ is

$$\text{HE} = N(B_t - B_{t+1}) + x_t(y_{t+1}^S - y_t^S). \tag{3.18}$$

Large hedging errors tend to increase the cost of rebalancing the hedge.

4 MINIMAX HEDGING STRATEGY

In this section, we present a strategy, based on Howe et al. (1994), to solve the hedging problem. This strategy is based on the concept of a worst-case scenario, which the hedger specifies in terms of movements in stock price, and it finds a hedge that minimizes the effect of such a scenario[15]. In Section 4.1, we formulate the minimax problem. In Section 4.2, we present the worst-case scenario and its two variants. In Section 4.3, we present the hedging error which is the underlying cost to be minimized. In Section 4.4, we present the objective function. In Section 4.5, we define the minimax hedging error. In Section 4.6, we discuss the treatment of transaction costs. In Section 4.7, we present the variants of the minimax hedging strategy. In Section 4.8, we discuss the minimax solution to the problem given in Section 4.1.

4.1 Minimax Problem Formulation

The problem that minimax addresses is that of minimizing an objective function under a worst-case scenario. In the context of the hedging problem, the

---

[15] In practice, the Black and Scholes assumption of constant instantaneous standard deviation of returns is rarely satisfied. The effect of volatility variations is frequently as bad, if not worse, than that of the variations of the underlying variable and neither delta hedging nor the minimax hedging approach in this chapter can hedge against these variations. We address this problem in Section 7.

minimizing variable is x_t and the maximizing variable is y_t^S, which is allowed to take any value within predefined bounds.

The minimax problem is

$$\min_{x_t} \max_{y_{t+1}^S} f(x_t, y_{t+1}^S) \tag{4.1}$$

subject to

$$y_{t+1}^{S,\text{lower}} \le y_{t+1}^S \le y_{t+1}^{S,\text{upper}} \tag{4.2}$$

where $f(x_t, y_{t+1}^S)$ is the objective function, presented in Section 4.4, and $y_{t+1}^{S,\text{lower}} \le y_{t+1}^S \le y_{t+1}^{S,\text{upper}}$ is either the range defined under Worst Case 1, presented in Section 4.2.1, or under Worst Case 2, presented in Section 4.2.2.

There are no constraints on x_t, the number of shares to be held at time t: nonnegative x_t implies a long position in shares[16]; negative x_t implies a short position in shares[17].

4.2 The Worst-case Scenario

4.2.1 Worst Case 1

In Worst Case 1, the hedger defines the worst case over extreme movements in stock price. The range of y_{t+1}^S has upper and lower bounds that describe the 95% confidence interval[18] of all possible values of the future stock price, that is, *two standard deviations* about the expected value of the stock price at time $t + 1$. This 95% confidence interval is based on an estimate of the volatility of the stock price and on the assumption that the stock price follows a lognormal distribution function. Worst Case 1 is hereafter referred to as the *95%-level*. However, it is not always the case that the worst case corresponds to the edges of this interval[19].

4.2.2 Worst Case 2

We define the *state* of the option as one of three possibilities: in-the-money[20],

---

[16] The hedger bought the shares.

[17] The hedger sold the shares. This situation is possible in the minimax context.

[18] We consider the 95% confidence interval as a reasonable range to consider under normal market conditions. Under abnormal market conditions, 99% confidence interval may be more appropriate.

[19] In examples at the end of the chapter, it is shown that the worst case may indeed occur in the middle of these ranges.

[20] If the current stock price is greater than the exercise price, then the option is said to be in-the-money.

at-the-money[21], and out-of-the-money[22]. Most of the business in exchange traded options is with respect to options that are at-the-money. The prices of the underlying stocks of these options usually oscillate about the exercise price[23]. That is, they move from sometimes being in-the-money to sometimes being out-of-the-money. In this scenario we focus on movements of the stock price which may result in a switch in the state of the option, that is, from being in-the-money at time t, to being out-of-the-money at time $t + 1$, and vice versa. This switch in the state of the option means that there is the danger of a higher hedging error being incurred in the interval between t and $t + 1$. In contrast to Worst Case 1, we define the range of y_{t+1}^S as the range whose upper and lower bounds describe the possible values of the future stock price within *one and three standard deviations (sd)* from the expected value of the stock price at time $t + 1$, in the direction of the exercise price X. This means that if $y_t^S > X$, the relevant range would be on the left side of the distribution of future stock price, with a lower bound of 3 sd and an upper bound of 1 sd; if $y_t^S \leq X$, the relevant range would be on the right side, with a lower bound of 1 sd and an upper bound of 3 sd. Worst Case 2 is hereafter referred to as the *Abrupt Change*.

4.3 The Hedging Error

From the discussion on delta hedging in Section 3, the hedging error is given by (3.18). When actual values of B_t, y_t^S, B_{t+1} and y_{t+1}^S are substituted into (3.18), we have the actual hedging error under delta hedging. When actual values of B_t and y_t^S and potential values of B_{t+1} and y_{t+1}^S are substituted into (3.18), we have the potential hedging error under delta hedging. The latter is the basis of the objective function in the minimax hedging strategy. In minimax, potential y_{t+1}^S is taken from a predefined range that maximizes the objective function; potential B_{t+1} is the value of the call option based on the pricing model[24] given potential y_{t+1}^S, that is, potential $B_{t+1} = B_{t+1}(y_{t+1}^S)$. The minimax strategy minimizes the maximum potential hedging error plus

---

[21] If the current stock price is equal to the exercise price, then the option is said to be at-the-money.

[22] If the current stock price is lower than the exercise price, then the option is said to be out-of-the-money.

[23] This is the region of greatest elasticity (curvature) and therefore the area about which market makers are mostly concerned. Formally, we are talking about the effects of the option's "gamma", the change in the option's delta for a unit change in the stock price.

[24] We use the Black and Scholes (1973) option pricing model, (3.10), with a modified volatility, (3.14).

interest payments on borrowed money[25]. In Section 4.5, we define the mini-max hedging error and give the definition of actual hedging error and potential hedging error in the context of minimax.

4.4 The Objective Function

In any dynamic hedging strategy, hedging errors are incurred; in order to correct for these errors, the hedge is rebalanced, with the cost of rebalancing being added to the cost of hedging. At time t, the hedger can aim to minimize the potential hedging error in the period between t and $t + 1$. Her decision at time t on x_t, the number of shares to hold, affects the actual hedging error between t and $t + 1$. The minimax hedging strategy aims to minimize the maximum potential hedging error between t and $t + 1$. The objective function used is thus the potential hedging error. In discrete delta hedging, where rebalancing is done at discrete intervals, we expect that the desirable proper-ties of delta hedging given in Section 3 will not be observed consistently in time. The hedger adopts a cautious strategy by minimizing the maximum potential hedging error plus interest payments on borrowed money, should the worst case occur. If the worst case does occur, she has effectively mini-mized its worst effect. If it does not, she will incur a hedging error more favorable than the worst case due to the noninferiority of the minimax solution (see Theorem 1.3.1).

This direct way of minimizing the potential hedging error is based on the no-arbitrage argument of Merton (1973) where a portfolio containing an option is considered. The underlying stock and a riskless bond (i.e., riskless in the sense of default) are suitably chosen such that the aggregate investment in the portfolio is zero. Merton (1973) demonstrates that there is a strategy of finding the mix of option, stock and bond that would ensure that the return on the portfolio would be nonstochastic. Because of the condition of zero aggre-gate investment, in order to avoid arbitrage profits, the return on this portfolio must be zero. In the case of a portfolio of written call options, underlying stock and bonds, given Merton's (1973) assumptions, the return on this particular portfolio must be zero. We consider such a portfolio and we call it the "ideal portfolio". This "ideal portfolio" is the benchmark used in the definition of the objective function. We derive basic properties of the minimax hedging strat-egy on the basis of a self-financing portfolio; conditional on these results, we add the effect of costs. We return to this when we discuss (4.6) below.

We define $U_1 : \Re^k \times \Re^k \to \Re^1$, $U_2 : \Re^k \to \Re^k$, $U : \Re^k \times \Re^k \to \Re^{k+1}$, $x_t \in \Re^k$, $y_{t+1}^S \in \Re^k$ and Q is a $(k + 1) \times (k + 1)$ positive definite weighting matrix. $U^d \in \Re^{k+1}$ is the vector of desired values for the potential hedging

---

[25] A rebalancing may involve borrowing of cash to cover any need to increase the stock holdings.

error and the transaction cost terms: we use a desired value of zero, that is, the desired hedging error is zero[26] and the desired transaction cost is zero.

The objective function is given by

$$f(x_t, y_{t+1}^S) = \frac{1}{2}\langle U - U^d, Q(U - U^d)\rangle \tag{4.3}$$

where

$$x_t = \begin{bmatrix} x_{1,t} \\ \vdots \\ x_{k,t} \end{bmatrix} \quad \text{and} \quad y_{t+1}^S = \begin{bmatrix} y_{1,t+1}^S \\ \vdots \\ y_{k,t+1}^S \end{bmatrix} \tag{4.4}$$

$$U(x_t, y_{t+1}^S) = \begin{bmatrix} U_1(x_t, y_{t+1}^S) \\ \cdots\cdots\cdots \\ U_2(x_t) \end{bmatrix} \quad \text{and} \quad U^d = \begin{bmatrix} U_1^d \\ \cdots \\ U_2^d \end{bmatrix} = \begin{bmatrix} 0 \\ \cdots \\ 0 \end{bmatrix} \tag{4.5}$$

$$U_1(x_t, y_{t+1}^S) = \sum_{i=1}^{k} x_{i,t}(y_{i,t+1}^S - y_t^S) + \sum_{i=1}^{k} N_i\left(B_{i,t} - B_{i,t+1}(y_{t+1}^S)\right)$$

$$+ \sum_{i=1}^{k} \left(-(x_{i,t} - x_{i,t-1})y_{i,t}^S + C_{i,t-1}(1 + r\Delta t)\right)r\Delta t \tag{4.6}$$

where

$$C_{i,t-1} = C_{i,t-2}(1 + r\Delta t) - (x_{i,t-1} - x_{i,t-2})y_{i,t-1}^S - \hat{K}\left|(x_{i,t-1} - x_{i,t-2})y_{i-1}^S\right|. \tag{4.7}$$

$$U_2(x_t) = \begin{bmatrix} U_{1,2}(x_{1,t}) \\ \vdots \\ U_{k,2}(x_{k,t}) \end{bmatrix} \tag{4.8}$$

where

$$U_{i,2}(x_{i,t}) = \hat{K}\left(x_{i,t} - x_{i,t-1}\right)y_{i,t}^S. \tag{4.9}$$

We identify all the variables in (4.4)–(4.9) and then give the economic interpretation of (4.6). $C_{i,t-1}$ is the cumulative value of *cash inflow minus cash outflow* at time $t - 1$. $C_{i,t-1}(1 + r\Delta t)$ is $C_{i,t-1}$ with interest payments.

---

[26] In minimax, we have the opportunity to adopt any desired value. We adopted a desired value of zero because in delta hedging, the expected value of the hedging error is zero.

The first term of (4.7) is the cumulative value of cash inflow minus cash outflow from the previous period with interest payments. The second term is a cash outflow if the $x_{i,t-1} > x_{i,t-2}$; otherwise, it is a cash inflow. The third term is always a cash outflow. We note that $C_{i,t-1}$ will normally be a negative number. At time t, $C_{i,t-1}$ is a constant: all the variables in (4.7) have actual values. We also note that the transaction cost term on (4.7) is the cost from the previous period and that this term, along with the other terms in (4.7), does not affect the optimization.

Because transaction costs for the current period introduce nondifferentiability into the equation, they do enter the objective function directly as part of U_1. Instead, we introduce a penalty term, $U_{i,2}$ to represent a penalty for transaction costs for each option i at time t. The treatment of transaction costs is discussed in detail in Section 4.6.

In the weighting matrix Q, we use diagonal weights q_i, $i = 1, ..., k + 1$, which are specified by the hedger, represent her preferences: with a high q_1 she prefers to minimize the potential hedging error that may be incurred from time t to time $t + 1$; with a high q_i, $i = 2, ..., k + 1$, she prefers to minimize the penalty term.

For each option $i = 1, ..., k$, $B_{i,t+1}(y_{i,t+1}^S)$ is determined using (3.10), (3.11) and (3.12) with the modified volatility estimate (3.14).

The economic interpretation of (4.6) is that U_1 represents the potential hedging error, including interest payments on borrowed money, between time t and time $t + 1$. It comprises the potential shift in the stock position, the potential shift in the option position and the potential interest payment. The first two terms of (4.6) give the return on a portfolio of written call options and underlying stocks. The third term represents the opportunity cost of money, that is, the interest payments on borrowed money, because the portfolio is not self-financing.

We wish to find the mix of options and stocks that minimizes the deviation of the return on the portfolio, including opportunity cost, from the return on the "ideal portfolio", the value of which is zero, based on Merton's (1973) conditions of zero aggregate investment and no-arbitrage.

4.5 The Minimax Hedging Error

In minimax, we distinguish actual from potential hedging error. *Actual hedging error, inclusive of interest payments on borrowed money*, is calculated when *actual B_t, y_t^S, B_{t+1} and y_{t+1}^S are used in (4.6). Potential hedging error, including interest payments on borrowed money*, is calculated when *actual* values of B_t and y_t^S and *potential* values of B_{t+1} and y_{t+1}^S are used in (4.6). Potential y_{t+1}^S is taken from a predefined range that maximizes the objective function. Potential B_{t+1} is the value of the call option based on the pricing model given potential y_{t+1}^S, that is, potential $B_{t+1} = B_{t+1}(y_{t+1}^S)$.

We define the *minimax hedging error* at time t as

$$\text{minimax hedging error} = U_1(x_t^*, y_{t+1}^{S^*}). \qquad (4.10)$$

The minimax hedging error is the worst-case potential hedging error, inclusive of interest payments on borrowed money, given the solution x_t^* and $y_{t+1}^{S^*}$.

4.6 Transaction Costs

In this section, we discuss the treatment of transaction costs in the minimax hedging strategy. The roundtrip transaction cost K is used in valuing the option and $\hat{K}$, with $\hat{K} = 1/2K$, is used as part of the cumulative value of cash inflow minus cash outflow and as part of the penalty term in the objective function. We first consider transaction costs as part of the cumulative value of cash inflow minus cash outflow. Subsequently we formulate transaction costs as part of the objective function.

The performance of delta hedging and the variants of minimax is measured by the final cumulative value of cash inflow minus cash outflow at the maturity of the option. After finding x_t by solving the minimax problem using (4.4)–(4.9), we can evaluate the actual cumulative value of cash inflow minus cash outflow at time t. This is given by

$$C_{i,t} = C_{i,t-1}(1 + r\Delta t) - (x_{i,t} - x_{i,t-1})y_{i,t}^S - \hat{K}|(x_{i,t} - x_{i,t-1})y_{i,t}^S|. \qquad (4.11)$$

The last term is the transaction cost at time t: this is always incurred and it is always a cash outflow. At time $t = 0$, the actual cumulative value of cash inflow minus cash outflow includes the option premium which is a cash inflow. This is given by

$$C_{i,0} = -x_{i,0}y_{i,0}^S + NB_{i,0} - \hat{K}|x_{i,0}y_{i,0}^S|. \qquad (4.12)$$

In all variants of the minimax hedging strategy we use (4.11) and (4.12) to compute the actual cumulative value of cash inflow minus cash outflow.

In the objective function, from (4.4), (4.8) and (4.9), the transaction cost term (TC) can be expressed as the penalty term

$$\text{TC} = \sum_{i=1}^{k} q_i(U_{i,2} - U_{i,2}^d)^2 = \sum_{i=1}^{k} q_i\left(\hat{K}(x_{i,t} - x_{i,t-1})y_{i,t}^S\right)^2. \qquad (4.13)$$

The right equality follows as $U_2^d = 0$ holds. The effect of this term on the solution is dependent on the level of transaction cost $\hat{K}$ and on the weights q_i. We adopt below a uniform weighting system: $q_i = q$, $i = 1, ..., k$, where q is a given constant.

For low values of $\hat{K}$, a high value of q is needed so that the transaction cost term TC is not dominated by other terms in the objective function. Conversely, for high values of $\hat{K}$, a low value of q is needed to ensure that TC does not dominate the objective function. For the simulation and empirical illustration below, the roundtrip transaction cost is set[27] to $K = 0.02$ and $q = 100$.

4.7 The Variants of the Minimax Hedging Strategy

We consider delta hedging[28] and five variants of minimax: A, B, B*, C, and D, where each variant has a specific objective function and definition of worst case. Transaction costs are included when computing the cumulative value of cash inflow minus cash outflow for all the strategies (Table 4.7).

Table 4.7   The hedging strategies to be used in the empirical illustration and simulation study.

Code	Strategy	Objective function	Condition of y_{t+1}^S	Transaction costs in objective function
	Delta	Delta neutrality	n.a	No
A	Minimax	Potential hedging error	95% level	No
B	Minimax	Potential hedging error	Abrupt change	No
B*	(Described in the next paragraph)			
C	As A	As A	As A	Yes
D	As B	As B	As B	Yes

B* is a weighted version of B in which the minimax recommendation on the number of shares to hold is weighted by a factor ranging from 0 to 1 representing the hedger's assessment of the information contained in changes in the underlying stock[29].

4.8 The Minimax Solution

The solution to (4.1)–(4.2) is obtained using the algorithm discussed in Chapter 4. The algorithm is based on generating successive directions of descent for

---

[27] This value is based on simulation results, reported in Howe (1994), showing the variation of K with q.

[28] Delta with gamma hedging is not considered here.

[29] This is a heuristic method of adjusting the hedge recommendations. The method is given in Appendix A.

$f(x, y^S)$ in x, while ensuring that the direction chosen maximizes $f(x, y^S)$ with respect to y^S. The direction chosen is therefore one that iteratively progresses towards the minimax solution.

Because the hedge recommendation under minimax is different from the hedge recommendation under delta hedging, following Black and Scholes (1973), the hedge recommendation under minimax is suboptimal. In the minimax hedging strategy, for any fixed x_t, we determine y_{t+1}^S, from the predefined range $y_{t+1}^{S,\text{lower}} \le y_{t+1}^S \le y_{t+1}^{S,\text{upper}}$, that maximizes the hedging error. We can, therefore, identify theoretically all the maxima corresponding to all possible values of x_t. The strategy calculates x_t that minimizes over these maxima. Although the number of shares x_t introduces some risk into the portfolio because it is not the same as the hedge recommendation under delta hedging, x_t ensures that if the *actual* y_{t+1}^S, as opposed to the minimax value, falls within the range $y_{t+1}^{S,\text{lower}} \le y_{t+1}^S \le y_{t+1}^{S,\text{upper}}$, the absolute value of the *actual* hedging error, inclusive of interest payments on borrowed money, will not be worse (higher) than the absolute value of the minimax hedging error. This is the *minimax robustness property* discussed in Theorem 1.3.1. The x_t value thus computed results in a robust strategy that is noninferior in performance for any stock price within the predefined range.

Given the 95% level as the definition of the worst-case scenario, $y_{t+1}^{S,\text{lower}} \le y_{t+1}^S \le y_{t+1}^{S,\text{upper}}$, the minimax algorithm ensures that either

- x_t is chosen such that, for extreme point maximizers, the objective function value is the same for all those upper and lower bounds corresponding to those maximizers, or

- the objective function value corresponds to a worst-case price that is in the middle of the range.

Given the abrupt change as the definition of the worst-case scenario, $y_{t+1}^{S,\text{lower}} \le y_{t+1}^S \le y_{t+1}^{S,\text{upper}}$, the minimax algorithm ensures that either

- x_t is chosen such that, for extreme point maximizers, the objective function value for the upper limit is as close as possible to the value for the lower limit, or

- the objective function value corresponds to a worst-case price that is in the middle of the range.

In all cases, it can be shown that the chosen x_t places an upper bound on the absolute value of the hedging error that can be incurred for any price in the given range.

5 SIMULATION

We describe the simulation of the performance of the minimax variants against that of delta hedging when they are used to hedge the risk of writing a European call option. Options with their underlying stock series are generated and then categorized under five general groups of options. The objective of the simulation is to identify which variants, for which groups of options, outperform delta hedging. Thus, we aim to establish the characteristics of options, in terms of crossovers and abrupt changes in the price of the underlying stock, for which the different variants of minimax are particularly suited.

5.1 Generation of Simulation Data

For the generation of the full data set for the simulation, five volatility levels were used, namely, 0.20, 0.30, 0.40, 0.50 and 0.60. Holding the volatility constant, 1250 sets of option and stock time series were generated which were then screened based on a selection procedure described below. Selection is based on whether a particular set of option and stock time series falls within any of the specified groups identified below.

For all sets having the same volatility, each set is placed into one of five option groups. These groups are defined based on two events: the crossover and the abrupt change (Table 5.1a).

Table 5.1a   The idealized groups.

	Crossovers	Abrupt changes
Group 1	None	None
Group 2	Several	None
Group 3	None	Several
Group 4	Several	Several
Group 5	Few	Few

- A *crossover* is an event such that $y_t^S \leq X$ and $y_{t+1}^S > X$.

- An *abrupt change* is an event such that $y_{t+1}^S \in$ Abrupt Change.

The allocation of a set (an option/stock time series) to any of the above groups is determined by the distribution of the total number of crossovers per series, the distribution of the total number of abrupt changes per series and the distribution of the total number of simultaneous crossovers and abrupt changes per series.

Allocating a Stock Price Series to a Group

Given the exercise price and the Abrupt Change range defined in Section 4, the following numbers are known for each stock series:

J_X	total number of crossovers
J_{AC}	total number of abrupt changes
$J_{X\&AC}$	total number of simultaneous crossovers and abrupt changes

The distributions of J_X, J_{AC} and $J_{X\&AC}$, each based on a sample of 1250 stock series (given a constant volatility σ) are ascertained. These distributions are used to allocate a time series of a specified volatility into one of the five option groups. Given the means, μ_{J_X}, $\mu_{J_{AC}}$, $\mu_{J_{X\&AC}}$, and standard deviations, σ_{J_X}, $\sigma_{J_{AC}}$, $\sigma_{J_{X\&AC}}$, of these distributions, a stock price series is allocated to a group if it has the properties given below for that group:

Group 1:

$$J_X < \mu_{J_X} - a\sigma_{J_X}, \quad J_{AC} < \mu_{J_{AC}} - b\sigma_{J_{AC}} \text{ and } J_{X\&AC} < \mu_{J_{X\&AC}} - c\sigma_{J_{X\&AC}}$$

Group 2:

$$J_X > \mu_{J_X} + a\sigma_{J_X}, \quad J_{AC} < \mu_{J_{AC}} - b\sigma_{J_{AC}} \text{ and } J_{X\&AC} < \mu_{J_{X\&AC}} - c\sigma_{J_{X\&AC}}$$

Group 3:

$$J_X < \mu_{J_X} - a\sigma_{J_X}, \quad J_{AC} > \mu_{J_{AC}} + b\sigma_{J_{AC}} \text{ and } J_{X\&AC} < \mu_{J_{X\&AC}} - c\sigma_{J_{X\&AC}}$$

Group 4:

$$J_{X\&AC} > \mu_{J_{X\&AC}} + c\sigma_{J_{X\&AC}}$$

Group 5:

$$\mu_{J_X} - a\sigma_{J_X} \le J_X \le \mu_{J_X} + a\sigma_{J_X} \text{ and } \mu_{J_{AC}} - b\sigma_{J_{AC}} \le J_{AC} \le \mu_{J_{AC}} + b\sigma_{J_{AC}}$$

In the simulation, the coefficients of the standard deviations are: $a = 1$, $b = 1$ and $c = 2$. These values are chosen to ensure that the groups are sufficiently differentiated and that there are a large number of elements within a group. We set $c = 2$ to ensure that all groups, except Group 4, have a low incidence of simultaneous crossovers and abrupt changes. Each group has a total of 250 options, representing 50 options for each of the five volatility levels. We refer to one simulation run for one option as a *replication*. A total of 1250 replications were done in the simulation.

In Table 5.1b, very roughly the allocation has generated actual groups of time series of the underlying stock with the following characteristics.

Table 5.1b The actual groups.

	Crossovers	Abrupt changes
Group 1	Very few	Very few
Group 2	Several	Very few
Group 3	Very few	Several
Group 4	Several	Several
Group 5	Few	Few

5.2 Setting Up and Winding Down the Hedge

In this section, we discuss the mechanics of hedging, from setting up to winding down of the hedge. We consider delta and five minimax hedging strategies. These strategies involve rebalancing the hedge at uniform intervals of time; in the simulation, the interval is 1 day. Daily data include the stock price, the option price, the risk-free interest rate, the time to maturity and the volatility. The risk-free interest rate[30] is preset to 0.10. Dividends are excluded from the analysis and N, the number of contracted shares, is 100.

At time 0, each strategy experiment is assumed to hold the same number of shares (x_0) based on delta, and the same initial cumulative value of cash inflow minus cash outflow given by (4.12). Every day through to the maturity date, the hedge is rebalanced according to the strategies' recommendation. The trajectory of the number of shares held at time t, x_t, varies with the hedging strategy used. The actual cumulative value of cash inflow minus cash outflow at time t is given by (4.11). At time T, if the holder does not exercise her option to buy the shares, each strategy disposes of its portfolio in the same way: selling any shares held, or buying any shares sold short, at time $T - 1$ at y_T^S.

5.3 Summary of Simulation Results

We present the results of this simulation mainly in terms of the performance and the relative performance of the minimax strategy. We define the performance of a strategy[31] as the final cumulative value[32] of cash inflow minus cash outflow in using that strategy on an option, standardized[33] as a percentage of

---

[30] This is the continuous rate.

[31] A strategy can either be a minimax variant or delta hedging.

[32] This is calculated after winding down the hedge.

[33] Following Samuelson (1965), all cells in Table 5.5.1 have been standardized by dividing the original profit by the exercise price. In the simulation study, we set the exercise price at $X = 1000$; this makes the original profit effectively standardized. In the empirical study, because of the differences in exercise prices, the standardization becomes relevant.

the notional[34] contract value of that option. We define relative performance of a minimax variant as the performance of that variant minus the performance of delta hedging (DH). A variant is said below to outperform DH if its performance is higher than that of DH and, for any group of options, the difference for that group is significant at the 1% level.

5.3.1 Performance of Delta Hedging and Minimax

The performance of a strategy is averaged over all options in a group. Each cell in the table gives the performance of a particular strategy, and the relative performance of that strategy. Table 5.3.1 summarizing the simulation results is accompanied by Figure 5.3.1 which gives a graphical representation of the relative performance of the minimax variants. The horizontal axis gives the average distance of a particular group from the exercise price. Each group consists of 250 options or stock price series, corresponding to 250 replications; each stock price series consists of 190 daily prices, corresponding to 38 weeks,

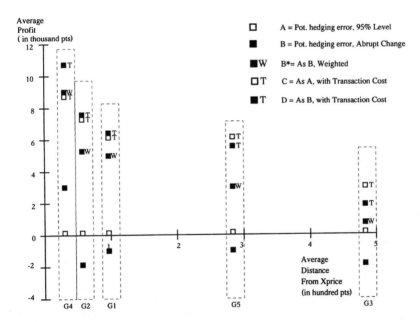

Figure 5.3.1 Relationship between relative performance and average distance from the exercise price.

---

[34] The notional value of the contract is the number of contracted shares multiplied by the exercise price. The total is the summation over all notional values.

Table 5.3.1 Performance and relative performance of the different strategies. Units: percentage points of final cumulative value of cash inflow minus cash outflow divided by the notional value of the contract

	Delta hedging	A	B	B*	C	D
		Potential hedge error		As B	As A	As B
		95% level	A.C.	Weighted	With TC	With TC
Group 1 (no crossovers, no abrupt changes)						
Average profit	0.7	0.8	−0.3	5.4	6.8	6.9
Average profit over DH	0.0	0.1	−1.0	4.7	6.1	6.2
Group 2 (crossovers only)						
Average profit	0.7	0.7	−1.2	6.2	7.7	8.1
Average profit over DH	0.0	0.0	−1.9	5.5	7.0	7.4
Group 3 (abrupt changes only)						
Average profit	−3.3	−3.3	−5.4	−2.6	−0.5	−1.4
Average profit over DH	0.0	0.0	−2.1	0.7	2.8	1.9
Group 4 (crossovers and abrupt changes)						
Average profit	−0.8	−0.6	2.4	8.1	7.7	9.9
Average profit over DH	0.0	0.2	3.2	8.9	8.5	10.7
Group 5 (general case)						
Average profit	−4.0	−4.0	−5.1	−0.7	2.0	1.4
Average profit over DH	0.0	0.0	−1.1	3.3	6.0	5.4

with 5 trading days per week, for a 9-month option. The average squared deviation from the exercise price over 47,500 (= 190 × 250) prices was calculated; the square root gives the "average distance from the exercise price" for that group. Hence, we have

$$\text{the average distance from the exercise price} = \sqrt{\frac{\sum_{i=1}^{250}\sum_{l=1}^{190}(y_l^{S,i} - X)^2}{190 \times 250}} \quad (5.1)$$

where i refers to a particular time series in a group and l refers to a particular price. The "average distance from the exercise price" represents the degree of *moneyness*[35] of the options in that group.

In Figure 5.3.1, the two groups closest to the origin consist of options which are at-the-money. The group farthest from the origin has the largest distance from the exercise price which indicates that it consists of options which are either deep-in-the-money or deep-out-of-the-money.

Hypothesis testing has shown that variants B*, C and D are robust: they perform better than delta hedging for the group for which they are designed to perform best and in general they do not perform worse than delta hedging. Variants B*, C and D are strategies designed to constrain transaction costs: variants C and D both have a transaction cost term in their objective function which is then minimized; Variant B* uses a heuristic method of constraining transaction costs based on short-term trends and distance from the exercise price.

Statistical hypothesis testing has also shown that:

1. Variants B*, C and D perform better than delta hedging for at-the-money options; in particular, when there are crossovers as well as abrupt changes.

2. The abrupt change variants, B and D, are not suitable for deeply-in-the-money or deeply-out-of-the-money options even if the corresponding price series are characterized by a large number of abrupt changes.

3. The abrupt change variants, B and D, are more suitable than their 95% level counterparts, A and C, for at-the-money options when the price series are characterized by a large number of abrupt changes.

4. The transaction-costs variants, C and D, perform better than their no-transaction-costs counterparts, A and B. The inclusion of a transaction cost term in the objective functions of variants C and D provides more cautious hedge recommendations and this helps constrain trading costs.

5. The weighted minimax variant B* performs better than its nonweighted variant, B. This implies that the proposed weighting system helps provide hedge recommendations that are conditioned by recent stock price levels. This conditioning leads to cautious trading and helps to constrain trading costs.

In the next section, we present the high performing variants of minimax. These are C, D and B*, the variants that explicitly constrained transaction costs.

---

[35] The amount that the option is in-the-money or out-of-the-money.

5.3.2 The High Performing Variants of Minimax

In Figures 5.3.2.1–5.3.2.3, respectively, we show the variation with volatility in the relative performance of variants C, D and B*. Variant C uses the 95% level to define the worst-case scenario. Variant D uses the abrupt change to define its worst-case scenario. Variant B* uses the abrupt change to define its worst-case scenario; it also uses a heuristic weighting system to give weighted hedge recommendations. In the figures, for each of the five volatility levels, a regression line has been fitted showing the relationship between relative performance and degree of moneyness which is represented by the average distance from the exercise price. All five regression lines have a negative slope. The analysis also shows that the higher the volatility, the more robust the performance of the minimax strategy. This is exhibited in the figures by the decreasing slope with increasing volatility. For volatility levels 0.2 and 0.3, the performances of the three variants are significantly better than that of delta hedging for distances relatively close to the exercise price. For volatility levels 0.5 and 0.6, the performances are significantly better than that of delta hedging for a wider range of distances from the exercise price.

Comparing the figures, we see a trend towards increasing parallelism between the regression lines from Figure 5.3.2.1 to Figure 5.3.2.3. This can

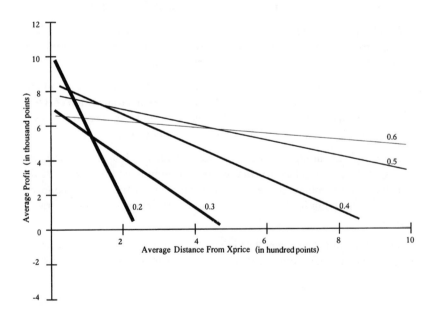

Figure 5.3.2.1 Performance of Variant C for different levels of sigma. C is the objective function with penalty for transaction costs, 95% Level.

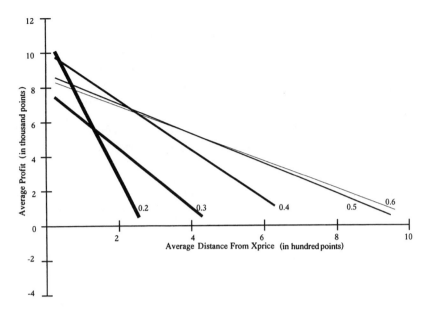

Figure 5.3.2.2   Performance of Variant D for different levels of sigma. D is the objective function with penalty for transaction costs, Abrupt Change.

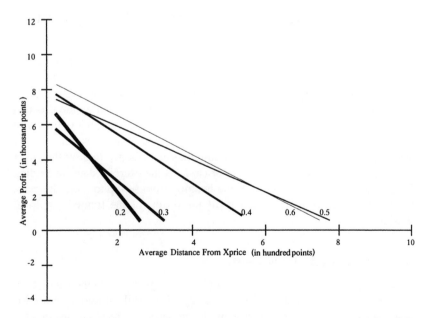

Figure 5.3.2.3   Performance of Variant B* for different levels of sigma. B* is the objective function without a penalty term for transaction costs, Abrupt Change, weighted.

be interpreted as follows: in Figure 5.3.2.3, the lines have a relatively high degree of parallelism which shows that the higher the volatility, the better the performance; however, variant B* starts to perform badly when the lines cross the x-axis. The point of intersection with the x-axis for each of the five volatility levels are closer to zero when compared with the other figures (Figures 5.3.2.1 and 5.3.2.2). This implies that variant B* performs best for a limited degree of moneyness.

In Figure 5.3.2.1, the lines have a relatively low degree of parallelism which shows that there is a trade-off between degree of moneyness and volatility. When the average distance is close to zero, variant C performed well for all volatility levels but has a slightly better performance for lower volatilities. The inverse relationship between relative performance and volatility can be observed when the average distance is large. This implies that variant C performs well for a wide degree of moneyness.

In Figure 5.3.2.2, the parallelism is intermediate between those in Figures 5.3.2.1 and 5.3.2.3. This strategy is the best performer of the three when the distance is close to zero. This implies that variant D is the most suitable strategy for at-the-money options. When the distance is large, variant D performs worse than variant C. This implies that variant D is not suitable for deeply-out or deeply-in-the money.

6 ILLUSTRATIVE HEDGING PROBLEM: A LIMITED EMPIRICAL STUDY

We consider the problem of hedging the risk of writing a European call option for a number of options available in the UK options market. We present a limited empirical study of 30 options: this is done as an illustration of the performance of the minimax hedging strategy when applied to real data. Actual market prices were used for both the stock and the option. The writer of the call incurs a potential liability in the case of exercise of the option by the buyer, and receives a premium; she is obliged to offer the buyer N shares of the stock at the exercise price, X, at the exercise date. In Section 6.1, we describe the mechanics of the hedging from set-up to wind-down. In Section 6.2, we present a summary of the results of the limited empirical study.

6.1 From Set-Up to Wind-Down

In Section 5.2, we discussed the mechanics of hedging used in the simulation, from setting up to winding down of the hedge. For the illustration, we use the same mechanics with a few changes in the parameters. The strategies involve rebalancing the hedge at uniform intervals of time; in the illustration, the interval is 1 week, that is, 5 trading days. Weekly data[36] include the market price of the stock, the market price of the option, the risk-free interest rate, the time to maturity and an estimate of the volatility of the stock price. The estimates of volatility used in the illustration are based on the most recent 100 days of stock price movement and on implied volatility[37]. The risk-free interest rate is based on the discount rate[38] provided by a Treasury Bill that expires at about the same time as the option. In the illustration, N, the number of contracted shares, is 1000.

6.2 The Hedging Strategies Applied to 30 Options: Summary of Results

The strategies defined in Section 4 were used to hedge the risk of writing one call option; 30 options were used in the study. Table 6.1 presents the performance of the different strategies relative to delta hedging. The final row of Table 6.1 indicates that the relative performance of four of the five variants of minimax averaged over 30 options is better than that of delta hedging, and that for variant B, for which the relative performance is worse, the weighted version B* performs better than delta hedging. Under variant B, the recommended change in the number of shares to hold is highly volatile, and when transaction costs are included, the cumulative cost of this variant can get very high. The effect of the weighting is to dampen this volatility, and so reduce transaction costs. The final row of Table 6.1 shows marked differences in the relative performance of different minimax variants. The columns of Table 6.1 are generally highly variable. The variants of minimax are essentially specific to different classes of options. This implies that as options change their moneyness, then the minimax approach must also change. In other words, the minimax approach to be implemented is a function of moneyness.

Because we have not established that the universe of calls from which the sample of 30 was selected is representative of the universe of all calls, the result that minimax performs better than delta hedging is not necessarily generalizable. However, the criteria for the universe from which we selected the 30 calls do not include requirements on the number or size of abrupt changes, nor the number of crossovers. Therefore, it is possible that minimax would perform (slightly) better than delta hedging for the universe of all calls written for UK stocks.

[36] Data were supplied by Datastream, International. All prices are mid-prices. The stock price series used were Datastream's Adjusted Stock Price Series: these are the original stock price series with dividend adjustments.

[37] Implied volatility is the volatility implied by an option price observed in the market.

[38] The discount rate was used to find the current value of the Treasury Bill based on a face value of 1 and the time to maturity. The risk-free interest rate is the continuous rate that discounts the face value to the computed current value.

Table 6.1 Gain over delta hedging for the different strategies[a] (standardized values, in hundred points)

	Xprice	A	B	B*	C	D
Tesco	220	−4	−116	−57	14	−26
Boots	280	−1	−67	−7	17	15
Sainsbury	360	−1	−131	−33	24	−32
BAir	220	−1	−39	−8	15	26
GEC	180	−2	−140	−92	2	−48
Allied	650	−1	−81	11	35	37
Ladbroke 1	240	−1	−65	42	−24	72
Ladbroke 2	260	−2	−80	−63	−23	−49
Cadbury 1	420	−3	−56	2	22	25
Cadbury 2	460	8	−17	60	78	93
Hanson 1	200	6	10	90	25	99
Hanson 2	180	−2	−62	−43	12	−4
P&O	600	7	0	66	45	53
Vodafone 1	390	−1	−101	−57	15	−33
Vodafone 2	360	0	−19	28	18	58
Prudential 1	240	15	26	131	76	142
Prudential 2	220	6	−41	97	53	113
Marks&Spencer	300	4	31	89	17	119
Shell 1	460	2	28	69	23	76
Shell 2	500	1	−115	−30	33	16
Eurotunnel	390	7	10	48	44	62
Glaxo	800	6	27	84	28	72
Guinness 1	500	7	−98	6	42	55
Guinness 2	550	9	62	113	52	107
Forte	200	−2	−04	14	−29	17
Thames Water	330	−1	30	10	14	20
U. Biscuits	390	−2	−20	15	−28	27
Btelecom 1	330	1	−44	18	53	63
Btelecom 2	360	5	−126	−28	78	52
Wellcome	900	4	−19	62	37	76
Average		2	−41	21	26	43

[a] See Table 4.7 (page 194).

7 MULTIPERIOD MINIMAX HEDGING STRATEGIES

In earlier sections, we consider several variants of a minimax strategy, hereafter called Basic minimax, that determines the number of shares that minimizes the worst-case potential hedging error for the next period. In this section and the next, we present strategies introduced in Howe et al. (1996). We consider the 95% level variant (variant C in Section 4.7) and explore extensions to this variant. The choice of variant C is due to its incorporation of transaction costs in its objective function and that it computes the minimax hedging error based on the most likely future values of the stock price, that is, the worst-case scenario chosen from the 95% level range.

The first extension is a two-period minimax strategy, hereafter called Two-Period minimax, where the worst case is defined over a two-period setting. In this extension, the objective function of Basic minimax is augmented to include the hedging error for the second period. In Basic minimax we have a one time period setting that is equal to the rebalancing interval. In Two-Period minimax we have a two time period setting, but the rebalancing interval remains equal to one period.

The second extension is a variable minimax strategy, hereafter called Variable minimax, where early rebalancing is triggered by the minimax hedging error. In Basic minimax, we preset a rebalancing interval that is constant throughout the life of the option and the hedger necessarily rebalances at the end of each interval, and may not rebalance within that interval. In Variable minimax we also preset a constant rebalancing interval, but the hedger may rebalance before the end of that interval. Under Variable minimax the hedger can monitor the actual hedging error within a rebalancing interval; if she finds the actual hedging error unacceptable, she can rebalance before the end of that interval.

In Section 7.1, we present Two-Period minimax, and in Section 7.2, Variable minimax. In Section 8, we summarize the results of a simulation study where the performances of Basic minimax, the two multiperiod extensions and delta hedging are compared.

7.1 Two-Period Minimax Strategy

This strategy provides the hedger with a tool for computing the minimax hedging error in two time periods. The strategy is designed for the hedger who wishes to have a constant rebalancing interval, that is, $\Delta t \to 0$. She decides on the number of shares to hold on the basis of the calculated minimax hedging error for the following two time periods, and she rebalances at the end of the first of the two periods. If she cannot rebalance at the end of the first period (say due to a shortage of stock), the two-period minimax yields a solution that ensures the worst case over both periods are adequately addressed. If the worst case in one or both periods is not realized, the noninferiority of minimax ensures that the hedging error will improve (see Theorem 1.3.1).

The difference between the worst case under Two-Period minimax and the worst case under Basic (single period) minimax is that the range of uncertainty for the future stock price is wider for the two-period case. This results in a higher minimax hedging error for the two periods compared to the summation of the errors over two periods when using Basic minimax. In this sense, Two-Period minimax is a more cautious strategy than Basic minimax. However, when applying Basic minimax separately for two periods in succession, transaction cost contributions may result in less favorable hedging error. Indeed, when Basic minimax is considered over two periods simultaneously, with separate decision variables over both time periods, the best hedging error may be improved in view of the inherent transaction cost minimization for the second period.

7.1.1 Minimax Problem Formulation

The minimax problem is given by

$$\min_{x_t} \max_{(y_{t+1}^S, y_{t+2}^S)} f(x_t, y_{t+1}^S, y_{t+2}^S) \tag{7.1}$$

subject to

$$y_{t+1}^{S,\text{lower}} \leq y_{t+1}^S \leq y_{t+1}^{S,\text{upper}}$$

$$y_{t+2}^{S,\text{lower}} \leq y_{t+2}^S \leq y_{t+2}^{S,\text{upper}} \tag{7.2}$$

where $f(x_t, y_{t+1}^S, y_{t+2}^S)$ is the objective function, presented in Section 5.2, $y_{t+1}^{S,\text{lower}} \leq y_{t+1}^S \leq y_{t+1}^{S,\text{upper}}$ and $y_{t+2}^{S,\text{lower}} \leq y_{t+2}^S \leq y_{t+2}^{S,\text{upper}}$ are ranges defined as the 95% level.

As with Basic minimax, there are no constraints on x_t, the number of shares to hold at time t: nonnegative x_t implies a long position in shares; negative x_t implies a short position in shares.

7.1.2 The Objective Function

We define $U_1 : \Re^k \times \Re^k \to \Re^1$, $U_2 : \Re^k \times \Re^k \times \Re^k \to \Re^1$, $U_3 : \Re^k \to \Re^k$, $x_t \in \Re^k$, $y_{t+1}^S \in \Re^k$, $y_{t+2}^S \in \Re^k$, $U : \Re^k \times \Re^k \times \Re^k \to \Re^{k+2}$ and Q as a $(k+2) \times (k+2)$ positive definite weighting matrix. U_1 refers to the potential hedging error for the first time period and U_2 refers to that for the second time period. U_3 refers to a penalty term for transaction costs associated with buying or selling of stocks. $U^d : \Re^k \times \Re^k \times \Re^k \to \Re^{k+2}$ is the vector of desired values for the potential hedging error for the two periods and the transaction cost terms: we use a desired value of zero, that is, the desired hedging error is zero and the desired transaction cost is zero. The objective function is given by

$$f(x_t, y_{t+1}^S, y_{t+2}^S) = \frac{1}{2}\langle U - U^d, Q(U - U^d)\rangle \qquad (7.3)$$

where

$$x_t = \begin{bmatrix} x_{1,t} \\ \vdots \\ x_{k,t} \end{bmatrix}, \quad y_{t+1}^S = \begin{bmatrix} y_{1,t+1}^S \\ \vdots \\ y_{k,t+1}^S \end{bmatrix}, \quad y_{t+2}^S = \begin{bmatrix} y_{1,t+2}^S \\ \vdots \\ y_{k,t+2}^S \end{bmatrix} \qquad (7.4)$$

$$U(x_t, y_{t+1}^S, y_{t+2}^S) = \begin{bmatrix} U_1(x_t, y_{t+1}^S) \\ U_2\left(x_t, \begin{bmatrix} y_{t+1}^S \\ y_{t+2}^S \end{bmatrix}\right) \\ U_3(x_t) \end{bmatrix} \quad \text{and} \quad U^d = \begin{bmatrix} U_1^d \\ U_2^d \\ U_3^d \end{bmatrix} = \begin{bmatrix} 0 \\ 0 \\ 0 \end{bmatrix} (7.5)$$

$$U_1(x_t, y_{t+1}^S) = \sum_{i=1}^{k} x_{i,t}(y_{i,t+1}^S - y_{i,t}^S) + \sum_{i=1}^{k} N_i(B_{i,t} - B_{i,t+1}(y_{i,t+1}^S))$$

$$+ \sum_{i=1}^{k}\left(-(x_{i,t} - x_{i,t-1})y_{i,t}^S + C_{i,t-1}(1 + r\Delta t)\right)r\Delta t \qquad (7.6)$$

(first period hedging error)

$$U_2\left(x_t, \begin{bmatrix} y_{t+1}^S \\ y_{t+2}^S \end{bmatrix}\right) = \sum_{i=1}^{k} x_{i,t}(y_{i,t+2}^S - y_{i,t+1}^S) + \sum_{i=1}^{k} N_i(B_{i,t+1}(y_{i,t+1}^S) - B_{i,t+2}(y_{i,t+2}^S))$$

$$+ \sum_{i=1}^{k}\left(-(x_{i,t} - x_{i,t-1})y_{i,t}^S + C_{i,t-1}(1 + r)\Delta t\right)(1 + r\Delta t)(r\Delta t) \qquad (7.7)$$

(second period hedging error)

where

$$C_{i,t-1} = C_{i,t-2}(1 + r\Delta t) - (x_{i,t-1} - x_{i,t-2})y_{i,t-1}^S - \hat{K}\left|(x_{i,t-1} - x_{i,t-2})y_{i,t-1}^S\right|$$

(cumulative value of cash inflow minus cash outflow with transaction cost)

$$(7.8)$$

$$U_3(x_t) = \begin{bmatrix} U_{1,3}(x_{1,t}) \\ \vdots \\ U_{k,3}(x_{k,t}) \end{bmatrix} \tag{7.9}$$

where

$$U_{i,3}(x_{i,t}) = \hat{K}(x_{i,t} - x_{i,t-1})y_{i,t}^S \quad \text{(transaction penalty)} \tag{7.10}$$

We first identify all the variables in (7.3)–(7.10) and then give an economic interpretation of (7.6) and (7.7). The variables and the treatment of transaction costs are as discussed in Section 4, suitably extended for the two-period strategy.

$C_{i,t-1}$ is the cumulative value of cash inflow minus cash outflow at time $t-1$. $C_{i,t-1}(1 + r\Delta t)$ is $C_{i,t-1}$ with interest payments. The first term of (7.8) is the cumulative value of cash inflow minus cash outflow from the previous period with interest payments. The second term is a cash outflow if $x_{i,t-1} > x_{i,t-2}$; otherwise, it is a cash inflow. The third term is always a cash outflow. We note that $C_{i,t-1}$ will normally be a negative number. At time t, $C_{i,t-1}$ is a constant: all the variables in (7.8) have actual values.

Transaction costs introduce nondifferentiability into the equation. Thus, they do not enter the objective function as part of U_1 nor U_2. Instead, we introduce $U_{i,3}$ to represent a penalty for transaction costs for each option i at time t.

In the weighting matrix Q, we adopt diagonal weights q_i, $i = 1, ..., k + 2$, which are specified by the hedger, represent her preferences: a high q_1 represents an emphasis on minimizing the potential hedging error that may be incurred from time t to time $t + 1$; a high q_2 represents an emphasis on minimizing the potential hedging error that may be incurred from time $t + 1$ to time $t + 2$. High q_i, $i = 3, ..., k + 2$, represents an emphasis on minimizing the corresponding transaction cost term.

For each option i, $i = 1, ..., k$, $B_{i,t+1}(y_{i,t+1}^S)$ and $B_{i,t+2}(y_{i,t+2}^S)$ is determined using the Black and Scholes (1973) option pricing model and using Leland's (1985) modified volatility.

In (7.6) and (7.7), U_1 represents the potential hedging error between time t and time $t + 1$; it comprises the potential shift in the stock position, the potential shift in the option position and the potential interest payment. U_2 represents the potential hedging error between time $t + 1$ and time $t + 2$. U_3 refers to a penalty term for transaction costs associated with buying or selling of stocks. We wish to minimize the potential hedging error, including interest payments on borrowed money, for two time periods. At the same time, we wish to constrain transaction costs.

7.1.3 The Two-Period Minimax Hedging Error

In contrast to the minimax hedging error for one time period given by (4.10) in Section 4.5, the minimax hedging error for two time periods is given by

$$\text{Two-period minimax hedging error} = \hat{U}_1(x_t^*, y_{t+1}^{S^*}) + \hat{U}_2(x_t^*, y_{t+1}^{S^*}, y_{t+2}^{S^*}).$$
$$(7.11)$$

7.2 Variable Minimax Strategy

This strategy is the same as Basic minimax in all respects except that under Variable minimax the hedger can rebalance within the preset interval, and in deciding when to rebalance she takes account of the actual hedging error. If the actual hedging error is unacceptable to her, she may wish to rebalance before the end of the preset interval; if the actual hedging error is not unacceptable, she would rebalance at the end of the preset interval.

At the start of each time period, the hedger specifies her worst-case scenario for that time period. One such scenario may be a large movement in the price of the underlying stock that results in an actual hedging error that is unacceptable to her. She uses Basic minimax to minimize the potential hedging error that corresponds to such a scenario. If the stock price moves in the direction that makes the hedging error unacceptable to her, she may wish to rebalance early. For example, if the preset rebalancing interval is 1 week, that is 5 trading days, she may rebalance on day 1, 2, 3 or 4, and the next preset interval will start on the following day.

Under Variable minimax, the hedger could use the minimax hedging error calculated by Basic minimax as a criterion in deciding whether to rebalance before the end of the preset interval. Because the minimax hedging error corresponds to the worst-case scenario for the next period, if the stock price is within the preset range, she knows that the absolute value of the actual hedging error would not be higher than the minimax hedging error. Despite this knowledge, she may consider such an actual hedging error to be unacceptable. If so, she can avoid the accumulation of unacceptable hedging errors by rebalancing early should the actual hedging error be worse than the threshold, which is defined as the proportion of the minimax hedging error acceptable to her. This threshold error serves as a trigger for early rebalancing.

If, in the preset interval, the hedger finds that the actual hedging error is worse than her threshold error, she may rebalance at that time. However, if, at any time within the preset interval, the actual error is not worse than her threshold error, she would not rebalance before the end of the preset interval. She can use the minimax hedging error as a criterion for deciding when to rebalance. The time at which she rebalances becomes the start of the next preset interval.

Because Variable minimax uses a system to monitor the actual hedging error and allows the hedger to rebalance early when the actual hedging error becomes unacceptable to her, Variable minimax is more responsive to unfavorable stock price movements than Basic minimax. In this sense, Variable minimax is a more aggressive strategy than Basic minimax.

7.2.1 Basic Minimax Problem Formulation

We recall the Basic minimax problem (4.1)–(4.2) given by

$$\min_{x_t} \max_{y^S_{t+1}} f(x_t, y^S_{t+1})$$

subject to

$$y^{S,\text{lower}}_{t+1} \leq y^S_{t+1} \leq y^{S,\text{upper}}_{t+1}$$

where $f(x_t, y^S_{t+1})$ is the objective function, presented in Section 4, and $y^{S,\text{lower}}_{t+1} \leq y^S_{t+1} \leq y^{S,\text{upper}}_{t+1}$ is the range defined under Worst Case 1, the 95% level.

The definition of variables is identical to that in Section 4. The minimax hedging error is the same as that in Section 4.5. Variable minimax is essentially the above Basic minimax, augmented by a system to monitor the actual hedging error.

7.2.2 The Monitoring System

The minimax potential hedging error, with the corresponding minimizing variable x_t, applies to the time period between t and $t + 1$; hereafter we refer to this time period as τ. We define a smaller interval Δt such that $m\Delta t = \tau$ where m is the number of intervals of length Δt. Here, we consider multi-periods within the period τ. On solving the minimax problem, we find the value of x_t that minimizes the maximum hedging error that could occur within the time period t to $t + 1$, given the preset range of y^S_{t+1}, that is, the maximum hedging error is the value that we have insured against, when using x_t. It should be noted that there may be combinations of time $t + m_0\Delta t$, $m_0 = 0, 1, ..., m$, and corresponding stock price $y^S_{t+m_0\Delta t}$, that give the same hedging error as the one defined by the minimax solution.

For each small interval Δt, we define a certain percentage $z\%$ of the absolute value of the minimax hedging error M as the threshold V, that is,

$$V = \frac{z}{100} M. \tag{7.14}$$

For each time period t to $t + m_0\Delta t$, $m_0 = 0, 1, ..., m$, we calculate the absolute value of the actual hedging error, A, and compare this with the threshold V. If the actual hedging error is negative and

$$A_{t+m_0\Delta t} \geq V. \qquad (7.15)$$

at time $t + m_0\Delta t$, with actual stock price $y^S_{t+m_0\Delta t}$, then we rebalance and solve the minimax problem again, and update t, that is, set to $t + m_0\Delta t$ (the current time). If condition (7.15) is not satisfied for any time $t + m_0\Delta t$, $m_0 = 0, 1, ..., m$, then we rebalance at time $t + m\Delta t$.

8 SIMULATION STUDY OF THE PERFORMANCE OF DIFFERENT MULTIPERIOD STRATEGIES

The simulation in this section is intended to serve as a feasibility study on potential extensions to Basic minimax. Towards this, the simulation is used to ascertain whether the two multiperiod extensions of Basic minimax outperform delta hedging (DH), to ascertain whether Basic minimax outperforms Two-Period minimax, which is designed to be a more cautious strategy than Basic minimax, and to ascertain whether Variable minimax, which is designed to be a more aggressive strategy than Basic minimax, outperforms Basic minimax.

8.1 The Simulation Structure

The method for generating stock price series and option price series is described in Section 5. Also in Section 5, we discuss the mechanics of hedging, from setting up to winding down the hedge. For this simulation, we use the same mechanics with a few changes in parameters. Delta hedging, Basic minimax, and Two-Period minimax involve rebalancing the hedge at uniform intervals of time; in this simulation, the interval is 1 week. Variable minimax involves the monitoring of the actual hedging error on a daily basis. Weekly and daily data include the price of the stock, the price of the option, the risk-free interest rate, the time to maturity and the volatility (sigma) of returns on the stock, given as one of five preset levels: 0.2, 0.3, 0.4, 0.5, and 0.6. The risk-free interest rate is preset at 0.10. Dividends are excluded from the analysis. In this simulation, N, the number of contracted shares, is 100. In addition, for Variable minimax, we set the threshold level at 10% of the minimax hedging error.

We apply delta hedging, Basic minimax, Two-Period minimax and Variable minimax to 1250 options; as in Section 5, these are subdivided into five levels of sigma: 0.2, 0.3, 0.4, 0.5 and 0.6. However, in contrast to the simulation in Section 5, we do not stratify in terms of groups because, as mentioned in Section 8.1, this simulation study is intended only to explore the feasibility of extensions to Basic minimax.

8.2 Results of the Simulation Study

All four strategies were used to hedge the risk of writing each of the 1250 European call options. The average performance in using a strategy is calculated for all options with a constant sigma level. In Table 8.2a, we summarize the relative performance of Basic minimax and the two multiperiod extensions. In Table 8.2b, we summarize the difference in relative performance between minimax strategies. These tables also contain, in parentheses, the absolute value of the t-statistics, followed by ** if the difference is significant at the 2% level, and * if it is significant at the 10% level.[39]

From Table 8.2a, each strategy outperforms DH by about 3 percentage points, with Variable minimax being slightly the better performer. Two-Period minimax is the worst performer, outperforming DH by just over 2 percentage points. For all three strategies, their relative performances fall with increasing sigma; the fall is most marked in Two-Period minimax.

From Table 8.2b, for low levels of sigma, Variable minimax outperforms Basic minimax, but for high levels of sigma, Variable minimax performs much the same as Basic minimax. For low levels of sigma, Two-Period minimax performs much the same as Basic minimax, but for high levels of sigma, Basic minimax outperforms Two-Period minimax.

Table 8.2a Relative performance of three minimax strategies (t-values in parentheses) (Units: percentage points of final cumulative value of cash inflow minus cash outflow divided by the notional value of the contract)

Sigma	Strategy		
	Basic	Two-Period	Variable
0.2	3.3 (20.1)**	3.3 (26.4)**	3.6 (30.6)**
0.3	2.9 (19.7)**	2.9 (23.0)**	3.4 (14.4)**
0.4	2.9 (12.3)**	2.4 (10.4)**	3.2 (13.5)**
0.5	2.8 (11.9)**	1.5 (6.8)**	2.9 (13.2)**
0.6	2.8 (7.0)**	1.7 (10.1)**	2.8 (11.7)**
Average	2.9	2.3	3.1

8.3 Rank Ordering

The simulation results suggest the following rank order of positive differences in performance. Table 8.3 gives the rank ordering of the strategies. All minimax strategies for the five levels of sigma outperform delta hedging. For low levels of sigma (sigma = 0.2 or 0.3), Basic minimax has the same rank as

---

[39] We use significance levels of 2% and 10% to express the results of a two-tailed test; however, we use a 1-tailed test for the sign of the difference at the 5% level.

Table 8.2b Difference in relative performance between minimax strategies (*t*-values in parentheses) (Units: percentage points of final cumulative value of cash inflow minus cash outflow divided by the notional value of the contract)

Sigma	Strategy		
	Basic minus Two-Period	Variable minus Basic	Variable minus Two-Period
0.2	0.0 (0.1)	0.3 (1.7)*	0.3 (2.5)**
0.3	0.0 (0.2)	0.5 (3.4)**	0.5 (3.6)**
0.4	0.5 (2.2)*	0.3 (1.7)*	0.8 (3.7)**
0.5	1.3 (4.7)**	0.1 (0.2)	1.4 (5.5)**
0.6	1.1 (2.7)**	0.0 (0.0)	1.1 (4.4)**
Average	0.6	0.2	0.8

Two-Period minimax while for a high level of sigma (sigma = 0.6), Basic minimax has the same rank as Variable minimax.

For the average rank order, Variable minimax outperforms Basic minimax, which is consistent with the view that Variable minimax is more responsive to the development of unacceptable hedging errors. Basic minimax outperforms Two-Period minimax, which is consistent with the view that Two-Period minimax is less suitable when the hedger rebalances at the end of one period.

Table 8.3 Rank order of positive significant differences in performance for each level of sigma

Strategy	Sigma level					
	0.2	0.3	0.4	0.5	0.6	Average
Variable Minimax	1	1	1	1	1.5	1.1
Basic Minimax	2.5	2.5	2	2	1.5	2.1
Two –period minimax	2.5	2.5	3	3	3	2.8
Delta Hedging	4	4	4	4	4	4.0

9 CAPM-BASED MINIMAX HEDGING STRATEGY

In Section 4 above, we discuss several variants of a minimax strategy, which we call Basic minimax, that determines for an individual option the number of shares that minimizes the worst-case potential hedging error for the next period. In Section 4.2 we consider the worst-case scenario in terms of movements in the price of the underlying stock, with the source of uncertainty being the total volatility of returns on the underlying stock. In this section, we present the strategy from Howe et al. (1998). We define the worst-case scenario in terms of the two components of total volatility: the market risk and the specific risk. We use the Capital Asset Pricing Model (CAPM) as a basis for a price determination function, and develop a CAPM-based minimax hedging strategy that uses both components of total volatility; we refer to this strategy as CAPM minimax.

The CAPM is an equilibrium model for the pricing of assets in capital markets. It is a one-factor model that attempts to explain an asset's expected return in terms of a multiple, the *beta*, of the risk premium for holding the asset market instead of holding risk-free assets, that is government bonds[40].

Consider the problem of hedging the risk of holding a portfolio of written call options. The risk of holding such a portfolio can be hedged by holding a portfolio of underlying stocks. We define the hedge portfolio as the combination of the portfolio of written call options and the portfolio of the underlying stocks.

We consider CAPM minimax because, as discussed in Section 10, when Basic minimax is used to hedge the risk of holding a portfolio of more than one written call option, the performance of Basic minimax is inferior compared to its performance when used to hedge the risk of writing the options individually. This inferior performance of Basic minimax when it is applied to a portfolio may be caused by a compounding of the worst-case scenarios the hedger has specified for each option, giving a worst-case scenario for the portfolio that may markedly overstate her view of the worst case for the portfolio. The hedger may consider that this compounded scenario has a very low likelihood of being realized, and so may choose a less severe worst-case scenario for the portfolio.

As noted above, CAPM minimax uses both components of total volatility. Therefore, the worst-case analysis incorporates information about these components separately in the solution. For an individual written call option, we express the worst-case scenario in terms of total volatility because we wish to define a range for future stock price movements. In this case it is not necessary to separate total volatility into its components to define the range. In contrast, for a portfolio of several written call options, we express the worst-case scenario in terms of the components of total volatility for the following reasons. As the number of written options increases, the specific risk of the portfolio of underlying stocks decreases. At the limit, where options are written for all the stocks in the market, the specific risk of the portfolio of underlying stocks reaches a minimum value. For such a portfolio of underlying stocks, where specific risk may be very small compared to the market risk,

---

[40] See Elton and Gruber (1991) for a comprehensive introduction to the Capital Asset Pricing Model.

the remaining source of uncertainty for the value of this portfolio becomes the market risk. Further, for a given number of underlying stocks in a portfolio, this portfolio's market risk will be more dominant than its specific risk if the underlying stocks in the portfolio have high market risks and low specific risks. Equally, the portfolio's market risk will be less dominant if the underlying stocks have low market risks and high specific risks. Furthermore, the market risk is a shared component among all the stocks in the portfolio. Thus, the worst case defined under the restriction of the CAPM model is less pessimistic than that defined over the unrestricted individual stocks. Where the portfolio's market risk is more dominant than its specific risk, the compounded scenario of individual worst cases based on total volatility would overstate the worst-case scenario the hedger would specify if she based her specification on market risk.

In Section 9.1, we present the Capital Asset Pricing Model and the Market Model, and in Sections 9.2–9.4, CAPM minimax and the definitions of two worst-case scenarios. In Section 10, we present the results from a simulation when two variants of Basic minimax and of CAPM minimax are applied to 150 portfolios of 5 options. In Section 11, we present the beta-risk profile of a hedge portfolio under CAPM minimax.

9.1 The Capital Asset Pricing Model

In this section we summarize the Capital Asset Pricing Model (CAPM) and develop from it a price determination function. Let:

β_i	beta of stock i
$R_{i,t}$	total return on stock i for the period up to time t
R_M	return on the market portfolio M for the period up to time t
ϵ_i	disturbance variable
r_f	risk-free rate
$y_{i,t}^S$	stock price i at time t
I_t	market index level at time t
R_t^I	return on the market index I for the period up to time t

The CAPM is an expectations theory where the expected return on a stock is a function of the expected return on the market. This is given by[41]

$$E(R_i) = r_f + \beta_i\left(E(R_M) - r_f\right)$$ (9.1)

---

[41] This is a simplified version of the CAPM given in Huang and Litzenberger (1988, Equation (10.6.1)), where the return on a portfolio having a zero covariance with respect to the market portfolio is represented by the riskless rate.

where M is the market portfolio and $E(\cdot)$ is the expectation or average over some historical time series. Because the CAPM is an expectations theory, it is frequently assumed that the underlying relationship between the realized prices of the assets and the market portfolio can be modeled by the Market Model.[42] First we assume that there is a market proxy, represented by an index I, having a beta equal to 1. The Market Model, based on the CAPM, uses the returns on the market index as proxy to the returns on the market portfolio and defines the return on stock i for the time interval t to $t + 1$ as

$$R_{i,t+1} = (1 - \beta_i)r_f + \beta_i R^I_{t+1} + \epsilon_{i,t+1} \qquad (9.2)$$

under the assumptions that $\text{Cov}(\epsilon_{i,t+1}, R^I_{t+1}) = 0$ and $E(\epsilon_{i,t+1}\epsilon_{j,t+1}) = 0$, $i = 1,...,k$, $j = 1,...,k$, $i \neq j$. The disturbance variable[43], $\epsilon_{i,t+1}$, has zero mean and variance equal to $(\sigma^{\epsilon_i})^2$.

We define the total return on stock i for the time interval t to $t + 1$ as price returns

$$R_{i,t+1} = \frac{y^S_{t+1} - y^S_t}{y^S_t}. \qquad (9.3)$$

We use a market index as proxy to the market portfolio and define the return on the market index for the time interval t to $t + 1$ as index returns

$$R^I_{t+1} = \frac{I_{t+1} - I_t}{I_t}. \qquad (9.4)$$

For the time interval t to $t + 1$, the disturbance variable is given by

$$\epsilon_{i,t+1}. \qquad (9.5)$$

Using (9.3)–(9.5), (9.2) becomes

$$y^S_{i,t+1} = y^S_{i,t}\left\{(1 - \beta_i)r_f + \beta_i \frac{I_{t+1} - I_t}{I_t} + \epsilon_{i,t+1}\right\} + y^S_{i,t}, \quad i = 1,...,k \quad (9.6)$$

This is the price determination function used by minimax. (9.6) is derived directly from the Market Model; we use the term *CAPM minimax* to indicate that the basic motivation is from the CAPM.

9.2 The CAPM-based Minimax Problem Formulation

In CAPM minimax, the minimizing variable is x_t and the maximizing variable is y^E_{t+1} (9.10 below) which is a vector of uncertainties on the market index and on the individual stocks. y^E_{t+1} is allowed to take any value within predefined

---

[42] This is the version of the Market Model, expressed in terms of (9.1). See Elton and Gruber (1991, p. 338).

[43] This is referred to as the disturbance term in Huang and Litzenberger (1988, p. 311).

bounds. Therefore, the problem is

$$\min_{x_t} \max_{y_{t+1}^E} f(x_t, y_{t+1}^E) \tag{9.7}$$

subject to

$$y_{t+1}^{E,\text{lower}} \leq y_{t+1}^E \leq y_{t+1}^{E,\text{upper}} \tag{9.8}$$

where $f(x_t, y_{t+1}^E)$ is the objective function, discussed in Section 9.3, and $y_{t+1}^{E,\text{lower}} \leq y_{t+1}^E \leq y_{t+1}^{E,\text{upper}}$ is the range that defines the worst case, discussed in Section 9.4. There are no constraints on x_t, the number of shares to hold at time t: nonnegative x_t implies a net holding of shares; negative x_t implies a net sale of shares.

9.3 The Objective Function

We define $\bar{U}_1 : \Re^k \times \Re^{k+1} \to \Re^1$, $U_2 : \Re^k \to \Re^k$, $\bar{U} : \Re^k \times \Re^{k+1} \to \Re^{k+1}$, $x_t \in \Re^k$, $y_{t+1}^E \in \Re^{k+1}$ and $\bar{Q}$ as a $(k+1) \times (k+1)$ positive definite weighting matrix. $\bar{U}_1$ refers to the potential hedging error for the next period. $\bar{U}_2$ refers to transaction costs associated with buying or selling of stocks. $\bar{U}^d : \Re^k \times \Re^{k+1} \to \Re^{k+1}$ is the vector of desired values for the potential hedging error for the two periods and the transaction cost terms. We adopt a desired value of zero, that is, it is desired that there are no hedging errors or transaction costs.

We consider the objective function

$$f(x_t, y_{t+1}^E) = \frac{1}{2} \langle \bar{U} - \bar{U}^d, \bar{Q}(\bar{U} - \bar{U}^d) \rangle \tag{9.9}$$

where

$$x_t = \begin{bmatrix} x_{1,t} \\ \vdots \\ x_{k,t} \end{bmatrix} \quad \text{and} \quad y_{t+1}^E = \begin{bmatrix} I_{t+1} \\ \epsilon_{i,t+1} \\ \vdots \\ \epsilon_{k,t+1} \end{bmatrix} \tag{9.10}$$

$$\bar{U}(x_t, y_{t+1}^E) = \begin{bmatrix} \bar{U}_1(x_t, y_{t+1}^E) \\ \cdots\cdots\cdots\cdots \\ \bar{U}_2(x_t) \end{bmatrix} \quad \text{and} \quad \bar{U}^d = \begin{bmatrix} \bar{U}_1^d \\ \cdots \\ \bar{U}_2^d \end{bmatrix} = \begin{bmatrix} 0 \\ \cdots \\ 0 \end{bmatrix} \tag{9.11}$$

$$\bar{U}_1(x_t, y_{t+1}^E) = \sum_{i=1}^{k} x_{i,t}\left(y_{i,t+1}^S(y_{t+1}^E) - y_{i,t}^S\right) + \sum_{i=1}^{k} N_i\left(B_{i,t} - B_{i,t+1}(y_{i,t+1}^S(y_{t+1}^E))\right)$$

$$+ \sum_{i=1}^{k} \left(-(x_{i,t} - x_{i,t-1})y_{i,t+1}^S + C_{i,t-1}(1 + r\Delta t)\right)r\Delta t$$

(9.12)

where

$$C_{i,t-1} = C_{i,t-2}(1 + r\Delta t) - (x_{i,t-1} - x_{i,t-2})y_{i,t-1}^S - \hat{K}\left|(x_{i,t-1} - x_{i,t-2})y_{i,t-1}^S\right|$$

(9.13)

$$\bar{U}_2(x_t) = \begin{bmatrix} \bar{U}_{1,2}(x_{1,t}) \\ \vdots \\ \bar{U}_{k,2}(x_{k,t}) \end{bmatrix}$$

(9.14)

where

$$\bar{U}_{i,2}(x_{i,t}) = \hat{K}(x_{i,t} - x_{i,t-1})y_{i,t}^S.$$

(9.15)

The variables and the treatment of transaction costs are as discussed in Section 4, suitably extended for the CAPM version of the minimax strategy. $C_{i,t-1}(1 + r\Delta t)$ is the value of $C_{i,t-1}$ with interest payments. At time t, $C_{i,t-1}$ is a constant: all the variables in (5.12b) have actual values. We note that $C_{i,t-1}$ will normally be a negative number.

Transaction costs do not come into the objective function directly as part of $\bar{U}_1$ because they introduce nondifferentiability into the equation. Instead, we introduce $\bar{U}_{i,2}$ to represent a penalty for transaction costs for each option i at time t.

In the weighting matrix $\bar{Q}$, we adopt diagonal weights $\bar{q}_i$, $i = 1, ..., k + 1$, specified by the hedger. They reflect her preferences: high $\bar{q}_1$ represents an emphasis on minimizing the potential hedging error. High $\bar{q}_i$, $i = 2, ..., k + 1$, represents an emphasis on minimizing the corresponding transaction cost term.

For each option $i = 1, ..., k$, $B_{i,t+1}(y_{i,t+1}^S(y_{t+1}^E))$ is valued using the Black and Scholes(1973) option pricing model and using Leland's(1985) modified volatility estimate. For each stock $i = 1, ..., k$, $y_{i,t+1}^S(y_{t+1}^E)$ is computed using the price determination function, (9.6).

$\bar{U}_1$ represents the potential hedging error between time t and time $t + 1$: it comprises the potential shift in the stock position, the potential shift in the option position and the potential interest payment. It is a function of the variable y_{t+1}^E which is a vector of specific error variables and the market risk. We wish to minimize the potential hedging error, including interest payments on borrowed money, using a worst-case scenario based on market risk and specific risk.

9.4 The Worst-case Scenario

The source of uncertainty y^E_{t+1} is defined in (9.10) in terms of I_{t+1}, the uncertainty due to market movements, and k variables $\epsilon_{i,t+1}$, the uncertainty specific to stock $i = 1, \ldots, k$. The uncertainty range, as defined by (9.8), can be reformulated as

$$I_l \leq I_{t+1} \leq I_u \qquad (9.16)$$

$$\epsilon^{\text{lower}}_{i,t+1} \leq \epsilon_{i,t+1} \leq \epsilon^{\text{upper}}_{i,t+1}, \quad i = 1, \ldots, k. \qquad (9.17)$$

In CAPM minimax we consider two worst-case scenarios, given in the following two sections.

9.4.1 Worst Case 1

Worst Case 1 relates to extreme movements in the market index and in specific error variables consistent with the most likely values that they may have within the 95% confidence interval. We define the range of y^E_{t+1}, where we evaluate the effect of the worst-case scenario, as the range whose upper and lower bounds delimit the 95% confidence interval of all possible values of the future market index and specific error variables, that is, *two standard deviations* about the expected value at time $t + 1$. For I_{t+1}, the 95% confidence interval will be based on market risk. For the specific error variables $\epsilon_{i,t+1}$, $i = 1, \ldots, k$, the 95% confidence interval will be based on the corresponding specific risks. Worst Case 1 is hereafter referred to as the *95% Level*.

9.4.2 Worst Case 2

Worst Case 2 relates to extreme movements in the market index that may result in a switch in the state of the option from in-the-money' to out-of-the-money, or vice versa. A switch in the state of the option may result in a higher hedging error. We define the range of I_{t+1} as the upper and lower bounds of the future market index within *one and three standard deviations (sd)* from the expected value of the market index at time $t + 1$. We choose the side of the distribution that may result in a stock price that is closer to the exercise price X. This means that if $y^S_t > X$, the relevant range for I_{t+1} would be on the left side of the distribution of future market index, with I_{t+1} having a lower bound of 3 sd and an upper bound of 1 sd. If $y^S_t \leq X$, the relevant range for I_{t+1} would be on the right side, with a lower bound of 1 sd and an upper bound of 3 sd.

For a portfolio of stocks, the range of I_{t+1} may be determined by several criteria such as the states of the options (whether in-the-money, at-the-money, or out-of-the-money) and the relative distance of each stock price from the exercise price weighted by the volatility. In this simulation we use the criterion that if the majority of the options are in-the-money, then we use the left side of the distribution of future market index. If the majority of the options are out-of-the-money, then we use the right side. *The specific error variables will take their 95% level ranges as defined in Worst Case 1.* Worst Case 2 is hereafter referred to as the Abrupt Change[44].

10 SIMULATION STUDY OF THE PERFORMANCE OF CAPM MINIMAX

We ascertain the performance of a minimax variant relative to that of delta hedging (DH). Two variants of CAPM minimax are considered. These are the 95% Level variant and the Abrupt Change variant. A variant is taken to outperform DH if its performance is higher than that of DH. Additionally, for any group of options, the difference for that group is significant at the 5% level. We simulate the relative performance of the two variants of CAPM minimax when they are used to hedge the risk of writing a European call option. The generation of the options and their underlying stock price series are described in Section 10.2. We select a set of options for this simulation on the basis of the correlation of the returns on their underlying stock with the returns on a market index. From this set we randomly select portfolios of 5 options. This simulation is intended to serve as a feasibility study on potential extensions to Basic minimax. Towards this, the simulation is used to ascertain whether CAPM minimax outperforms Basic minimax for individual options and for portfolios of options.

10.1 Generation of Simulation Data

The method used for generating stock price and option price series is the same as that discussed in Section 5. A market index series is generated with which we compute the beta of the stocks; 150 portfolios of 5 options are then constructed by randomly selecting from all options whose underlying stocks have betas within 0.5 to 1.5. We refer to a *simulation run* on a portfolio as a *replication*.

For this simulation, we use the same mechanics as in Section 5 with a few changes in parameters. The strategies involve rebalancing the hedge at uniform intervals of time. In the simulation, the interval is 1 week. Weekly data include the price of the stock, the price of the option, the risk-free interest

---

[44] Abrupt Change refers to the 1 to 3 standard deviation region of the distribution of future levels. For Basic minimax, we consider stock price levels and for CAPM minimax, we consider the market index level.

rate, the time to maturity and the volatility (sigma) of the returns on the stock, given as one of five preset levels: 0.2, 0.3, 0.4, 0.5, and 0.6. The risk-free interest rate is preset at 10%. Dividends are excluded from the analysis. In the simulation, N, the number of contracted shares, is 100. In addition, for CAPM minimax, the beta of the stocks varied between 0.5 and 1.5.

In the simulation we apply Basic minimax and CAPM minimax, to sets of 5 options. First, we apply the strategies to the options individually, and then to a portfolio of the 5 options. We also apply delta hedging to the options individually. Delta hedging is not considered for a portfolio of the options because its performance when applied to a portfolio is the summation over its performance when applied to individual options in the portfolio. Table 10.1 summarizes the stratification according to the treatment of options.

Within the minimax strategies, we use two variants that correspond to two worst-case scenarios: the 95% Level and the Abrupt Change variants. For Basic minimax, these are the same as, respectively, variants C and D in Section 4. For CAPM minimax they are as described in Section 9.5.

10.2 Summary of Simulation Results

Table 10.1 Stratification of the strategies according to the treatment of options

Name	Strategy	Applied to	Measure of overall performance
DH	Delta Hedging	Individual options	Sum of all the individual performances
Basic(I)	Basic Minimax	Individual options	Sum of all the individual performances
CAPM(I)	CAPM Minimax	Individual options	Sum of all the individual performances
Basic(P)	Basic Minimax	Portfolio of options	Performance for the portfolio
CAPM(P)	CAPM Minimax	Portfolio of options	Performance for the portfolio

In Table 10.2a, we summarize the relative performance of the strategies applied to different treatments of options when the worst-case scenario is defined by the 95% Level and by Abrupt Change. In Table 10.2b, we summarize the difference in performance between minimax strategies. The tables also contain, in parentheses, the absolute value of the t-statistics, followed by ** if the difference is significant at the 2% level, and * if it is significant at the 10% level.[45]

---

[45] We use significance levels of 2% and 10% to express the results of a 2-tailed test; however, we use a 1-tailed test for the sign of the difference at the 5% level.

Table 10.2a Relative performance of strategies (*t*-values in parentheses) (Units: percentage points of final cumulative value of cash inflow minus cash outflow divided by the notional value of the contract)

Worst-case scenario	Strategy			
	Basic(I)	Basic(P)	CAPM(I)	CAPM(P)
95% Level	1.9 (23.0)**	0.8 (8.8)**	1.7 (15.9)**	2.1 (24.1)**
Abrupt Change	2.1 (23.6)**	1.9 (21.0)**	1.8 (18.8)**	2.1 (23.9**)

Table 10.2b Difference in performance between minimax strategies (*t*-values in parentheses) (Units: percentage points of final cumulative value of cash inflow minus cash outflow divided by the notional value of the contract)

Worst-case scenario	Strategy					
	CAPM(P) Minus Basic(P)	CAPM(P) minus CAPM(I)	CAPM(P) minus Basic(I)	CAPM(I) minus Basic(I)	CAPM(I) minus Basic(P)	Basic(P) minus Basic(I)
95% Level	1.3	0.4	0.2	−0.2	0.9	−1.1
	(16.9)**	(3.6)**	(1.8)*	(2.4)**	(10.4)**	(18.6)**
Abrupt Change	0.2	0.3	0.0	−0.3	−0.1	−0.2
	(2.0)*	(2.6)**	(0.4)	(2.3)**	(0.0)	(1.8)*

From Table 10.2a, for the Abrupt Change scenario, the relative performance of the strategies is much the same. Each strategy outperforms DH by about 2 percentage points, with CAPM(P) being joined by Basic(I) as the best performers. For the 95% Level scenario, although all strategies outperform DH, there is a greater variation in relative performance, with CAPM(P) being the best performer, and with Basic(P) being the worst performer.

From Table 10.2b, the biggest differences in performance occurred under the 95% Level scenario, with the biggest positive difference occurring for CAPM(P) minus Basic (P), and the largest negative difference occurring for Basic(P) minus Basic(I).

10.3 Rank Ordering

The simulation results suggest the following rank order of positive differences in performance. Table 10.3a gives the rank order for an individual option, and

Table 10.3b gives the rank order for a portfolio. These tables are based on Tables 10.2a and 10.2b.

Table 10.3a Rank order of positive significant differences in performance for an individual option

Strategy	Worst-case scenario		
	95% Level	Abrupt Change	Average
Basic (Individual)	1	1	1
CAPM (Individual)	2	2	2
Delta Hedging	3	3	3

Table 10.3b Rank order of positive significant differences in performance for a portfolio of options

Strategy	Worst-case scenario		
	95% Level	Abrupt Change	Average
CAPM (Portfolio)	1	1.5	1.25
Basic (Individual)	2	1.5	1.75
CAPM (Individual)	3	3.5	3.25
Basic (Portfolio)	4	3.5	3.75
Delta Hedging	5	5	5.00

For individual options and for portfolios, all minimax strategies for the two worst-case scenarios outperform delta hedging. For individual options, Basic(I) outperforms CAPM(I) as expected since there is nothing to be gained by using CAPM(I) to isolate the risks for an individual stock. Indeed, imposing CAPM(I) does introduce further modeling errors which deteriorate the performance. In contrast, for portfolios, CAPM(P) outperforms Basic(I) for the 95% Level scenario; however, the difference between them is not significant at the 1% level. This difference suggests that CAPM(P) is sensitive to market index levels and may be the most suitable strategy for portfolios. Whereas CAPM(P) outperforms CAPM(I), Basic(I) outperforms Basic(P), especially under the 95% level scenario. These results suggest that CAPM(P) has succeeded in dealing with the problem of the compounding of worst-case scenarios that occurs under Basic(P). However, our results are not adequate to establish that CAPM(P) is a superior strategy to Basic(I).

11 THE BETA OF THE HEDGE PORTFOLIO FOR CAPM MINIMAX

The *beta of a stock* is a coefficient that relates market returns and stock returns. This relationship is given by (5.4). The true beta of a stock can be estimated using regression analysis. The *beta of an option*[46] is a coefficient that relates market returns and option returns. Because the value of an option is dependent on the value of the stock, the beta of the option is expressed in terms of the stock price.

$$\beta^B = \frac{\Theta(d_1)y^S\beta^S}{B} \qquad (11.1)$$

where β^B is the beta of the option and β^S is the beta of the stock. This equation shows that the returns on the option are affected by the returns on the market via the stock and the stock beta. The *beta of the hedge portfolio* is the sum of the betas of its components (stocks and options) weighted by the proportion of the total value of the hedge portfolio contributed by the individual components.

We illustrate the performance of CAPM minimax for a randomly selected replication from the simulation. Figure 11 shows the variation with time of the beta of the hedge portfolio. The graph also shows a horizontal line that is the hedge portfolio's beta-risk profile under delta hedging, and the market index level. The figure also shows that, unlike delta hedging, CAPM minimax gives a nonzero-beta hedge portfolio. This results in an increasing beta when the market is falling and in a decreasing beta when the market is rising, with the highest beta occurring at the time when the market is at its lowest (see arrow). This pattern illustrates the performance of the strategy over time. In a falling market, CAPM minimax gives a hedge recommendation in anticipation of a rise in the market index, and in a rising market, in anticipation of a fall in the market index.

12 HEDGING BOND OPTIONS

12.1 European Bond Options

The discussions in Sections 1–11 focus on the hedging of stock options. In this section, we present a similar application of minimax as applied to the hedging of bond options[47]. The discussion below gives an overview of how the formulations for stock options are adapted for bond options.

In this section we apply minimax to the problem of hedging the risk of writing a bond option. We illustrate using a call option on a discount bond but

---

[46] See Black and Scholes (1973), and Sears and Trennepohl (1982).

[47] A bond option is a type of interest rate derivative. For a comprehensive discussion on interest rate derivatives, see Hull (1997) and references therein.

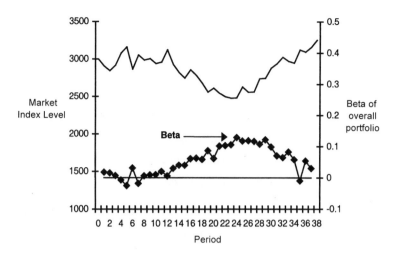

Figure 11 Beta-risk profile using CAPM minimax.

the principle remains applicable to other interest rate derivatives. As in the writing of stock options, for bond options, the writer receives a premium and incurs a potential liability. We concentrate initially on European bond options, present the formulation for this type of option, and later discuss the formulation for American options.

We use the notation given in Section 1, replacing S, the stock, with G, to represent a bond. Let y_t^G be the bond price at time t. There are, however, a few alternative concepts used. The overall notation is given by

$B_t^G = B_t^G(y_t^G)$	call price
y_t^G	bond price at time t
$y_t^{G,\text{lower}}$	the lower bound on y_t^G
$y_t^{G,\text{upper}}$	the upper bound on y_t^G
X^G	exercise price of the bond option
r	risk-free interest rate
t	current date
T	expiration date
$T - t$	time to maturity
σ	volatility

$\Theta(d)$	the cumulative normal distribution function
Δt	hedging interval
N	the contracted number of bonds
x	number of bonds to hold

The subscripts:

0	time 0, the initiation date of the contract
t	time t, any time such that $0 < t < T$
T	time T, the expiration date
i	refers to bond i or option i

At any time t, the value of the call option using Black's (1976) Model is

$$B^G = e^{-rT}[F\Theta(d_1) - X^G\Theta(d_2)] \qquad (12.1)$$

where

$$d_1 = \frac{\ln(F/X^G) + \sigma^2 T/2}{\sigma\sqrt{T}}$$

$$d_2 = d_1 - \sigma\sqrt{T}$$

where F is the bond futures[48] price at the maturity of the option. This is calculated as $F = (y_t^G - I^r)e^{rT}$ where y_t^G is the bond's spot cash price[49] and I^r is the present value of the coupons up to the maturity of the option. The exercise price X^G is also in terms of the cash price of the bond. Because of the assumptions underlying Black's model, that is, mainly the assumptions that the volatility of F is constant and that the bond price at time T has a lognormal distribution, then the hedging of European bond options is similar to delta hedging for stock options.

For a portfolio of one European bond option and a share of the bond held long, the change in portfolio value, that is, the hedging error (HE) is

$$\text{HE} = N(B_t^G - B_{t+1}^G) + x_t(y_{t+1}^G - y_t^G). \qquad (12.2)$$

The minimax hedge ratio x solves

$$\min_{x_t} \max_{y_{t+1}^G} f(x_t, y_{t+1}^G) \qquad (12.3)$$

subject to

[48] A futures contract, normally exchange-traded, is an agreement to buy or sell an asset for a specified price at a specified time in the future.

[49] A bond's cash price, also referred to as the dirty price, is the sum of the quoted, clean, price and the accrued interest.

$$y_{t+1}^{G,\text{lower}} \leq y_{t+1}^{G} \leq y_{t+1}^{G,\text{upper}}$$

Here, $f(x_t, y_{t+1}^{G})$ is given by (4.3)–(4.9) where references to stock, S, are replaced by bonds, G, and the option value is calculated using (12.1). The upper and lower bounds are defined by the 95% Level or the Abrupt Change worst-case scenario, given in Section 4.2.

12.2 American Bond Options

For American options, the option can be exercised at any time prior to the maturity of the option. A tree structure, say a Binomial Tree or a Trinomial Tree, can be used to help simulate yield curve scenarios, or the equivalent scenarios of the term structure of interest rates, that vary through time. Any node of a tree represents a future yield curve that can be used for pricing the bond at that future point in time. The option, at the same node, can then be categorized as in-the-money, out-of-the-money, or at-the-money.

In-the-money options are analyzed for their optimality of exercise. An American option has two potential values at any node: its expected value calculated as the probability-weighted sum of its future values, or the payoff from exercise of the option calculated as the difference between the exercise price and the spot price at that node. If the expected value is greater than the payoff, then it is not optimal to exercise, and the option value at that node equals the expected value, and vice versa.

The consistency of a tree is important in terms of creating a plausible evolution of interest rates. The analysis of American-style interest rate derivatives requires the use of an interest-rate model (or a yield curve model) that specifies the evolutionary nature of interest rates. For discussion, we present a model of the term structure of interest rates, the Hull and White (1990) model[50], that defines the behavior of the short rate r[51]. This incorporates mean reversion to an average level at a rate a. In the Hull and White (1990) model, the price $P(t, T)$ of a discount bond, maturing at T, at a future time t is given by

$$P(t, T) = A(t, T)e^{-\beta(t,T)r(t)} \tag{12.4}$$

where

$$\beta(t, T) = \frac{1 - e^{-a(T-t)}}{a}$$

---

[50] See Hull (1997). Other yield curve models can be used such as the models developed by Ho and Lee (1986), Vasicek (1977), Cox et al. (1985).

[51] In this type of model, the short rate is the instantaneous interest rate which is approximated by the overnight rate.

$$\ln A(t, T) = \ln \frac{P(0, T)}{P(0, t)} - \beta(t, T) \frac{\partial \ln P(0, t)}{\partial t} - \frac{1}{4a^3} \sigma^2 (e^{-aT} - e^{-at})^2 (e^{2at} - 1).$$

The Hull and White (1990) model is used for generating a tree whose nodes represent a future term structure of interest rates. We do not use (12.4) directly in the formulation below but mention it as a basis for scenario generation. The bond pricing given by (12.4) applies to discount bonds that are used as building blocks for propagating the instantaneous interest rate. For details on term structure tree construction, see Hull (1997). Once the tree has been generated, then the information on each node can be used for modeling American options.

Figure 12.1 illustrates a trinomial tree where each node corresponds to one of three possibilities at any point in time. At a node, a particular yield curve characterizes the bond market, and corresponding bond and option values. Similarly, at any node, the short rate r read off from the yield curve is used for discounting future values of the option, from some future nodes, back to that node. The whole yield curve is used to value the bond at that node. If the bond is coupon-bearing, then it is valued based on the yield curve at that node, with all coupon payments plus principal payments considered. The figure illustrates a one-period trinomial tree; for the pricing of options, the life span of the option is subdivided into several periods, perhaps 30 periods, in order to refine the calculations that can be done using a tree structure[52].

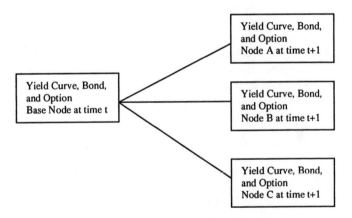

Figure 12.1   A one-period trinomial tree starting at time t.

---

[52] A step-by-step procedure for using tree structures for the pricing of American options is described in Hull (1997).

The price of a bond at any node is determined by the yield curve at that node. The option is then exercised, or not, depending on whether the payoff on exercise is higher than the expected value of the option. This means that, at Node A in Figure 12.1, the value of the option is the higher of two calculations: (1) the expected value of the option based on the propagation of the tree past Node A, and (2) the difference between the bond price and the exercise price at Node A, that is, the payoff.

Creating a robust hedge using the minimax framework for an American-style option involves a similar framework for (12.3) adjusted for computational efficiency. In (12.3), the future bond price is allowed to vary within predefined bounds that represent a continuum of future bond prices. Each point in this continuum represents a bond price that, in turn, implies an underlying yield curve. One can associate with each point in this continuum a Trinomial Tree with its first node sitting at time $t + 1$. Due to the computational complexity implied by the continuous framework, the discretization of the maximizing variable, that is, the future bond price, permits a more feasible formulation. Consider therefore the discrete minimax problem:

$$\min_{x_t} \max_{j \in [1, \ldots, m^{sce}]} f(x_t)^j \tag{12.5}$$

where j is an element of the discrete set of m^{sce} scenarios characterizing the possible yield curves that may occur at the horizon. This is a one-stage discrete minimax strategy, where the bounds constraint on the future values of y_{t+1}^G in the continuous minimax formulation have been replaced by a finite number of scenarios. The objective function $f(x_t)^j$, parameterized by the scenario j, is the potential hedging error if the yield curve were to move from its current state to the yield curve defined under scenario j. The scenarios in this formulation are clearly individual yield curve scenarios. Within the context of the worst-case scenarios defined in Section 4.2, one possible discretization of (12.3) would be scenarios corresponding to the three nodes at time $t + 1$, as shown in Figure 12.1.

The objective function $f(x_t)^j$ is similar to (4.3)–(4.9), where the future bond price y_{t+1}^G is replaced by its discrete scenario counterpart $(y_{t+1}^G)^j$, and the future option price $B_{t+1}^G(y_{t+1}^G)$ is replaced by the scenario $B_{t+1}^G((y_{t+1}^G)^j)$.

We define $U_1^j : \mathfrak{R}^k \times \mathfrak{R}^k \to \mathfrak{R}^1$, $U_2^j : \mathfrak{R}^k \to \mathfrak{R}^k$, $U^j : \mathfrak{R}^k \times \mathfrak{R}^k \to \mathfrak{R}^{k+1}$, $x_t \in \mathfrak{R}^k$, $y_{t+1}^G \in \mathfrak{R}^k$ and Q is a $(k + 1) \times (k + 1)$ positive definite weighting matrix. $U^d : \mathfrak{R}^k \times \mathfrak{R}^k \to \mathfrak{R}^{k+1}$ is the vector of desired values for the potential hedging error and the transaction cost terms: we use a desired value of zero, that is, the desired hedging error is zero[53] and the desired transaction cost is zero.

---

[53] As in the previous sections, we use a desired value of zero because in delta hedging, the expected value of the hedging error is zero.

The objective function is given by

$$f(x_t)^j = \frac{1}{2}\langle U^j - U^d, Q(U^j - U^d)\rangle \tag{12.6}$$

where

$$x_t = \begin{bmatrix} x_{1,t} \\ \vdots \\ x_{k,t} \end{bmatrix} \quad \text{and} \quad (y_{t+1}^G)^j = \begin{bmatrix} (y_{1,t+1}^G)^j \\ \vdots \\ (y_{k,t+1}^G)^j \end{bmatrix} \tag{12.7}$$

$$U^j(x_t, y_{t+1}^G) = \begin{bmatrix} U_1^j(x_t, (y_{t+1}^G)^j) \\ \cdots\cdots\cdots\cdots\cdots \\ U_2(x_t) \end{bmatrix} \quad \text{and} \quad U^d = \begin{bmatrix} U_1^d \\ \cdots \\ U_2^d \end{bmatrix} = \begin{bmatrix} 0 \\ \cdots \\ 0 \end{bmatrix} \tag{12.8}$$

$$U_1^j\left(x_t, (y_{t+1}^G)^j\right) = \sum_{i=1}^{k} x_{i,t}((y_{i,t+1}^G)^j - y_{i,t}^G) + \sum_{i=1}^{k} N_i\left(B_{i,t}^G - B_{i,t+1}^G((y_{i,t+1}^G)^j)\right)$$

$$+ \sum_{i=1}^{k}\left(-(x_{i,t} - x_{i,t-1})y_{i,t}^G + C_{i,t-1}(1 + r\Delta t)\right)r\Delta t \tag{12.9}$$

where

$$C_{i,t-1} = C_{i,t-2}(1 + r\Delta t) - (x_{i,t-1} - x_{i,t-2})y_{i,t-1}^G - \hat{K}\left|(x_{i,t-1} - x_{i,t-2})y_{i,t-1}^G\right|. \tag{12.10}$$

$$U_2(n_t) = \begin{bmatrix} U_{i,2}(x_{1,t}) \\ \vdots \\ U_{k,2}(x_{k,t}) \end{bmatrix} \tag{12.11}$$

where

$$U_{i,2}(x_{i,t}) = \hat{K}(x_{i,t} - x_{i,t-1})y_{i,t}^G. \tag{12.12}$$

The interpretation of these equations is similar to that in Section 4. In (12.9), $B_{i,t+1}^{G,j}((y_{i,t+1}^G)^j)$ is the value of the option that corresponds to the bond price $(y_{i,t+1}^G)^j$ and to the yield curve scenario j. This calculation is associated with a particular trinomial tree whose starting node holds the yield curve scenario j at time $t + 1$. The tree is used for valuing an American option with a life span that starts from $t + 1$ and ends at time T. Figure 12.2 is a representation of the discrete minimax formulation, illustrating the relationship between the yield curve scenarios and the trees that emanate from nodes at time $t + 1$.

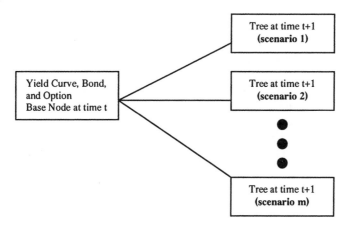

Figure 12.2   A multiscenario diagram where each scenario is represented by a trinomial tree.

The minimax formulation presented here applies to general types of interest rate derivative whose value is determined by the level of interest rates from a segment or from the whole term structure of interest rates. The above formulation is a single-period minimax optimization. Scenarios are defined at the end of the period, and the expectations framework is used for the valuation of the options.

13 CONCLUDING REMARKS

We have developed a dynamic hedging strategy that minimizes the effect of the worst-case scenario. It requires a range for the future stock price to be specified. Minimax differs from delta hedging in that it allows the hedger to incorporate information or her beliefs about the future level of prices. The results of the simulation study suggest that minimax is robust in the sense that it performs better than delta hedging for the set of options for which it is explicitly designed, and in general, it does not perform worse than delta hedging. The simulation results also suggest that three variants of minimax are suitable for hedging the risk of writing an option when the price of the underlying stock is both highly volatile and crosses over the exercise price frequently (at-the-money options). This problem is of particular interest to market makers, investors, as well as speculators. The results of the limited empirical study are mainly consistent with the simulation study. The results also show that the minimax strategy is robust and that the worst case is not always characterized by the upper or lower bound of the stock price, as initial intuition might suggest.

We have also presented two multiperiod extensions of the basic minimax hedging strategy that the hedger can use under specific situations. Two-Period minimax is designed for the hedger who wishes to consider the possibility that she may fail to rebalance at the end of a preset, one-period, interval. Variable minimax is designed for the hedger who wishes to actively avoid negative hedging errors. The results of the simulation suggest the following rank ordering of the strategies: Variable minimax, Basic minimax, Two-Period minimax, delta hedging.

Variable minimax is a more aggressive strategy in the sense that the hedger regularly monitors the development of the actual hedging error and rebalances early in response to an undesirable event. The strength of Variable minimax is that it gives the hedger a criterion that provides the opportunity to rebalance early in order to limit accumulated hedging errors to some acceptable level. However, to the extent that more frequent rebalancing increases transaction costs, any benefits from early rebalancing may be offset by such an increase, even though the Variable minimax constrains these costs.

Two-Period minimax is a more cautious strategy and, as such, it is less suitable than Basic minimax when the hedger can rebalance at the end of the first period. The strength of the Two-Period minimax is that it provides the hedger with a buffer that can, to some extent, absorb the negative effects of an undesirable event that may occur during the second period, if she fails to rebalance at the end of the first period.

The two multiperiod extensions have been shown to perform well in the circumstances for which they have been designed. The results of the simulation are such that further advances would be expected through studies of Variable minimax under different degrees of moneyness, and studies of Two-Period minimax under changing levels of volatility. However, we do not consider that Variable minimax could usefully be extended to an n-period setting.

We have presented an extension of the basic minimax hedging strategy that uses information about the movements in the market index in the calculation of the potential hedging error; we have also simulated its performance. The results of the simulation suggest that CAPM(P) is sensitive to movements in the market index, and has succeeded in dealing with a major difficulty in the Basic(P) strategy (described in Section 9), a difficulty that was most conspicuous under the 95% level worst-case scenario.

The variants of CAPM minimax have been shown to perform well in the circumstances for which they have been designed. The results of the simulation suggest that CAPM minimax may be suitable for other hedging problems where the securities are highly sensitive to market movements.

Finally, we presented an outline of the application of minimax to bond options, both European and American, and showed how a tree-based model for pricing American options can be used in the context of a scenario-based minimax model.

References

Black, F. (1976). "The Pricing of Commodity Contracts", *Journal of Financial Economics*, 3, 167–179.

Black, F. and M. Scholes (1973). "The Pricing of Options and Corporate Liabilities", *Journal of Political Economy*, 81, 637–659.

Boyle, P. and D. Emanuel (1980). "Discretely Adjusted Option Hedges", *Journal of Financial Economics*, 8, 259–282.

Cox, J., J.E. Ingersoll and S. Ross (1985). "A Theory of the Term Structure of Interest Rates", *Econometrica*, 53, 385–407.

Cox, J. and S. Ross (1976). "The Valuation of Options for Alternative Stochastic Processes" *Journal of Financial Economics*, 3, 145-166.

Cox, J. and M. Rubinstein (1984). *Option Markets*, Prentice Hall, Englewood Cliffs, NJ.

Davis, M.H.A. and A.R. Norman (1990). "Portfolio Selection with Transaction Costs", *Mathematics of Operational Research*, 15, 676–713.

Elton, E.J. and M.J. Gruber (1991). *Modern Portfolio Theory and Investment Analysis*, 4th edition, John Wiley & Sons, New York.

Gilster, J. and W. Lee (1984). "The Effects of Transaction Costs and Different Borrowing and Lending Rates on the Option Pricing Model: A Note", *Journal of Finance*, 39, 1215–1222.

Ho, T.S.Y. and S.B. Lee (1986). "Term Structure Movements and Pricing Interest Rate Contingent Claims", *Journal of Finance*, 41, 1011–1029.

Hodges, S.D. and A. Neuberger (1989). "Optimal Replication of Contingent Claims under Transaction Costs", *Review of Futures Markets*, 8, 222–239.

Howe, M. (1994). "A Quasi-Newton Algorithm for Continuous Minimax with Applications to Risk Management in Finance", PhD Dissertation, Imperial College London.

Howe, M., B. Rustem and M.J.P. Selby (1994). "Minimax Hedging Strategy", *Computational Economics*, 7, 245–275.

Howe, M., B. Rustem and M.J.P. Selby (1996). "Multi-Period Minimax Hedging Strategies", *European Journal of Operations Research*, 93, 185–204.

Howe, M., B. Rustem and M.J.P. Selby (1998). "CAPM-Based Minimax Hedging Strategy", Working Paper, Department of Computing, Imperial College, London.

Huang, C. and R.H. Litzenberger (1988). *Foundations for Financial Economics*, North-Holland, New York.

Hull, J. (1997) *Options, Futures and Other Derivatives*, 3rd edition, Prentice Hall, Englewood Cliffs, NJ.

Hull, J. and A. White (1990). "Pricing Interest Rate Derivative Securities", *Review of Financial Studies*, 3–4, 573–592.

Leland, H. (1985) "Option Pricing and Replication with Transaction Costs", *Journal of Finance*, 5, 1283–1301.

Merton, R. (1973). "Theory of Rational Option Pricing", *Bell Journal of Economics and Management Science*, 4, 141–183.

Neuhaus, H. (1990). "Option Valuation and Hedging under Transaction Costs", PhD Dissertation, London Business School.

Panas, V.G. (1993). "Option Pricing with Transaction Costs", PhD Dissertation, Imperial College, London.

Rosenberg, J.M. (1993). *Dictionary of Investing*, John Wiley & Sons, New York.

Samuelson, P.A. (1965). "Rational Theory of Warrant Pricing", *Industrial Management Review*, 6, 13–31.

Sears, R.S. and G.L. Trennepohl (1982). "Measuring Portfolio Risk in Options", *Journal of Financial and Quantitative Analysis*, 17, 391–409.

Vasicek, O. (1977). "An Equilibrium Characterisation of the Term Structure", *Journal of Financial Economics*, 5, 177–188.

APPENDIX A: WEIGHTING HEDGE RECOMMENDATIONS, VARIANT B*

Changes in stock price often contain an element of noise; the hedger may wish to react only to the signal element in the change. The probability that the hedger will consider that a given Δy_t^S has a large noise element is higher, for high values of the standard deviation of y_t^S. Some crossovers leave y_t^S close to X. Others leave y_t^S far from X. The probability of another crossover occurring is higher when y_t^S is close to X. The recommendation on n_t, the number of shares to hold, has a higher probability of being reversed, when y_t^S is close to X. The hedger is less likely to accept the transaction costs in realizing the recommended x_t when she considers that another crossover is likely to occur. In this case, there may be a possibly large and countervailing change in recommended x_t with corresponding large transaction costs. We give below an expression that the hedger can use to weight the number of shares recommended by a strategy to reflect her perception of the amount of noise in Δy_t^S and the potential reversibility of the crossover. The hedger can give any weights to a and b to reflect the importance she attaches to the standard deviation, denoted by sd, of y_t^S. The general expression is

$$k1 = \frac{\left[w1 \dfrac{|\Delta y_t^S|}{a^* sd} + w2 \dfrac{|y_t^S - X|}{b^* sd} \right]}{(w1 + w2)} \tag{A1}$$

She may give a nonzero value to $w2$ when there is a crossover from $t - 1$ to t; there are no restrictions on the value she gives to $w2$.

In Section 4.5, we apply the expression to minimax variant B, to give a weighted version of that variant, B*.

APPENDIX B: NUMERICAL EXAMPLES

We present illustrations of the performance of minimax from an algorithmic point of view. The quasi-Newton algorithm and Kiwiel's algorithm have been implemented as discussed in Chapter 5. Because both algorithms require a maximization problem to be solved in Step 1, they are both relatively computationally more expensive compared to most nonlinear programming algorithms. For the maximization subproblem, we used the comprehensive NAG[54] optimization routine E04VDF which can handle both linear and nonlinear constraints of a nonlinear programming problem. All the illustrations were solved by using both algorithms. Kiwiel's algorithm has been implemented as a check to the solutions found by the quasi-Newton algorithm. In the implementation of Kiwiel's algorithm, we set the linear approximation parameter m to a constant for all examples: $m = 2.0e^{-03}$.

Output Variables

x_0	initial value of x
y_0	initial value of y
$\Phi(x^*)$	objective function value at the solution
x^*	final value of x
y^*	final value of y
$k_\alpha \vert \alpha_k = 1, \ \forall k \geq k_\alpha$	iteration number where the stepsize equals 1 for all succeeding iterations
No. of iterations	total number of iterations the algorithm took to solve the problem
Time	total computer time using a Sun SparcStation ELC running at 7 million instructions per second

The Stopping Criterion

The stopping criterion for both algorithms is the condition that the approximate directional derivative is sufficiently close to zero, that is,

$$\text{If } \Psi \geq -\epsilon, \text{ then terminate.} \tag{A2}$$

The values of the stopping parameter ϵ used in the examples are within the range $[1.0e^{-6}, 1.0e^{-14}]$. The values reported here are for the quasi-Newton and Kiwiel's algorithms, as in the chapter containing numerical results.

[54] Numerical Algorithms Group, Oxford, UK.

In the illustrations, we use data from the empirical study in Section 4. The rebalancing date is December 4, 1991. The expiration date for all options is August 1992. Their exercise prices, as well as other parameters, are given in the simulation sections. The first three illustrations show the performance of Basic minimax. Illustration 1 shows the more likely situation when the minimax algorithm is applied to the hedging strategy. When the minimax hedging error is negative, which is the more likely event, the solution is an extreme point. Illustration 2 shows a case where the minimax hedging error is positive. In this case, the solution is strictly within the upper and lower bounds (i.e., a *mid-range* solution). Illustration 3 shows the performance of the algorithms when applied to a portfolio of 5 options. The next two illustrations show the performance of Two-Period minimax. Illustration 4 also shows a mid-range solution. Illustration 5 shows the performance for a portfolio of three options. Finally, Illustration 6 shows the performance for a portfolio of 5 options under CAPM Minimax.

We give the data[55] used for the examples in this chapter. All prices are in pence.

Stock	Stock price	Option price	Exercise price	Option's maturity
British Telecom	347	39.5	330	August 1992
Prudential Corp	226	16.5	240	August 1992
Guinness	499	46.0	500	August 1992
Thames Water	336	31.5	330	August 1992
Tesco	237	33.0	220	August 1992
Cadbury Schweppes	457	63.0	420	August 1992

The Concavity of the Maximization Subproblem

If $f(\cdot, y)$ is convex with respect to y, then the solution to minimax lies at the upper or lower bound of y. There may be multiple maxima, all of which are at the boundary. If, on the other hand, $f(\cdot, y)$ is concave with respect to y, the solution may lie anywhere within the feasible region for y.

The potential of finding a nonextreme point solution, that is, a solution not at the boundary, to the minimax formulation is exemplified by Illustrations 2 and 5 below. In this case, the minimax hedging error is positive thereby contributing to the concavity of the maximization subproblem. To see this, consider the second derivative of (4.6) when applied to a single-option minimax problem:

---

[55] Data were obtained from a financial information provider, DataStream International.

$$\frac{\partial^2 U_1}{\partial (y_{t+1}^S)^2}$$

$$= \left(x_t - \frac{\partial B_{t+1}(y_{t+1}^S)}{\partial y_{t+1}^S} \right)^2 + \left(2 \left(\begin{array}{c} x_t \left(y_{t+1}^S - y_t^S \right) + N \left(B_t - B_{t+1}(y_{t+1}^S) \right) \\ + \left(-(x_t - x_{t-1}) y_t^S + C_{t-1}(1 + r\Delta t) r\Delta t \right) \end{array} \right) \right)$$

$$\times \left(-\frac{\partial^2 B_{t+1}(y_{t+1}^S)}{\partial (y_{t+1}^S)^2} \right)$$

The concavity of the problem is determined by the second term: because the problem refers to the writing of a call option, the second derivative of the option price is negative. When this is premultiplied by a positive hedging error, the second term's potential to dominate the first term may shift the problem towards a concave maximization subproblem. Examples 2, 4 and 5 below are illustrations of this case.

Example 1 (Extreme Point Solution) Hedging the Risk of Writing an Option on British Telecom Stock

Minimize $f(x_t, y_{t+1}^S)$ subject to

$$3.289920e+02 \le y_{t+1}^S \le 3.686350e+02$$

Initial point:

$$x_t = (6.185360e+01), y_{t+1}^S = (3.470000e+02)$$

	Quasi-Newton algorithm	Kiwiel's algorithm
$\Phi(x_t^*)$	1.508524e+04	1.508910e+04
x^*	6.303667e+01	6.299798e+01
y_{t+1}^{S*}	3.289920e+02	3.289920e+02
$k_\alpha \| \alpha_k = 1; \forall k > k_\alpha$	1	1
No. of iterations (time)	5 (0.3 s)	1036 (39.8 s)
Minimax hedging error	−168 points	−168 points

Example 2 (Mid-range Solution) Hedging the Risk of Writing an Option on the Cadbury Stock

Minimize $f(x_t, y_{t+1}^S)$ subject to

$$4.354752e+02 \leq y_{t+1}^S \leq 4.831153e+02$$

Initial point:

$$x_t = (6.554278e+01), \quad y_{t+1}^S = (4.570000e+02)$$

	Quasi-Newton algorithm	Kiwiel's algorithm
$\Phi(x_t^*)$	1.053631e+05	1.057401e+05
x^*	6.763951e+01	6.755767e+01
y_{t+1}^{S*}	4.602877e+02	4.601593e+02
$k_\alpha \vert \alpha_k = 1;\ \forall k > k_\alpha$	2	1
No. of iterations (time)	3 (0.1 s)	511 (72.1 s)
Minimax hedging error	459 points	459 points

Example 3 (Extreme Point Solution) Hedging the Risk of Writing Five Options: Guinness, Prudential, British Telecom, Thames Water and Tesco

Minimize $f(x_t, y_{t+1}^S)$ subject to

$$4.731038e+02 \leq y_{1,t+1}^S \leq 5.301120e+02$$

$$2.142715e+02 \leq y_{2,t+1}^S \leq 2.400908e+02$$

$$3.289920e+02 \leq y_{3,t+1}^S \leq 3.686350e+02$$

$$3.185629e+02 \leq y_{4,t+1}^S \leq 3.569492e+02$$

$$2.247006e+02 \leq y_{5,t+1}^S \leq 2.517767e+02$$

Initial point:

$$(x_t)^T = (4.8076e+01, 7.4848e+01, 6.1853e+01, 7.3183e+01, 8.2752e+01)$$

$$(y_{t+1}^S)^T = (4.9900e+02, 2.2600e+02, 3.4700e+02, 3.3600e+02, 2.3700e+02)$$

	Quasi-Newton algorithm	Kiwiel's algorithm
$\Phi(x_t^*)$	4.747561e+05	4.910554e+05
x^*	4.723223e+01	4.725500e+01
	7.593698e+01	7.536454e+01
	6.266327e+01	6.266036e+01
	7.369503e+01	7.369543e+01
	7.657575e+01	8.004109e+01
y_{t+1}^{S*}	5.301120e+02	5.301120e+02
	2.142715e+02	2.142715e+02
	3.289920e+02	3.686350e+02
	3.569492e+02	3.569492e+02
	2.517767e+02	2.517767e+02
$k_\alpha \vert \alpha_k = 1; \ \forall k > k_\alpha$	$\alpha = 1$ not attained	$\alpha = 1$ not attained
No. of iterations (time)	36 (4.3 s)	749 (115.6 s)
Minimax hedging error	-876 points	-818 points

Example 4 (Mid-range Solution) Hedging the Risk of Writing an Pption on the Guinness Stock

Minimize $f(x_t, y_{t+1}^S)$ subject to

$$4.731038e+02 \leq y_{t+1}^S \leq 5.301120e + 02$$

$$4.637506e+02 \leq y_{t+2}^S \leq 5.447065e+02$$

Initial point:

$$(x_t)^{\mathrm{T}} = (4.807649e+01), \quad (y_{t+1}^S, y_{t+2}^S)^{\mathrm{T}} = (4.990000e+02, 4.990000e+02)$$

	Quasi-Newton algorithm	Kiwiel's algorithm
$\Phi(x_t^*)$	1.800928e+05	1.800939e+05
x^*	4.723716e+01	4.726027e+01
$(y_{t+1}^{S*}, y_{t+2}^{S*})^T$	4.971921e+02	4.972337e+02
	5.447065e+02	5.447065e+02
$k_\alpha \vert \alpha_k = 1; \ \forall k > k_\alpha$	1	1
No. of iterations (time)	3 (0.2 s)	607 (43.4 s)
Minimax hedging error	-585 points	-585 points

Example 5 (Mid-range Solution) Hedging the Risk of Three Options: Guinness, Prudential and British Telecom

Minimize $f(x_t, y_{t+1}^S)$ subject to

$$4.731038e+02 \leq y_{1,t+1}^S \leq 5.301120e+02$$

$$2.142715e+02 \leq y_{2,t+1}^S \leq 2.400908e+02$$

$$3.289920e+02 \leq y_{3,t+1}^S \leq 3.686350e+02$$

$$4.637506e+02 \leq y_{1,t+2}^S \leq 5.447065e+02$$

$$2.100345e+02 \leq y_{2,t+2}^S \leq 2.467007e+02$$

$$3.224879e+02 \leq y_{3,t+2}^S \leq 3.787839e+02$$

Initial point:

$$(x_t)^{\mathrm{T}} = (4.807640e+01, 7.484878e+01, 6.185360e+01)$$

$$(y_{t+1}^S, y_{t+2}^S)^{\mathrm{T}} = (4.990000e+02, 2.260000e+02, 3.470000e+02,$$

$$4.990000e+02, 2.260000e+02, 3.470000e+02)$$

	Quasi-Newton algorithm	Kiwiel's algorithm
$\Phi(x_t^*)$	5.280679e+05	5.286092e+05
x^*	4.867006e+01	4.842086e+01
	7.576715e+01	7.576387e+01
	6.343740e+01	6.343626e+01
$(y_{t+1}^{S*}, y_{t+2}^{S*})^T$	5.002876e+02	4.997487e+02
	2.271383e+02	2.271341e+02
	3.495103e+02	3.495293e+02
	5.447065e+02	5.447065e+02
	2.100354e+02	2.100354e+02
	3.224879e+02	3.224879e+02
$k_\alpha \vert \alpha_k = 1; \ \forall k > k_\alpha$	8	$\alpha = 1$ not attained
No. of iterations (time)	9 (5.5 s)	112 (103.2 s)
Minimax hedging error	-1031 points	-1042 points

Example 6 (Extreme Point Solution) Hedging the Risk of Writing Five Options: Guinness, Prudential, British Telecom, Thames Water and Tesco

Minimize $f(x_t, y_{t+1}^E)$ subject to

$$2.972739e+03 \leq I_{t+1} \leq 3.087715e+03$$

$$-2.360554e-02 \leq \epsilon_{1,t+1} \leq 2.360554e-02$$

$$-4.770382e-02 \leq \epsilon_{2,t+1} \leq 4.770382e-02$$

$$-3.302775e-02 \leq \epsilon_{3,t+1} \leq 3.302775e-02$$

$$-3.302775e-02 \leq \epsilon_{4,t+1} \leq 3.302775e-02$$

$$-4.475385e-02 \leq \epsilon_{5,t+1} \leq 4.475385e-02$$

Initial point:

$$(x_t)^{\mathrm{T}} = (3.8939e+01, 6.7017e+01, 3.7248e+01, 3.0025e+01, 3.5765e+01)$$

$$(y_{t+1}^E)^{\mathrm{T}} = (2.9150e+03, 0.0, 0.0, 0.0, 0.0, 0.0)$$

	Quasi-Newton Algorithm	Kiwiel's Algorithm
$\Phi(x_t^*)$	2.211716e+06	2.211741z+06
x^*	5.433818e+01	5.432885e+01
	7.048898e+01	7.046786e+01
	4.677316e+01	4.678775e+01
	4.887535e+01	4.883134e+01
	4.885359e+01	4.887019e+01
y_{t+1}^{E*}	3.036498e+02	3.035329e+02
	-2.360554e-02	-2.360554e-02
	-1.189297e-03	-9.524208e-04
	-1.854275e-02	-1.825807e-02
	3.302775e-02	3.302775e-02
	3.462539e-02	3.500561e-02
$k_\alpha\|\alpha_k = 1; \ \forall k > k_\alpha$	1	1
No. of iterations (time)	53 (32.6 s)	268 (122.4 s)
Minimax hedging error	1629 points	1626 points

COMMENTS AND NOTES

CN 1: Brownian Motion

The term Brownian Motion has been used to describe the motion of a particle that is subject to a large number of small molecular shocks. In financial calculus, Brownian Motion is used interchangeably with the term Wiener Process which is a particular type of Markov stochastic process. Markov stochastic processes are processes where only the present value of a variable is relevant for predicting the future. Models of stock price behavior are usually expressed as a Wiener Process or Brownian Motion (see Hull, 1997).

CN 2: Ito's Lemma

The price of a stock option is a function of the underlying stock's price and time. More generally, we can say that the price of any derivative is a function of the stochastic variables underlying the derivative and time. Ito's lemma states that if a variable x follows an Ito Process, that is, that

$$dx = a(x,t)\ dt + b(x,t)\ dz$$

where dz is a Wiener Process and a and b are functions of x and t, then another function G of x and t follows the process

$$dG = \left(\frac{\partial G}{\partial x} a + \frac{\partial G}{\partial t} + \frac{1}{2} \frac{\partial^2 G}{\partial x^2} b^2 \right) dt + \frac{\partial G}{\partial x} b\ dz$$

where dz is the same Wiener Process. Thus, G also follows a Wiener Process.

CN 3: Cumulative Normal Distribution Function

To solve the Black and Scholes formula, one needs to calculate the cumulative normal distribution function, Θ. The function can be evaluated directly using numerical procedures. Alternatively, a polynomial approximation can be used that provides values for $\Theta(d)$ with a six-decimal-place accuracy. The following have been extracted from Hull (1997):

$$\Theta(d) = \begin{cases} 1 - \Theta'(d)(a_1 k + a_2 k^2 + a_3 k^3 + a_4 k^4 + a_5 k^5) & \text{when } x \geq 0 \\ 1 - \Theta(-d) & \text{when } x < 0 \end{cases}$$

where

$$k = \frac{1}{1 + \xi d}$$

$$\xi = 0.2316419$$

$$a_1 = 0.319381530$$

$$a_2 = -0.356563782$$

$$a_3 = 1.781477937$$

$$a_4 = -1.821255978$$

$$a_5 = 1.330274429$$

and

$$\Theta'(d) = \frac{1}{\sqrt{2\pi}} e^{-d^2/2}.$$

Chapter 9

Minimax and asset allocation problems

In this chapter, we consider potential uses of minimax in the context of portfolio asset allocation, with specific illustrations for bond portfolios. We demonstrate that the issue of mis-forecasting can be appropriately addressed within the minimax framework. An asset allocation based on minimax has robustness properties that cushion the performance of the portfolio against the occurrence of predefined worst-case scenarios. There is a guaranteed performance which improves when the worst-case scenario fails to materialize. A number of minimax asset allocation techniques are discussed, all applicable to stocks, bonds or currencies. These are considered in the context of mean-variance optimization and benchmark tracking. Additionally, a minimax formulation for a multistage asset allocation problem is presented. Lastly, the complementary use of both minimax and options for portfolio management is explored.

1 INTRODUCTION

In this chapter, we address the issue of asset allocation and present allocation strategies based on minimax that enable the investor to better assess the potential performance of a chosen portfolio. We initially consider two standard investment tools: the mean-variance and the benchmark-tracking approaches to portfolio selection. The extensions of these strategies to a minimax framework are explored with illustrations on how an investor can benefit from these extensions. Minimax index tracking is considered, showing how standard tracking techniques can be adapted to form robust trackers. A multistage framework for minimax portfolio selection is presented for portfolio rebalancing. Finally, the simultaneous use of minimax and options is studied in a complementary manner that allows a fund manager to benefit from lower insurance premia.

Starting with the work of Markowitz (1952) on portfolio selection, considerable work has been devoted to the issue of asset allocation. Markowitz proposed the idea that a sensible investing strategy does not look solely at return-maximization, but also at the interplay between return-maximization and risk-minimization, and the trade-off between these two. Hence, a balance can be found between return-maximization and risk-minimization that may be more consistent with the utility function of an investor. A follow-up work by Levy and Markowitz (1979) looks at approximating an expected utility function by a function of mean and variance. Sharpe (1994) discusses the use of his reward-to-variability ratio, now called the Sharpe ratio, in guiding the investment process. Thus, by looking at the expected return per unit of risk, an investor has a more informed measure of the trade-off between risk and return.

The introduction of constraints to the asset allocation process leads to some interesting variations on the mean-variance theme. These works include Leibowitz and Kogelman (1991b) where the focus is on the balance between risky and risk-free assets. A simple model of quantifying risk tolerance is proposed and then used to determine the maximum investment in the risky assets. Downside risk (CN 2) is measured by the shortfall probability relative to a minimum return threshold. By specifying both this threshold and a shortfall probability, a shortfall constraint is established to determine the maximum allocation to risky assets. Leibowitz and Kogelman (1991a) also consider the issue of diversification and how the introduction of a foreign asset could improve the risk performance of a portfolio, again, in the context of a shortfall constraint. The application of constraints was also studied by Don Ezra (1991), in the context of surplus optimization where provisions are made for any liabilities that the portfolio might be supporting. By looking at the surplus, or the net between the value of the portfolio's assets and those of the liabilities, the resulting asset allocation is argued to be more robust in terms of meeting the portfolio's liability obligations when compared to an allocation done on an asset basis only.

Research into the use of scenarios in asset allocation, as compared to the covariance matrix framework of Markowitz's mean-variance analysis, have been carried out by Dembo (1991), Clarke and Silva (1998), and Rustem et al. (2000), where model mis-specification is raised as a critical issue. Similarly, optimal dynamic portfolio decision in a continuous time framework in view of parameter uncertainty, mainly in the area of model mis-specification, has been explored by Maenhout (1999).

Benchmarking, or referencing relative to an index or another portfolio, has been a common method of creating a portfolio. This is due to the apparent ease of assessing the performance of the portfolio relative to the chosen benchmark. The assumption is that the performance of the benchmark is a suitable reference point for policy decision making and that its performance is easy to assess. Roll (1992) presents a formulation of benchmark-tracking in the mean-variance context and discusses the performance of benchmark-tracking efficient portfolios relative to their mean-variance efficient counterparts. Worzel et al. (1994) explores the creation of fixed income portfolios that track their chosen index. Optimization models are presented that penalize the downside deviations of the portfolio returns from the index. Lipman (1990) looks at the issue of maximizing utility and the incorporation of benchmarks in the utility-maximization process.

Both the absolute performance, in terms of returns and the volatility of returns, and the relative performance, that is, relative to a benchmark, of a portfolio are standard measures used by the investment community. To the extent that more robust methods of asset allocation complement existing decision tools, we explore the use of minimax as a computational strategy for enhancing current methods. In Section 2 we present robust asset allocation models based on minimax. In Section 3 we discuss the performance of a minimax portfolio selection process. In Section 4 we consider an enhanced version of benchmarking where we look at two benchmarks, that is, dual benchmarking. In Section 5 we discuss the applicability of the minimax approach to address other asset allocation objectives such as index tracking and downside risk optimization. In Section 6 we present a multistage framework for asset allocation that describes how the various minimax strategies, previously discussed for a single stage problem, can be cast to address stochastic multistage problems. In Section 7, we consider an integrated approach to options and worst-case analysis.

2 MODELS FOR ASSET ALLOCATION BASED ON MINIMAX

Rustem et al. (2000) explore various models of discrete minimax for asset allocation. The issue of inaccuracy in asset return forecasting and risk estimation is addressed. Minimax formulations are proposed for robust solutions in view of return and risk inaccuracies. Rival return forecasts as well as rival risk estimates are considered and these are cast as input scenarios in the minimax formulation. It is argued that, by using minimax to take account of all rival scenarios, the models proposed ensure a basic guaranteed return, in view of multiple scenarios. Consider the mean-variance framework for a given $\alpha \in [0, 1]$

$$\min_{x} \{J_{\alpha}(x) | x \in X\} \tag{2.1}$$

where $J_{\alpha}(x)$ is the quadratic objective function

$$J_{\alpha}(x) = -\alpha \langle E(r), x \rangle + (1 - \alpha) \langle x - \bar{x}, C(x - \bar{x}) \rangle \tag{2.2}$$

$E(r) \in \Re^{n}$ is the expected return vector of the set of assets being considered with

$$r = E(r) + \epsilon \tag{2.3}$$

ϵ is a random error from $N(0, C)$, $C \in \Re^{n \times n}$ is the covariance matrix of returns, $x \in \Re^{n}$ is the vector of portfolio weights to be optimally determined,

$\bar{x} \in \Re^n$ denotes benchmark[1] weights that x should follow closely, and X is the convex feasible set of weights including the budget constraint and restrictions specified by the investor. We note that the *minimization* of $-\langle E(r), x \rangle$ is equivalent to the *maximization* of $\langle E(r), x \rangle$. The main weakness of (2.1) is the inherent inaccuracy of the risk and return estimates.

2.1 Model 1: Rival Return Scenarios with Fixed Risk

Consider the case when, instead of a single return forecast, we are given a number of rival forecasts which represent plausible scenarios of the future. The ith scenario given by $E(r^i) \in \Re^n$ and

$$J_\alpha^i(x) = -\alpha \langle E(r^i), x \rangle + (1 - \alpha)\langle x - \bar{x}, C(x - \bar{x}) \rangle \qquad (2.4)$$

with m^{sce} scenarios and

$$r^i = E(r^i) + \epsilon, \quad i = 1, ..., m^{sce}. \qquad (2.5)$$

In this model, the risk estimate is assumed to be common across all scenarios.

For (2.4), the minimax formulation is given by

$$\min_x \max_{i=1,...,m^{sce}} \left\{ -\alpha \langle E(r^i), x \rangle + (1 - \alpha)\langle x - \bar{x}, C(x - \bar{x}) \rangle \mid x \in X \right\} \qquad (2.6)$$

which can be expressed as a quadratic programming problem

$$\min_{x, v \in \Re^{n+1}} \left\{ -\alpha v + (1 - \alpha)\langle x - \bar{x}, C(x - \bar{x}) \rangle \mid x \in X, \ \langle E(r^i), x \rangle \geq v, \ \forall i \right\} \qquad (2.7)$$

where $v \in \Re^1$.

2.2 Model 2: Rival Return with Risk Scenarios

The second minimax formulation takes into account rival return forecasts with each having an *associated* risk. Consider, therefore, the objective

$$J_\alpha^i(x) = -\alpha \langle E(r^i), x \rangle + (1 - \alpha)\langle x - \bar{x}, C^i(x - \bar{x}) \rangle \qquad (2.8)$$

and

$$r^i = E(r^i) + \epsilon^i, \quad i = 1, ..., m^{sce} \qquad (2.9)$$

where ϵ^i is a random error from $N(0, C^i)$, $C^i \in \Re^{n \times n}$.

For (2.8), the minimax formulation is

$$\min_x \max_i \left\{ -\alpha \langle E(r^i), x \rangle + (1 - \alpha)\langle x - \bar{x}, C^i(x - \bar{x}) \rangle \mid i = 1, ..., m^{sce}, \ x \in X \right\}$$
$$(2.10)$$

---

[1] The term benchmark is used extensively in this chapter. A broad discussion on benchmarks and benchmarking is given in CN 1.

and its nonlinear programming equivalent is given by

$$\min_{x,v \in \Re^{n+1}} \{v \mid -\alpha \langle E(r^i), x \rangle + (1-\alpha)\langle x - \bar{x}, C^i(x - \bar{x}) \rangle \leq v,$$

$$i = 1, \dots, m^{sce}, \ x \in X\} \qquad (2.11)$$

where $v \in \Re^1$.

2.3 Model 3: Rival Return Scenarios with Independent Rival Risk Scenarios

Consider a set of rival forecast scenarios specified independently from a set of rival risk scenarios. The minimax strategy looks at the compounding effect of risk and return scenarios and computes the worst-case solution.

The corresponding minimax formulation is

$$\min_x \max_{i^r, i^C} \big\{ -\alpha \langle E(r^{i^r}), x \rangle + (1-\alpha)\langle x - \bar{x}, C^{i^C}(x - \bar{x}) \rangle \mid i^r = 1, \dots, m^{sce^r},$$

$$i^C = 1, \dots, m^{sce^C}, \ x \in X\} \qquad (2.12)$$

where m^{sce^r} is the number of return scenarios and m^{sce^C} is the number of risk scenarios.

The equivalent nonlinear programming formulation is given by

$$\min_{x,v,z \in \Re^{n+2}} \{ -\alpha v + (1-\alpha)\, z \mid \langle E(r^{i^r}), x \rangle \geq v, \forall i^r,$$

$$\langle x - \bar{x}, C^{i^C}(x - \bar{x}) \rangle \leq z, \ \forall i^C, \ x \in X\} \qquad (2.13)$$

where $z \in \Re^1$ and $v \in \Re^1$.

2.4 Model 4: Fixed Return with Rival Benchmark Risk Scenarios

The final model concerns rival benchmark weights and rival risk scenarios. The minimax problem is given by

$$\min_x \max_{i^B, i^C} \big\{ -\alpha \langle E(r), x \rangle + (1-\alpha)\langle x - \bar{x}^{i^B}, C^{i^C}(x - \bar{x}^{i^B}) \rangle \mid i^C = 1, \dots, m^{sce^C},$$

$$i^B = 1, \dots, m^{sce^B}, \ x \in X\} \qquad (2.14)$$

where $\bar{x}^{i^B}$ is the (i^B)th rival benchmark and C^{i^C} is the (i^C)th rival variance. The equivalent nonlinear programming formulation is given by

$$\min_{x,z \in \Re^{n+1}} \{-\alpha \langle E(r), x \rangle + (1-\alpha)z \mid \langle x - x^{i^B}, C^{i^C}(x - x^{i^B}) \rangle \le z,$$

$$i^B = 1, ..., m^{sce^B}, \ i^C = 1, ..., m^{sce^C}, \ x \in X\}. \tag{2.15}$$

2.5 Efficiency

Problem (2.1)–(2.2) represents the simultaneous maximization of portfolio return and minimization of risk. Thus, the solution achieves the best return at minimal risk. The emphasis on risk is represented by the parameter α. As α increases from zero, the investment strategy becomes increasingly risk averse. It is important to note here the role of convexity of the underlying problem which ensures that all solutions are global optima (see Lemma 1.1.3). Problem (2.1)–(2.2) always has the best global portfolio return with least risk. Hence, at the given level of risk, a better return is not achievable and thus the solution is efficient for every value of α. This concept of efficiency also applies to Models 1, 3, and 4 due to the convexity of the underlying problem. The only exception is Model 2 where the solution of (2.11) is not necessarily the best return at minimal risk. It should be noted that (2.11) is also a convex problem with a global optimum. The trade-off between risk and return, represented by α, is no longer the determining factor for computing the best return versus risk unless the worst-case is defined by a single scenario (i.e., the unique maximizer case). A close examination of Model 3 also reveals that Model 2 is a restricted version with index equality $i^r = i^C$ and this further constraint complicates matters.

3 MINIMAX BOND PORTFOLIO SELECTION

In this section, we illustrate the use of minimax in managing a bond portfolio. Bond portfolio managers involved in forecasting for asset allocation face the issue of whether they need to come up with a commonly agreed set of forecasts. Creating consensus forecasts becomes problematic when individual managers involved in forecasting have widely conflicting views with no managerial framework for them to resolve the conflicting forecasts into one consensus forecast. We present a framework partly based on Model 3 in Section 2 and emphasize the worst-case risk for a given performance over all return scenarios. We demonstrate that an optimal bond portfolio can be created within this framework to generate a compromise at the asset allocation level instead of the forecast level. We further demonstrate the sub-optimality due to mis-forecasting by showing that an asset allocation decision based on any individual forecast may result in a severely poor performance if another forecast turns out to be a better estimator of the assets' return distributions. The proposed framework attempts to minimize the suboptimality across all forecasts and comes up with a compromise portfolio that may be easier for managers to accept than one that has been created based on any individual forecast. The most important characteristic of the minimax framework, however, is robustness and the guaranteed performance level provided. The robustness is due to the fact that the minimax strategy is chosen in view of the worst case scenario and that *if any other scenario is realized then performance is guaranteed to improve* (see Lemma 6.3.1).

The most common forecasting issue for asset allocation relates to the estimation of the asset return distributions. We consider two common estimation techniques[2]: the use of historical time series and the use of future scenarios. In using the former, one normally makes the assumption that the historical performance of an asset is a good estimator of its performance in the future. We refer to this as Model **H** below. In future scenarios, forward-looking fundamental and technical analysis is used to model the future performance of assets. For example, the forward-looking analysis of an equity market may involve dividend growth, capitalization, investment, as well as econometric considerations. This will provide an alternative estimate of the performance of the asset in the future. We refer to this as Model **S** below. We will use these two methods of estimating the asset return distributions to illustrate the danger of mis-forecasting and the benefits from using the proposed framework for asset allocation.

We first present in Section 3.1 the formulation of the single model problem and in Section 3.2 its application to two usd-based[3] bond portfolio managers who wish to generate mean-variance efficient portfolios, where one investor uses a model based on historical time series and another investor uses future scenarios. In Section 3.3, we give the minimax formulation for the compromise between the two models at the asset allocation level, and show the benefits of this formulation in an application in Sections 3.4 and 3.5.

3.1 The Single Model Problem

In the Markowitz[4] framework, the returns covariance matrix is an essential set of information that is required in order to optimally choose a portfolio. An optimal portfolio is selected by minimizing the variance of returns of the

---

[2] There are many ways of estimating an asset return distribution, both within the historical framework and the scenario framework. For the purpose of illustration we concentrate on these two broad estimation techniques, without going into the detail on the possible variations that one can use to enhance the estimation.

[3] US dollar-based.

[4] We use the Markowitz (1952) mean-variance framework for portfolio selection as the basic platform to illustrate the proposed approach, as the mean-variance framework is the most commonly used asset allocation technique.

resulting portfolio for a given level of expected portfolio return. Consider a universe of n assets for portfolio construction. As in Section 2, the portfolio is defined by the weights $x \in \mathfrak{R}^n$ and the expected returns $E(r) \in \mathfrak{R}^n$, with corresponding elements x_j and r_j associated with each asset j. Let $C \in \mathfrak{R}^{n \times n}$ be the $(n \times n)$ covariance matrix of returns, generated from the asset return distributions, which in turn has been generated using a particular model or estimation method.

Given a required level of expected portfolio return $\bar{R}$, an optimal portfolio is one that solves the following:

$$\min_{x \in \mathfrak{R}^n} \{\langle x, Cx \rangle\} \tag{3.1}$$

subject to

$$\langle x, E(r) \rangle \geq \bar{R}$$

$$\langle 1, x \rangle = 1$$

$$x \geq 0$$

where 1 is the vector with unit elements. The covariance matrix is used to estimate the volatility of the resulting portfolio. Having determined the portfolio, we can then monitor its performance by estimating its volatility and its actual return up to the end of the holding period.

Many implementations of the Markowitz framework use a historical covariance matrix suitably based on an assumed horizon period to hold the portfolio. For example, if a portfolio is to be determined today with a 1-month holding period, the covariance matrix to use is the one that estimates the 1-month volatility of returns of the risky assets.

A number of implementations are based on a different method of estimating the covariance matrix. Whereas historical data are ordinarily used for estimating the covariance matrix, an alternative approach may utilize forward-looking scenarios. We use historical data to generate a historical covariance matrix. Alternatively, a sufficient number of forward-looking scenarios are used for generating a scenario-based covariance matrix. We denote the historical covariance matrix by C^H and the scenario covariance matrix by C^S and consider these matrices as the product of two not necessarily conflicting *models* of asset returns in the application below.

3.2 Application: Two Asset Allocations Using Different Models

Consider two usd-based bond[5] portfolio managers who wish to adopt different models for their conjoint asset allocation decision. One uses historical time

---

[5] The applications considered in this chapter are for bond portfolios; however, the framework equally applies to other asset classes.

series while the other uses forward-looking scenarios. They wish to create a portfolio of international bonds from the following countries: Germany, Belgium, Spain, France, Italy, Japan, the Netherlands, Sweden, and the United Kingdom. Although both managers wish to attain an expected return of say 10% (we assume this target performance for illustration purposes), they differ in the model that they use for estimating returns. Any optimal asset allocation decision based on one model would yield a suboptimal return based on the other model. The manager who adopts the historical-time-series-based model of asset returns (hereafter Model **H**) wishes to create a portfolio that is optimal under that model and at the same time would not severely underperform under the scenario-based model (hereafter Model **S**). Similarly, the manager who adopts Model **S** wishes to create a portfolio that is optimal under her model and at the same time would not severely underperform under Model **H**.

The managers wish to know the risk/return trade-offs involved in using their models separately. Accordingly, we generate two efficient frontiers: one frontier, **H**, shown as an efficient frontier in Figure 3.1a, is generated using Model **H**; the other frontier, **S**, shown as an efficient frontier in Figure 3.1b, is generated using Model **S**. **H** and **S** trace the return and the risk in the context of Model **H** and Model **S**, respectively.

To see the performance of the portfolios on **S** if Model **H** happens to be a better representation of asset return distributions, we take the portfolios on **S** and identify in Figure 3.1a the location of this set of portfolios that traces the return and risk combinations in the context of Model **H**. Similarly, we take the portfolios on **H** and identify in Figure 3.1b the location of this set of portfolios

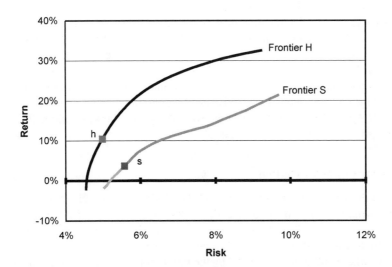

Figure 3.1a   Frontiers under the historical-time-series-based model, Model **H**.

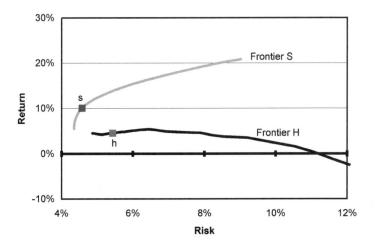

Figure 3.1b   Frontiers under the scenario-based model, Model **S**.

that traces out their return and risk combinations in the context of Model **S**. Figures 3.1a and 3.1b indicate that portfolios that are mean-variance efficient under one model are suboptimal under another model. Because no portfolio is simultaneously optimal under both models, the managers will seek a portfolio between these two efficient frontiers.

As noted above, the required portfolio must deliver the expected performance of 10% annual return. Accordingly we consider in Figure 3.1a a portfolio **h** that yields the expected performance of 10% under Model **H**. Similarly, we consider in Figure 3.1b a portfolio **s** that yields the expected performance of 10% under Model **S**. Next, we consider in Figure 3.1a the return and risk co-ordinates of **h** and **s** in the context of Model **H**. Relative to Model **H**,**h** has an expected performance of 10%. However, when located on Figure 3.1a, portfolio **s** is not on the efficient frontier under Model **H**. Because of the wide gap between **h** and **s**, evaluated under either model, the manager would want an optimal compromise that would minimize the suboptimality in the two contexts. In the next section, we present a method for minimizing the suboptimality across the two models.

3.3 Two-model Problem

The optimization of the two-model problem involves the selection of assets that would attempt to minimize the suboptimality across the two models, thereby creating the minimax-optimal bond portfolio. We can generate an efficient frontier that plots the risk and return for any minimax-optimal portfolio, where optimality is defined by the minimum risk, when optimized under the two models, for a required expected return.

We present the minimax formulation; we use the notation from Section 3.1, and augment the notation for the model from that section to distinguish between the two models used in the minimax formulation. As before, we use **H** and **S**, respectively, to refer to the historical-time-series-based model and the scenario-based model.

Consider a universe of n assets from which a portfolio could be constructed, where the portfolio is defined by the weights $x \in \Re^n$ associated with each asset. The expected returns $E(r^H) \in \Re^n$ under model **H** and $E(r^S) \in \Re^n$ under model **S** are respectively the expected values for the return on each asset based on historical time series and the expected value for the return on each asset based on multiple forward-looking scenarios. Let $C^H \in \Re^{n \times n}$ and $C^S \in \Re^{n \times n}$ be the $(n \times n)$ covariance matrix of returns under model **H** and model **S**, respectively. Given a required level of expected portfolio return $\bar{R}$, a minimax-optimal portfolio is one that solves the following:

$$\min_{x \in \Re^n} \max_{H,S} \left\{ \langle x, C^H x \rangle, \langle x, C^S x \rangle \right\} \tag{3.2}$$

subject to

$$\langle x, E(r^H) \rangle \geq \bar{R}$$

$$\langle x, E(r^S) \rangle \geq \bar{R}$$

$$\langle 1, x \rangle = 1$$

$$x \geq 0.$$

3.4 Application: Simultaneous Optimization across Two Models

To identify the optimal compromise when presented with two different models of return, we generate a frontier in which all portfolios are the result of simultaneously minimizing the suboptimality across two models, while maintaining the required level of expected return of 10%. We call this frontier the minimax frontier, **M**. While generating the efficient frontiers **H** and **S** is routine, generating the minimax frontier, **M**, is not: it requires specialized optimization software. We plot **M** in both Figures 3.2a and 3.2b which are replicas of Figures 3.1a and 3.1b with **M** overlaid. Under Model **H**, any portfolio on **M** will be inferior to any portfolio on **H**, and superior to any portfolio on **S**. Similarly, under Model **S**, any portfolio on **M** will be inferior to any portfolio on **S**, and superior to any portfolio on **H**. However, **M** provides a consistent level of performance under both models, and the managers know that, for the same expected return, any portfolio on **M** will provide the more acceptable risk than any portfolio on **H** and **S**. Minimax performance is the

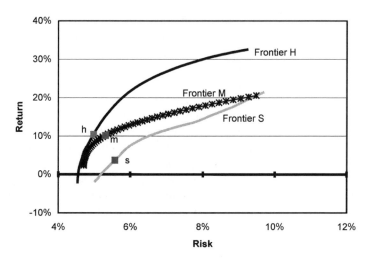

Figure 3.2a   Frontiers under Model **H**.

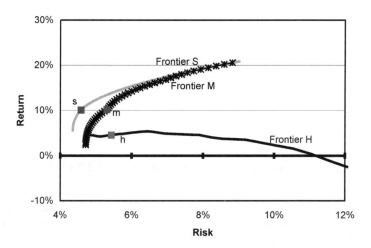

Figure 3.2b   Frontiers under Model **S**.

best worst-case performance, and hence the best lower bound given the scenarios.

3.5 Backtesting the Performance of a Portfolio on the Minimax Frontier

Investors may choose any portfolio on **M**; their choice will reflect their risk and return preferences. To illustrate the performance enhancement via mini-

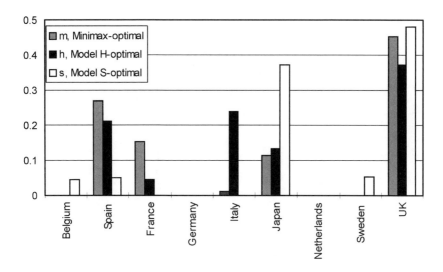

Figure 3.3   Asset allocations for the three portfolios.

max we arbitrarily select portfolio **m**, which we call the minimax-optimal portfolio, and show how the suboptimality discussed in Section 3.2 has been simultaneously minimized. Portfolio **m** simultaneously minimizes the gap between itself and **h**, and the gap between itself and **s**. We show in Figure 3.3 the asset allocations for the three portfolios that have been determined with a view to realizing a 10% return.

To illustrate the performance over time of the minimax-optimal portfolio **m**, we calculate the cumulative daily returns of **m** against **h** and **s** under Model **H**, and the cumulative daily returns of **m** against **h** and **s** under Model **S**. In Figures 3.4a and 3.4b, respectively, we plot the performance of **m** against **h** and **s** under Model **H**, and that of **m** against **h** and **s** under Model **S**. The cumulative returns are plotted on an in-sample and out-of-sample basis[6]; all cumulative returns start at zero returns for both in-sample and out-of-sample. The in-sample data for selecting a portfolio are specific to the model being used, whether Model **H** or Model **S**, whereas the out-of-sample data are the realized returns for all the assets considered in the analysis. In this illustration, the realized returns have been correctly modeled by Model **S**. From Figure 3.4a, for the out-of-sample period, **s** outperformed **h**: this demonstrates the effect of mis-forecasting. It is clear from the figures that **m** generally outperformed **s** under Model **H**, and also generally outperformed **h** under Model **S**, for both the in-sample period and the out-of-sample period.

---

[6] In-sample analysis shows the performance or fit of a model on the data used in creating the model whereas out-of-sample analysis shows its performance or fit on new data. Out-of-sample analysis is an acid test of the power of a model.

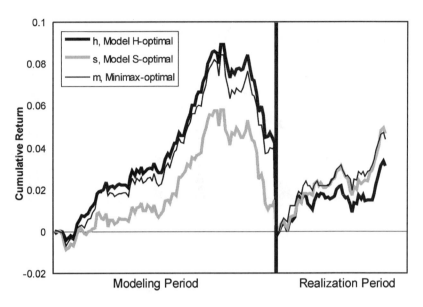

Figure 3.4a Cumulative returns in the context of Model **H**, and in the context of realized actual returns.

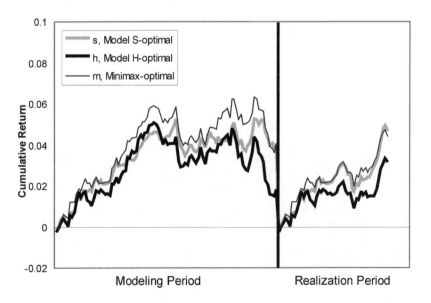

Figure 3.4b Cumulative returns in the context of Model **S**, and in the context of realized actual returns.

4 DUAL BENCHMARKING

Many portfolio managers may not consider any one benchmark appropriate for their purpose. Given their multiple objectives of maximizing returns and minimizing tracking error[7] and downside risk, they may wish to track multiple benchmarks simultaneously. We present a dual benchmarking method based on minimax to compute an optimized solution to the problem of benchmarking with multiple objectives. Managers may find it useful in developing a portfolio strategy that appropriately balances return and risk.

The portfolio manager who is certain she has selected the most appropriate benchmark to track and/or who is not particularly interested in avoiding downside risk may remain with single benchmarking. Other managers may consider dual benchmarking, under which they simultaneously track any two benchmarks. While any number of benchmarks can be tracked, most managers would perhaps be most interested in tracking those that meet different objectives. For example, one benchmark may be chosen because it performs well and is not easy to beat; the other, although low-performing, being relatively risk-free such as LIBOR[8], may be chosen because it helps immunize against downside risk. Accordingly, we illustrate dual benchmarking by describing how it could help a usd-based portfolio manager ensure her portfolio delivers both the minimum of LIBOR and the conditional maximum on excess returns by simultaneously tracking a global bond benchmark and the USD LIBOR benchmark. We then illustrate the benefits from such tracking.

We first present in Section 4.1 the formulation of the single benchmark tracking problem and in Section 4.2 its application to a usd-based portfolio manager looking at tracking a global benchmark and LIBOR separately. In Section 4.3, we give the minimax formulation of the dual benchmarking problem and show the benefits from using dual benchmarking in an application in Sections 4.4 and 4.5.

4.1 Single Benchmark Tracking

The single benchmark tracking problem involves the selection of assets that attempt to replicate the performance, in terms of tracking error and excess return trade-offs, of a chosen benchmark. We can generate an efficient frontier that plots the tracking risk and return for any optimal portfolio. Optimality is defined by the minimum tracking error for a given tracking return (i.e., excess return).

As in earlier sections, consider a universe of n assets from which a portfolio may be created. The portfolio is defined by the weights $x^P \in \Re^n$ and the vector

---

[7] Tracking error is the standard deviation of excess returns over a benchmark.

[8] London Inter-Bank Offered Rate, the benchmark borrowing rate between banking institutions in London, where the quotation is period-specific such as 3-month LIBOR.

of expected returns for n assets $E(r^P) \in \Re^n$, with corresponding elements x_j^P and $E(r_j^P)$ attached to each asset j. Further, consider a universe of m assets from which a benchmark is defined. The benchmark is identified by the weights $x^B \in \Re^m$ and the vector of expected returns for m assets $E(r^B) \in \Re^m$, with corresponding elements x_k^B and $E(r_k^B)$ associated with each asset k. The benchmark weights x^B are fixed for our purpose and are determined by the particular asset class being analyzed; in the case of bonds, the benchmark may be a composite of all outstanding investment-grade bonds.

In the case of bond portfolios, for example, we assume that the expected return for each asset is derived based on an estimation procedure using yield curve scenarios[9] where a scenario defines the future value of an asset, and therefore its return over the period[10]. Furthermore, we assume that each scenario is associated with a weight that defines the importance, or perhaps likelihood, of the scenario. Let $\theta \in \Re^{m^{sce}}$ be the vector of scenario weights for m^{sce} scenarios where each element θ^i represents the weight for scenario i. Associated with each of the n assets in the portfolio is a vector of scenario returns whose element $r_j^{P,i}$ defines the return for portfolio asset j under scenario i. Similarly, we define $r_j^{B,i}$ as the return for the benchmark asset j under scenario i.

The portfolio asset expected returns vector $E(r^P) \in \Re^n$ is a vector of expected returns where each element represents the expected return for each asset in the portfolio. In turn, the expected return for each asset j, $E(r_j^P) \in \Re^1$, represents the expectation over all scenarios for that particular asset. The return vector $r^{P,i} \in \Re^n$ under scenario i is a vector whose elements are the asset returns for that particular scenario. Hence we have

$$E(r^P) = \begin{bmatrix} E(r_1^P) \\ \vdots \\ E(r_n^P) \end{bmatrix} = \begin{bmatrix} \sum_{i=1}^{m^{sce}} \theta^i r_1^{P,i} \\ \vdots \\ \sum_{i=1}^{m^{sce}} \theta^i r_n^{P,i} \end{bmatrix}, \quad r^{P,i} = \begin{bmatrix} r_1^{P,i} \\ \vdots \\ r_n^{P,i} \end{bmatrix}.$$

Similarly, the expected returns of the assets in the benchmark, vector $E(r^B) \in \Re^m$, and the returns vector, $r^{B,i} \in \Re^m$, for scenario i are given by

---

[9] The estimation procedure using yield curve scenarios is only one of a number of ways of estimating a return distribution for an asset. An alternative procedure using historical time series of period returns, say daily or monthly, can be viewed as a series of period yield curve scenarios that would fit in the framework described in this section.

[10] Although the discussion in this particular section refers to the use of scenarios for estimating expected returns, this does not preclude the use of historical time series for estimating expected returns. In the case of historical time series, each time stamp can be regarded as a scenario.

$$E(r^B) = \begin{bmatrix} E(r_1^B) \\ \vdots \\ E(r_m^B) \end{bmatrix} = \begin{bmatrix} \sum_{i=1}^{m^{sce}} \theta^i r_1^{B,i} \\ \vdots \\ \sum_{i=1}^{m^{sce}} \theta^i r_m^{B,i} \end{bmatrix}, \quad r^{B,i} = \begin{bmatrix} r_1^{B,i} \\ \vdots \\ r_m^{B,i} \end{bmatrix}.$$

The portfolio (or benchmark) return is the weighted sum of the expected returns, for the assets in that portfolio (or benchmark). We can also identify the portfolio (or benchmark) return per scenario as the weighted sum of the return for a particular scenario, for the assets in that portfolio (or benchmark).

The expected portfolio return P and its return per scenario P^i are given by

$$P = \langle (x^P), E(r^P) \rangle$$

$$P^i = \langle (x^P), r^{P,i} \rangle.$$

The expected benchmark return B and its return per scenario B^i are given by

$$B = \langle (x^B), E(r^B) \rangle$$

$$B^i = \langle (x^B), r^{B,i} \rangle$$

where x^B is the weight of the asset in the benchmark. It is assumed to be constant at this stage and is specified in advance. Given a required level of expected excess return $\bar{T}$ over the benchmark, an optimal tracking portfolio is one that solves the following:

$$\min_{x^P \in \Re^n} \left\{ \sum_{i=1}^{m^{sce}} \theta^i \left((P^i - B^i) - (P - B) \right)^2 \right\} \tag{4.1}$$

subject to

$$\langle (x^P), E(r^P) \rangle - \langle (x^B), E(r^B) \rangle \geq \bar{T}$$

$$\langle 1, x^P \rangle = 1$$

$$x^P \geq 0.$$

The objective in (4.1) is to minimize the deviation of the excess return under each scenario, given by $(P^i - B^i)$, relative to the expected excess return, given by $(P - B)$. The first constraint specifies that the expected excess return, essentially $(P - B)$, is at least the same as some predefined level $\bar{T}$. Repeatedly solving (4.1) with varying levels of expected excess return $\bar{T}$ will generate a frontier of minimal risk for a given level of return. The frontier is efficient in the sense that it characterizes the best risk versus return trade-off that can be achieved and yields optimal tracking portfolios with minimal tracking error for any level of expected excess return.

4.2 Application: Tracking a Global Benchmark against Tracking LIBOR

A usd-based bond portfolio manager wishes to create a portfolio that tracks a global benchmark and simultaneously delivers a performance that is noninferior to LIBOR. The manager considers LIBOR as the minimum-return benchmark: the return on LIBOR is a good indicator of the return on an equivalent cash portfolio of relatively low risk. The manager has no incentive to create an asset portfolio, of relatively high risk, that delivers less than LIBOR. The manager will want to know the risk/return trade-offs involved in tracking a market-capitalization-weighted[11] global benchmark, shown in Figure 4.1, and in tracking LIBOR separately. Accordingly, we generate two efficient frontiers: one, **G**, shown in Figure 4.2a, which tracks the global benchmark; the other, **L**, shown in Figure 4.2b, which tracks LIBOR. **G** and **L** trace the tracking return and the tracking error with respect to the global benchmark and LIBOR, respectively.

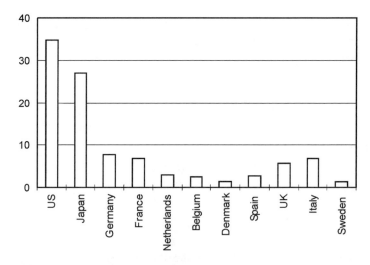

Figure 4.1 The market-capitalization-weighted global benchmark used by the bond portfolio manager.

---

[11] This means that the weightings in the global benchmark reflect the capitalization of equity markets in the countries that constitute the global benchmark.

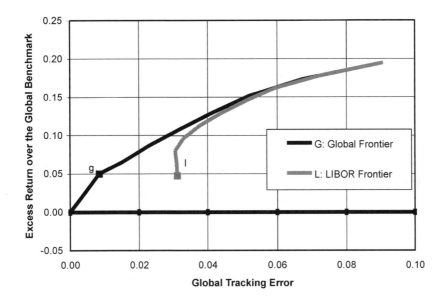

Figure 4.2a   Global benchmark tracking frontiers.

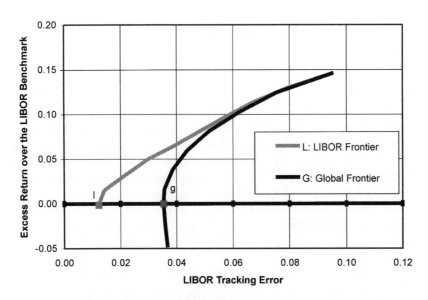

Figure 4.2b   USD LIBOR benchmark tracking frontiers.

Tracking error may be unacceptably high towards the right region of the frontier. Hence, the manager is likely to focus on the left region, where the tracking error is low. To see the performance of the portfolios corresponding to **L** in tracking the global benchmark **G**, we identify this performance of **L** in Figure 4.2a by tracing their excess return and tracking error combinations relative to the global benchmark **G**. Similarly, we evaluate the performance of the portfolios based on **G** and identify in Figure 4.2b by tracing excess return and tracking error combinations relative to LIBOR benchmarking **L**. Figures 4.2a and 4.2b indicate that portfolios that efficiently track the global benchmark are suboptimal in the LIBOR context, while portfolios that efficiently track LIBOR are suboptimal in the global benchmarking context. Because no portfolio is simultaneously optimal in both contexts, the manager will seek a portfolio between these two efficient frontiers.

As noted above, the required portfolio delivers the minimum performance target of LIBOR. Accordingly, we consider in Figure 4.2b the two portfolios **g** and **l** that yield zero tracking returns over LIBOR. We next consider in Figure 4.2a the return and risk coordinates of **g** and **l** in the global benchmarking context. When located on Figure 4.2a portfolio **g** is on the efficient frontier for global benchmarking. Relative to global benchmarking it has an expected outperformance of 5%. We refer to portfolio **g** as the 5% global-optimal portfolio when we illustrate the relative performance of a dual[12] portfolio. Because of the wide gap between **g** and **l**, in either figure, the investor may wish an optimal compromise that minimizes the suboptimality in the two benchmarking contexts. The suboptimality arises when measuring the performance of a portfolio, optimally tracking one benchmark, against a second benchmark. In Section 4.3, we present a method for minimizing this suboptimality.

4.3 Dual Benchmark Tracking

The dual benchmark tracking problem involves the selection of assets that optimally track a benchmark and simultaneously minimize the suboptimality in tracking another benchmark. We generate an efficient frontier that computes the tracking error and excess return for any dual-optimal bond portfolio. Optimality is defined by the minimum of the worst-case tracking error, when tracking both benchmarks, for given excess returns.

We augment the notation for the benchmark in Section 4.1 to distinguish between the two benchmarks used in the dual benchmark tracking formulation. We use $B1$ and $B2$, respectively to refer to the first and the second benchmark.

Given the required level of expected tracking returns $\bar{T}1$ and $\bar{T}2$, an optimal

---

[12] The dual portfolio is an optimal portfolio generated using the dual benchmarking framework described in Section 4.3.

tracking portfolio is one that solves the following:

$$\min_{x^P \in \Re^n} \max_{k=1,2} \left\{ \begin{array}{l} f_{k=1} = \left(\sum_{i=1}^{m^{sce}} \theta^i \big((P^i - B1^i) - (P - B1) \big)^2 \right) \\ \\ f_{k=2} = \left(\sum_{i=1}^{m^{sce}} \theta^i \big((P^i - B2^i) - (P - B2) \big)^2 \right) \end{array} \right\}$$

subject to

$$\langle (x^P), E(r^P) \rangle - \langle (x^{B1}), E(r^{B1}) \rangle \geq \bar{T}1 \qquad (4.2)$$

$$\langle (x^P), E(r^P) \rangle - \langle (x^{B2}), E(r^{B2}) \rangle \geq \bar{T}2$$

$$\langle 1, x^P \rangle = 1$$

$$x^P \geq 0.$$

The objective in (4.2) is to minimize the maximum, or worst-case, deviation relative to the two benchmarks. Relative to each benchmark, the excess return under each scenario, given by $(P^i - B1^i)$ and $(P^i - B2^i)$, relative to their corresponding expected excess return, given by $(P - B1)$ and $(P - B2)$. The first constraint specifies that the expected excess return relative to the first benchmark, essentially $(P - B1)$, is at least the same as some predefined level $\bar{T}1$. The second constraint specifies that the expected excess return relative to the second benchmark, essentially $(P - B2)$, is similarly required to be at least the same as some other predefined level $\bar{T}2$. These two constraints define the minimum expected excess returns for the portfolio whichever benchmark is relevant. We assume $\bar{T}1 = \bar{T}2$. Repeatedly solving (4.2) with varying levels of expected tracking returns $\bar{T}1$ ($= \bar{T}2$), we generate an efficient frontier of dual-optimal tracking portfolios with minimal tracking errors, when tracking both benchmarks, for any level of excess return.

4.4 Application: Simultaneously Tracking the Global Benchmark and LIBOR

We illustrate the optimal compromise when tracking two benchmarks. A frontier is generated in which all portfolios are the result of simultaneously minimizing the worst-case tracking errors under both global benchmarking and LIBOR benchmarking, while maximizing excess returns over both global benchmark and LIBOR. We call this frontier the dual benchmark frontier, **D**. While generating the efficient frontiers **G** and **L** (the global frontier and the LIBOR frontier) is routine, generating the dual benchmarking frontier, **D**, is not: it requires specialized optimization software. We illustrate **D** in both

Figures 4.3a and 4.3b which are replicas of Figures 4.2a and 4.2b with **D** overlayed. In Figure 4.3a, frontier **G**, generated from tracking a global benchmark, is more efficient than either **D** or **L**. Similarly, in Figure 4.3b, frontier **L**,

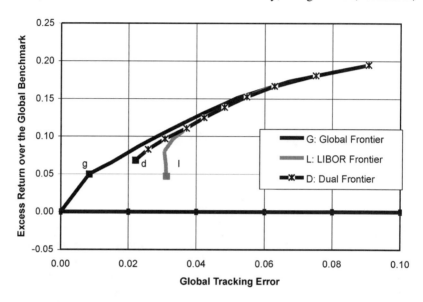

Figure 4.3a Global benchmark tracking frontiers.

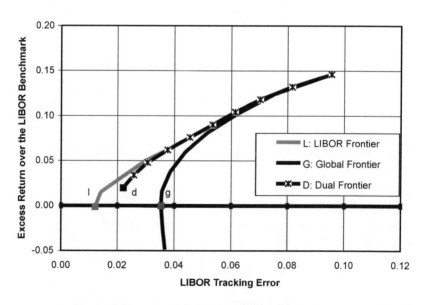

Figure 4.3b USD LIBOR benchmark tracking frontiers.

generated from tracking a LIBOR benchmark, is more efficient than either **D** or **G**. However, **D** provides a guaranteed level of performance over both benchmarks. For the same excess return, any portfolio on **D** will provide a more acceptable downside risk than any portfolio on either frontier **G** or **L**.

4.5 Performance of a Portfolio on the Dual Frontier

Any portfolio on **D** is chosen to reflect risk and return preferences. To illustrate the performance enhancement via dual benchmarking we arbitrarily select portfolio **d**, which we call the dual portfolio, and show how the suboptimality discussed in Section 4.2 has been simultaneously minimized. Portfolio **d** simultaneously minimizes the gap between itself and **g**, as shown in Figure 4.3a, and the gap between itself and **l**, as shown in Figure 4.3b. We show in Figure 4.4 the asset allocations for the 5% global-optimal portfolio (**g**) and for the dual portfolio (**d**).

To illustrate the performance over time of the dual portfolio **d**, we compute

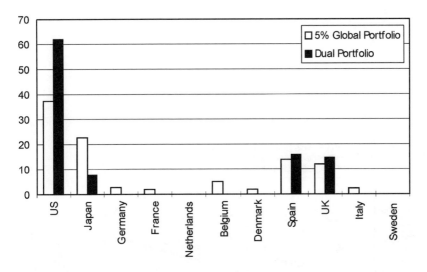

Figure 4.4   Asset allocations.

the cumulative daily returns of the dual portfolio (**d**) against the 5% global-optimal portfolio (**g**). In Figures 4.5a and 4.5b, respectively, we plot the tracking performance of the 5% global-optimal portfolio (**g**) against that of the dual portfolio (**d**) on a cumulative return basis. The cumulative returns are plotted in-sample and out-of-sample; all cumulative returns start at zero returns for both in-sample and out-of-sample. LIBOR returns are based on

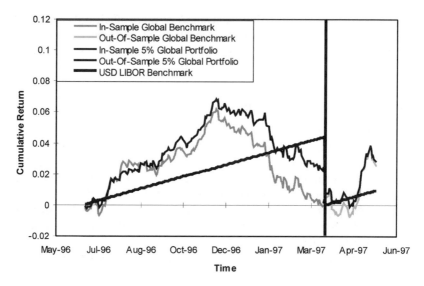

Figure 4.5a Cumulative returns of the global benchmark, USD LIBOR and the 5% global-optimal portfolio.

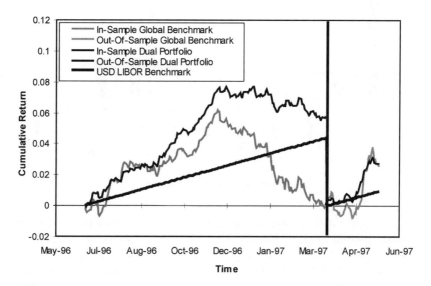

Figure 4.5b Cumulative returns of the global benchmark, USD LIBOR and the dual portfolio.

the average USD LIBOR rate for the in-sample and out-of-sample periods, with a simple linear cumulative return indicating the USD LIBOR performance over the periods. Looking at the figures we see that the 5% global-optimal portfolio (**g**) generally outperformed the global benchmark but underperformed LIBOR during the last quarter of the in-sample period, and during the first half of the out-of-sample period. The dual portfolio (**d**) also generally outperformed the global benchmark throughout the in-sample and out-of-sample periods, additionally, it also generally outperformed LIBOR.

5 OTHER MINIMAX STRATEGIES FOR ASSET ALLOCATION

Threshold returns (or minimum returns) are considered relevant to pension fund management where the gap between assets and liabilities defines a required minimum return. This leads to the broad area of asset management where benchmark tracking strategies define the required minimum outperformance. Furthermore, we consider index tracking, downside risk and the use of continuous minimax for range forecasts.

5.1 Threshold Returns and Downside Risk

The applications of Value-at-Risk models (Morgan Guaranty Trust Co. and Reuters Ltd., 1996) for estimating the potential loss in terms of portfolio value for adverse market movements have highlighted the need for downside risk optimization models. In this section we explore a minimax extension of downside risk modeling; for a comprehensive review of downside risk modeling in asset allocation (see CN 2 and Duarte, 1994).

We define the threshold return as the minimum acceptable or desired level of return that would satisfy the objectives of a fund. For a pension fund, the gap between the present value of assets and liabilities, as well as the gap between their duration levels (see CN 10.1), determine the threshold level of return that would maintain a high probability of meeting future liabilities as they fall due. The threshold return should provide cover against any potential widening of the gap between assets and liabilities. It should also cover the cost of maintaining the fund. For benchmark tracking strategies in asset management, where the performance of assets is assessed relative to the performance of a benchmark, the threshold return is the required outperformance net of management fees and transaction costs. For endowment funds, where maintaining the purchasing power of the assets may be the main objective, the threshold return should cover operating costs as well as expected inflation. In these types of funds where the investors' objectives are defined mainly in terms of a reference variable, for example, liabilities for pension funds and inflation for endowment funds, the investor's view of risk and return may deviate from the common measures in the mean-variance context, where risk is defined by the standard deviation of returns. This measure of risk is perhaps not the most appropriate for the illustrations given above, where the investors are more concerned about the downside risk, or the risk that the actual return on their portfolio is below their threshold level. We define downside risk[13] as

$$\text{downside risk} = \sqrt{\sum_{i=1}^{m^{sce}} \theta^i \left(\min(0, \langle x, r^i \rangle - TR))^2 \right)}$$

where x is the asset mix, TR is threshold return, r^i is the asset return vector under scenario i, and θ^i is the probability of occurrence of scenario i.

This definition of downside risk uses a number of scenarios over which the shortfall below the threshold return is evaluated. This definition depends on the existence of a sufficiently large set of scenarios in order to arrive at a meaningful figure. Another approach to downside risk is given in Section 5.2.1.

Downside risk optimization can be formulated as

$$\min_{x_p \in \Re^n} \left\{ \sum_{i=1}^{m^{sce}} \theta^i \left(\min \left(0, \langle x, r^i \rangle - TR \right) \right)^2 \right\} \tag{5.1}$$

subject to

$$\langle x, E(r) \rangle \geq \hat{R}$$

$$\langle 1, x \rangle = 1$$

$$x \geq 0$$

where $\hat{R}$ is a desired level of expected return, normally higher than the threshold return, and $E(r)$ is the expected return across all scenarios. An alternative way to estimate downside risk is to use historical data where a historical time series of returns for each asset is used in place of scenarios. In this case, the subscript i can be changed into a subscript t and can be interpreted as the time step over the time series, and m^{sce} can be interpreted as the number of time steps. When using historical data, (5.1) takes the following form:

$$\min_{x_p \in \Re^n} \left\{ \sum_{t=1}^{T} \theta^t \left(\min(0, \langle x, r_t \rangle - TR) \right)^2 \right\}$$

subject to

$$\langle x, \bar{E}(r) \rangle \geq \hat{R}$$

---

[13] Downside risk can be defined in a number of ways (see Duarte, 1994). A generalized formulation is provided in CN 2. The definition used here is one of the simplest in order to focus the discussion on the equivalent minimax formulation.

$$\langle 1, x \rangle = 1$$

$$x \geq 0$$

$\hat{R}$ is a desired level of expected return, normally higher than the threshold return, and $\bar{E}(r)$ is the historical average return.

In (5.1), the we take the minimum between the performance relative to the threshold and zero, that is, only the negative relative performance is taken into account. This is now formulated into a minimax framework as

$$\min_{x \in \mathfrak{R}^n} \max_i \left\{ \left(\min(0, \langle x, r^i \rangle - TR) \right)^2 \right\} \tag{5.2}$$

subject to

$$\langle x, E(r) \rangle \geq \hat{R}$$

$$\langle 1, x \rangle = 1$$

$$x \geq 0.$$

We note that with this formulation, the optimization of downside risk would not ensure that the standard deviation of returns is equally optimal. The solution generated by this formulation may correspond to a high-return high-volatility portfolio. Clearly, if these properties concern the investors then they should further modify their formulations to more accurately reflect their objectives. It is assumed that the investors interested in optimizing downside risk are more concerned by the performance of their portfolios relative to their reference variables, for example, liabilities or inflation concerns. Another downside risk formulation is presented in Section 5.2.1 for an application in robust index tracking.

5.2 Further Minimax Index Tracking and Range Forecasts

The problem presented in this section requires a framework for the investment manager to consider beating an index used as performance benchmark while feeling, to some extent, secure about the return forecast used. In other words, how erroneous can a forecast get (in terms of downside effects) while beating the index and ensure that the corresponding performance is secure, provided reality turns out to be within the downside bound used. If anything better than this bound is realized, the performance should improve (see Theorem 1.3.1).

Consider the classical mean-variance framework for a given $\alpha \in [0, 1]$

$$\min_x \{ J_\alpha(x) | x \in X \} \tag{5.3}$$

where $J_\alpha(x)$ is the quadratic objective function

$$J_\alpha(x) = -\alpha\langle E(r), x - \bar{x}\rangle + (1 - \alpha)\langle x - \bar{x}, C(x - \bar{x})\rangle. \qquad (5.4)$$

$E(r) \in \Re^n$ is the expected return vector of the set of assets being considered with

$$r = E(r) + \epsilon, \qquad (5.5)$$

$\epsilon \sim N(0, C)$ is the random error, $C \in \Re^{n \times n}$ is the covariance matrix of the returns, $x \in \Re^n$ is the portfolio weights to be optimally determined, and $\bar{x}$ denotes the benchmark weights for x to follow closely. X is the feasible set, which includes the budget constraint and the restrictions imposed by the investor. The vector ϵ is either the error of historical returns from the historical mean or the error between the historical actual return and corresponding forecast.

Ordinarily, the presence of $\bar{x}$ in the linear term in (5.4) does not influence the solution, as it is a constant shift of the objective. However, this may make a difference in a minimax strategy (Rustem et al., 2000).

5.2.1 Downside Risk Around the Lower Bound

An alternative approach is to consider a downside risk model (CN 2). A one-sided risk framework is considered below based on historical performance and where the one-sided risk is minimized while simultaneously maximizing the worst-case return.

The formulation entails the observation of historical returns on n assets over T periods. We thus have an observation matrix $R \in \Re^{T \times n}$. With a given benchmark and portfolio weights, the portfolio relative return is defined as

$$\varsigma = R(x - \bar{x}) \in \Re^T. \qquad (5.6)$$

We define downside risk as negative values of ς. This represents underperformance of the portfolio in view of historical return observations. We wish to minimize the mean square of such underperformance. Therefore, we consider

$$\varsigma^+ - \varsigma^- = \varsigma, \quad \varsigma^+, \varsigma^- \geq 0, \quad \varsigma^+ \varsigma^- = 0 \qquad (5.7)$$

and minimize

$$\frac{\langle \varsigma^-, \varsigma^- \rangle}{T - 1}. \qquad (5.8)$$

Hence, the objective of the classical portfolio problem becomes

$$J_\alpha(x) = -\alpha\langle E(r), x - \bar{x}\rangle + (1 - \alpha)\frac{\langle \varsigma^-, \varsigma^- \rangle}{T - 1} + c\langle \varsigma^+, \varsigma^- \rangle \qquad (5.9)$$

where $\varsigma^+ - \varsigma^- = R(x - \bar{x})$ and $\varsigma^+, \varsigma^- \geq 0$ are additional constraints. In (5.9), c is chosen large enough to ensure that $\langle \varsigma^+, \varsigma^- \rangle = 0$. We note that there are infinite $s^+, s^- \geq 0$ satisfying $s^+ - s^- = R(x - \bar{x})$ for any given x. However, there is a unique solution that minimizes (5.9). Indeed if $(R(x - \bar{x}))^i > 0$, that is, the ith term of the vector, then this can be achieved by $(s^-)^i = 0$, which minimizes the quadratic. If $(R(x - \bar{x}))^i < 0$ then the smallest $(s^-)^i$ that minimizes the quadratic is obtained by $(s^+)^i = 0$. The extension of this formulation to minimax with rival return scenarios is straightforward.

5.2.2 Range Forecasts: Continuous Minimax with Upper and Lower Bounds

The solution to the linear minimax robust portfolio problem

$$\min_x \max_r \left\{ -\langle r, x - \bar{x} \rangle \mid x \in X, \ r^{\text{lower}} \leq r \leq r^{\text{upper}} \right\} \qquad (5.10)$$

can be solved by using

$$s^+ - s^- = x - \bar{x}, \quad s^+, s^- \geq 0, \quad \langle s^+, s^- \rangle = 0$$

where $s^+, s^- \geq 0$ are buy and sell decisions, and solving the linear programming problem

$$\min_{s^-, s^+} \left\{ -\langle r^{\text{upper}}, s^- \rangle - \langle r^{\text{lower}}, s^+ \rangle \mid s^+ - s^- + \bar{x} \in X; s^+, s^- \geq 0 \right\}. \quad (5.11)$$

We note that the equality $\langle s^+, s^- \rangle = 0$ is ensured in the linear programming formulation. This corresponds to the risk-seeking end of the efficient frontier. It yields the optimal worst-case portfolio within this range. As always, the solution does not provide any protection over the variation of r outside the given ranges. For that, we need to introduce a certain degree of robustness to ensure that the solution is less sensitive to changes in r.

The risk associated with r in (5.10) is either the covariance of each r, $C(r)$, approximated with a single matrix C, corresponding to the central forecast error, or downside risk as in (5.2.1), or the historical covariance of r. In the former case, the risk term is consistent with the forecast but its inevitable approximation with a single matrix is not a desirable feature. Furthermore, the forecaster's risk is represented by both the forecast range and the covariance. The historical covariance is not consistent with the forecast but it seems to fit better with the interpretation that while the forecast error is represented by the forecast range, the historical return series does exhibit a certain variation, captured by the historical covariance. If the range is indeed such that any value in the range is possible, the historical covariance would represent the variation of any realization.

The down-sided risk approach penalizes appropriately the correct direction only. Especially when the worst case solution is at the upper or lower bound return value, the downside risk seems more realistic than a symmetric risk measure.

Let ς be given by

$$\varsigma = x - \bar{x}.$$

We reformulate the minimax problem (5.10) with the risk term as

$$\min_{\varsigma}\left\{\max_r\left(-\alpha\langle r, \varsigma\rangle \mid r^{\text{lower}} \leq r \leq r^{\text{upper}}\right) + (1 - \alpha)\langle \varsigma, C\varsigma\rangle \mid \varsigma + \bar{x} \in X\right\}.$$

$$(5.12)$$

As the objective is to maximize return in view of the worst case scenario, we may utilize the formulation in (5.11). In (5.12) the trade-off is between the worst-case portfolio and the traditional risk-averse policy. This problem can be solved by

$$\min_{\zeta^+, \zeta^-}\left\{\begin{array}{l} -\alpha\left(\langle r^{\text{lower}}, \zeta^-\rangle + \langle r^{\text{upper}}, \zeta^+\rangle\right) \\ +(1 - \alpha)\langle(\zeta^+ - \zeta^- + \bar{x}), C(\zeta^+ - \zeta^- + \bar{x})\rangle \\ \mid \zeta^+ - \zeta^- + \bar{x} \in X; \zeta^+, \zeta^- \geq 0 \end{array}\right\} \qquad (5.13)$$

The quadratic term in (5.12) minimizes the sensitivity of $-\langle r, \varsigma\rangle$ to changes in r. This is desirable as the actual performance of r may exhibit variation beyond the bounds of r. In that case, some degree of increased robustness might be traded off with a degree of deterioration in expected performance. Clearly, if only $-\langle r, \varsigma\rangle$ is important, then the risk-seeking solution of (5.12) is the desired strategy. That provides the best worst-case performance within the bounds and guarantees that actual performance will improve if any scenario, other than the worst case, is realized. However, it does not provide protection if the actual r transgresses the bounds.

The worst-case r for (5.12) is determined recognizing that the decision maker cares about the quadratic portfolio variance. The worst-case r is chosen to minimize the portfolio return; on the other hand, the portfolio investment decision, x, is computed ahead of r. Let ς_* be the solution of (5.12) and r be determined to create the most adverse condition in view of the sensitivity term and the linear term. After the event, the worst-case r is given by the linear programming problem

$$\max_r\left\{-\langle r, \varsigma_*\rangle \mid r^{\text{lower}} \leq r \leq r^{\text{upper}}\right\} \qquad (5.14)$$

which provides the lower bound performance of ς_*, provided r remains in the given bounds. Although a decision based on (5.10) would have better expected performance than (5.12), and consequently (5.14), if the bounds were not transgressed, (5.12) possesses the added robustness mentioned above. The lower bound given in (5.14) is calculated after ς_*. Therefore it does not influence ς_*, since the latter is determined by (5.12).

6 MULTISTAGE MINIMAX PORTFOLIO SELECTION

In this section, we present a multistage framework to reformulate the single-stage asset allocation problem as an adaptive multistage decision process. In the single-stage decision process the investor decides on the asset allocation based on expectations of returns at the horizon in the mean-variance sense, based on some predefined scenarios at the horizon in the minimax sense. In the multistage case, the investor decides based on expectations and/or scenarios up to some intermediate times prior to the horizon. These intermediate times may correspond to rebalancing or restructuring periods: we propose a system of deciding on the asset allocation with predefined rebalancing or restructuring times prior to the horizon. This assumption is not unreasonable in investment management where the horizon could be as long as 5 years and the restructuring may occur every year.

We present a two-stage process for simplicity. An asset allocation decision is to be made now, a restructuring to be made in 1 year's time and the liquidation of the portfolio to be made at the horizon which is in 2 years' time. We extend a single-period minimax model to a two-stage formulation. For ease of exposition, we show below the formulation for a one-stage problem as the starting point for our discussion. Other single-period minimax models may be adapted in a similar fashion. The model applies to a situation where there are rival return forecasts with fixed risk estimates, and with scenarios denoted by i.

$$\min_x \max_i \left\{ -\alpha \langle r^i, x \rangle + (1 - \alpha)\langle x - \bar{x}, C(x - \bar{x}) \rangle \right\}. \qquad (6.1)$$

In this formulation the value of r^i is essentially the expected return for each scenario i at the horizon of all the assets in the portfolio. In Figure 6.1, where a one-stage problem is considered, the branches emanating from the original node at time $t = 0$ represent scenarios. In this figure, there are five scenarios. At the tip of each branch is a terminal node whose return is r^i.[14]

We now extend (6.1) to accommodate a two-stage problem. For ease of exposition, we give in Figure 6.2 an extended version of the scenario tree of Figure 6.1. In Figure 6.2, the first stage is the same as the single stage of Figure 6.1. The second stage gives an expansion of each of the five nodes of the first

---

[14] Problem (6.1) can be re-formulated using $E(r^i)$ instead of r^i; in this case, (6.1) is the same formulation as (2.6) given in Section 2. For a formulation that uses an expected return estimate, there is no restriction imposed on how the expectation is generated. This implies that the return associated with each scenario is some expected value, perhaps provided by an economist's subjective forecast or by a forecasting model.

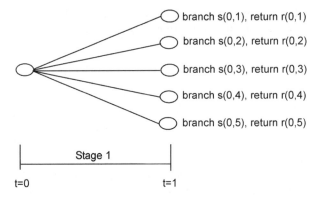

Figure 6.1   Tree structure showing a one-stage framework with 5 scenarios.

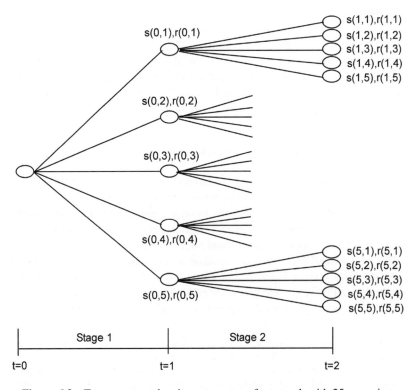

Figure 6.2   Tree structure showing a two-stage framework with 25 scenarios.

stage. This expansion shows that the return for each scenario of the first stage has been generated by taking the expectation over the five scenarios of the second stage that emanate from the relevant node. In the figure, $s(0, 1)$ is the first branch emanating from node 0, and its associated return is $r(0, 1)$. Similarly, $s(1, 2)$ is the second branch emanating from node 1, and its associated return is $r(1, 2)$.

We note that the nodes at the end of the second stage have return values that may also be based on expectation. In this case, the return associated with each second-stage scenario is some expected value, perhaps provided by an economist's subjective forecast or by a forecasting model. We shall not consider the detail of the nature of returns at Stage 2.

The use of a two-stage optimization formulation is more consistent with the nature of economic forecasting. The scenarios that forecasters may be able to generate for a one-year time frame (i.e., the return at the terminal nodes in Figure 6.1) may be more robust than those that the same forecasters may be able to generate for a 2-year horizon (i.e., the return at the terminal nodes in Figure 6.2). At the same time, forecasters are concerned about future events, perhaps as far forward as 2 years. A two-stage minimax formulation caters for this need for consistency.

We now explain the use of the scenario tree given in Figure 6.2. Let $s(a, b)$ represent a branch of a scenario tree from state a to state b, for example, $s(0, 1)$ is the branch emanating from state 0 to state 1. From Figure 6.2, we consider 5 branches in Stage 1, each with a return at the restructuring time $t = 1$. Each of the return, for each branch at time $t = 1$, is based on the 5 branches for Stage 2. For example, $r(0, 1)$ when is the return based on $\{s(1, 1), s(1, 2), s(1, 3), s(1, 4), s(1, 5)\}$; each of these branches emanating from $s(0, 1)$ has an associated return at the end of Stage 2, that is, at the horizon. The branches $\{s(1, 1), s(1, 2), s(1, 3), s(1, 4), s(1, 5)\}$ are the only relevant branches if, after Stage 1, state 1 is realized.

We define a scenario as a particular path through the tree, from state 0 to any one of the horizon states. For example, scenario S1 is the path $\{s(0, 1), s(1, 1)\}$. Similarly, scenario S2 is the path $\{s(0, 1), s(1, 2)\}$. The tree in Figure 6.1 has 25 scenarios. Therefore in the minimax formulation that we introduce below, the term scenario refers to any one of the 25 paths in Figure 6.2.

We introduce the concept of an ancestor state (Nielsen, 1997). Let $a_{t-1}(s)$ represent the ancestor state, that is, the state from which the current state at time t emanates. For scenario S1, the ancestor state of $s(1, 1)$ is $s(0, 1)$. Similarly, for scenario S2, the ancestor state of $s(1, 2)$ is $s(0, 1)$. With both $s(1, 1)$ and $s(1, 2)$, as well as $s(1, 3)$, $s(1, 4)$ and $s(1, 5)$, having a common ancestor $s(0, 1)$, it is clear that any decision to be made at Stage 2 would have to consider the conjoint effect of any previous asset allocation decision and the ancestor state on the portfolio.

At the end of Stage 1, the investor may wish to re-optimize to update her allocations in view of the relevant branches that she faces at that point. We note that by evaluating (6.1) up to time $t = 1$, that is, the restructuring time, the decision variable x may not be the overall optimal solution when the analysis is carried through to the horizon. In other words, by solving up to $t = 1$, we are not certain about the performance of the portfolio past that time. We now present a full multistage framework that determines a solution based on an analysis of all stages up to the horizon.

The discretization using scenarios approximates future expectations. Yet, each such approximation has an associated error which is the variability of the scenario which needs to be considered in addition to the variability over all scenarios. Let $x_t \in \Re^n$ be the asset allocation decision at time t, $x_t^i \in \Re^n$ be the vector of portfolio weights to be determined for scenario i and time t, and $x_2^+ \in \Re^n$ and $x_2^- \in \Re^n$, respectively, be the buy and sell decisions for Stage 2. Let $r(0, i)$ be the return at the end of Stage 1 based on branch i; similarly, let $r(i, j)$ be the expected return at the end of Stage 2 based on branch i in Stage 1 and branch j in Stage 2. Further, let m^b represent the number of branches that emanate from a node, while m^{sce} is the number of scenarios or paths in the tree. In Figure 6.2, $m^b = 5$ and $m^{sce} = 25$.

Recall that in the example above there are five branches emanating from a node and that there are 25 scenarios in total. In this formulation, the benchmark is assumed to be unchanged from Stage 1 to Stage 2. A varying benchmark can be used but this is not explored here.

The mean value of a multistage portfolio at time 1 is given by

$$\sum_{i=1}^{m^b} \theta^i r(0, i)^T (x_1^i - \bar{x}) \tag{6.1}$$

where θ^i is the associated probability function. The variance is given by

$$\sum_{i=1}^{m^b} \theta^i (x_1^i - \bar{x})^T \left(C^i + r(0, i) r(0, i)^T \right) (x_1^i - \bar{x})$$

(see Gulpinar et al., 2000). The mean and variance at Stage 2 are expressed similarly.

For one particular scenario s, the vectors $r(0, i)$ and $r(i, j)$ are unique, with parameters i and j both instantiated. When scenarios are used, the parameters i and j are both dependent on the scenario or path taken through the tree; hence the superscript s is used for identifying the return. Let $r^s(0, i^s)$ be the return at the end of Stage 1 based on branch i dependent on scenario s; similarly, let $r^s(i^s, j^s)$ be the return at the end of Stage 2 based on branch i in Stage 1 and branch j in Stage 2, where both parameters are dependent on scenario s. Suppose we consider only one scenario or path of the tree. We pick a unique

Stage 1 node and a unique Stage 2 node to yield

$$\min_{x_1^s, x_2^s, x_2^{s,+}, x_2^{s,-} \in \Re^n} \left\{ \begin{array}{l} -\alpha \langle r^s(0,i^s), x_1^s - \bar{x} \rangle \\ +(1-\alpha)\langle x_1^s - \bar{x}, (C + r^s(0,i^s)r^s(0,i^s)^T)(x_1^s - \bar{x}) \rangle \\ -\alpha \langle r^s(i^s,j^s), x_2^s - \bar{x} \rangle \\ +(1-\alpha)\langle x_2^s - \bar{x}, (C + r^s(i^s,j^s)r^s(i^s,j^s)^T)(x_2^s - \bar{x}) \rangle \end{array} \right\} \quad 6.2)$$

subject to generalized bounds constraints

$$l_1 \le x_1^s \le u_1$$

$$l_2 \le x_2^s \le u_2$$

$$\langle 1, x_1^s \rangle = 1$$

$$\langle 1, x_2^s \rangle = 1 + \langle (r^s(0,i^s)), x_1^s \rangle$$

$$x_1^s \ge 0$$

$$x_2^s \ge 0$$

$$x_2^s = x_1^s + x_2^{s,+} - x_2^{s,-}$$

$$x_2^{s,+} \ge 0$$

$$x_2^{s,-} \ge 0$$

$$\langle x_2^{s,+}, x_2^{s,-} \rangle = 0, \quad s = 1.$$

The last constraint simply ensures either $x_2^{s,+} > 0$ or $x_2^{s,-} > 0$ but not both. The constraint may be imposed as a penalty term in the objective. The rank-one terms $r^s(0,i^s)r^s(0,i^s)^T$ and $r^s(i^s,j^s)r^s(i^s,j^s)^T$ in the objective are simply intended to underline the influence of multiple scenarios discussed below. Assuming that the particular scenario s is realized, then (6.2) is an adequate formulation to find the optimal asset allocation. The presence of multiple scenarios makes (6.2) unrealistic but it can be used as a basis for accommo-dating other scenarios by weighting each scenario using its probability of occurrence. Additionally, we have to ensure that at the second stage we identify the path the scenario has taken in order to set a consistent system of accounting for the stage-one decision x_1^s. Recall that $a_{t-1}(s)$ denote the ancestor state; we introduce a new constraint called the nonanticipativity constraint. Let θ^s be the probability of scenario s. We have

$$\min_{x_1^s, x_2^s, x_2^{s,+}, x_2^{s,-} \in \Re^n} \left\{ \sum_{s=1}^{m^{sce}} \theta^s \begin{pmatrix} -\alpha \langle r^s(0, i^s), x_1^s - \bar{x} \rangle \\ +(1-\alpha)\langle x_1^s - \bar{x}, (C + r^s(0, i^s)r^s(0, i^s)^T)(x_1^s - \bar{x}) \rangle \\ -\alpha \langle r^s(i^s, j^s), x_2^s - \bar{x} \rangle \\ +(1-\alpha)\langle x_2^s - \bar{x}, (C + r^s(i^s, j^s)r^s(i^s, j^s)^T)(x_2^s - \bar{x}) \rangle \end{pmatrix} \right\}$$

(6.3)

subject to generalized bounds constraints

$$l_1 \leq x_1^s \leq u_1$$

$$l_2 \leq x_2^s \leq u_2$$

$$\langle 1, x_1^s \rangle = 1$$

$$\langle 1, x_2^s \rangle = 1 + \langle (r^s(0, i^s)), x_1^s \rangle$$

$$x_1^s \geq 0$$

$$x_2^s \geq 0$$

$$x_2^s = x_1^s + x_2^{s,+} - x_2^{s,-}$$

$$x_2^{s,+} \geq 0$$

$$x_2^{s,-} \geq 0$$

$$\langle x_2^{s,+}, x_2^{s,-} \rangle = 0$$

$$x_t^s = x_t^{s+1} \text{ for } t = 1, 2 \text{ and } a_{t-1}(s) = a_{t-1}(s+1), \quad \forall s.$$

The minimax equivalent of (6.3) is given by

$$\min_{x_1, x_2, x_2^+, x_2^- \in \Re^n} \max_s \left\{ \begin{matrix} -\alpha \langle r^s(0, i^s), x_1 - \bar{x} \rangle \\ +(1-\alpha)\langle x_1 - \bar{x}, (C + r^s(0, i^s)r^s(0, i_s)^T)(x_1 - \bar{x}) \rangle \\ -\alpha \langle r^s(i^s, j^s), x_2 \rangle \\ +(1-\alpha)\langle x_2 - \bar{x}, (C + r^s(i^s, j^s)r^s(i^s, j^s)^T)(x_2 - \bar{x}) \rangle \end{matrix} \right\}$$

(6.4)

subject to generalized bounds constraints

$$l_1 \leq x_1 \leq u_1$$

$$l_2 \leq x_2 \leq u_2$$

$$\langle 1, x_1 \rangle = 1$$

$$\langle 1, x_2 \rangle \leq 1 + \langle (r^s(0, i^s)), x_1 \rangle, \quad \forall s$$

$$x_1 \geq 0$$

$$x_2 \geq 0$$

$$x_2 = x_1 + x_2^+ - x_2^-$$

$$x_2^+ \geq 0$$

$$x_2^- \geq 0$$

$$\langle x_2^+, x_2^- \rangle = 0.$$

Problem (6.4) can be reformulated as the nonlinear program given by

$$\min_{Z \in \mathfrak{R}^1, x_1, x_2, x_2^+, x_2^- \in \mathfrak{R}^n} \{Z\} \tag{6.5}$$

subject to

$$-\alpha \langle r^s(0, i^s), x_1 - \bar{x} \rangle + (1 - \alpha) \langle x_1 - \bar{x}, (C + r^s(0, i^s) r^s(0, i^s)^T)(x_1 - \bar{x}) \rangle \leq Z$$

$$-\alpha \langle r^s(i^s, j^s), x_2 - \bar{x} \rangle + (1 - \alpha) \langle x_2 - \bar{x}, (C + r^s(i^s, j^s) r^s(i^s, j^s)^T)(x_2 - \bar{x}) \rangle \leq Z$$

$$l_1 \leq x_1 \leq u_1$$

$$l_2 \leq x_2 \leq u_2$$

$$\langle 1, x_1 \rangle = 1$$

$$\langle 1, x_2 \rangle \leq 1 + \langle (r^s(0, i^s)), x_1 \rangle, \quad \forall s$$

$$x_1 \geq 0$$

$$x_2 \geq 0$$

$$x_2 = x_1 + x_2^+ - x_2^-$$

$$x_2^+ \geq 0$$

$$x_2^- \geq 0$$

$$\langle x_2^+, x_2^- \rangle = 0.$$

Problem (6.5) is a flat minimax structure that looks at all possible scenarios that span the two stages. Clearly, other formulations are also possible where minimax and expected value optimization are combined. We discuss such strategies further in Chapter 10.

7 PORTFOLIO MANAGEMENT USING MINIMAX AND OPTIONS

The widespread use of options can be attributed partly to their insurance capability, where a portfolio's value is preserved by buying an option to sell a declining asset at a favorable price. There are many other perceived benefits from dealing with options, but from the point of view of portfolio management, the insurance capability remains to be perceived as the main benefit.

The use of options has entered the mainstream investment strategies of fund managers who operate in the alternative investments domain. These include hedge funds, arbitrage funds and commodity trading advisors who devise relatively more complex investment strategies compared to fund managers of traditional portfolios. For fund managers offering alternative investments, the use of options as a component part of their strategy provide the needed insurance to guard against extreme volatility, particularly on the downside.

An example of a hedge fund strategy that employs options is that of simultaneously buying a basket of stocks and selling another basket of stocks such that the net market exposure achieves a desired level and that both long and short sides of the portfolio are expected to generate positive returns. Options are then bought to protect the net portfolio performance and maintain the performance within some bounds. The long positions are insured against an unexpected fall in prices, and the short positions are also insured against an unexpected rise in prices. In order to keep a tight control on the bounds on performance, the fund manager may have to purchase expensive options whose strike price is consistent with the manager's desired cut-off point on poor performance. As the fund manager goes down the spectrum of available strike prices, the premium she pays goes down, depending on how far out-of-the-money she is prepared to go. The further out-of-the-money the option is, the weaker the insurance it provides. Because the decision to buy or sell an option, and at what strike price, is an integral part of the manager's investment strategy, she has to be able to assess the trade-off between premium levels and insurance benefit from the option.

An interesting question to raise in portfolio management is: How far out-of-the-money can an option be and still remain attractive as an insurance provider, if such use of an option is complemented by an active portfolio management via minimax. In view of the property of minimax strategies to provide a minimum guaranteed performance with respect to some predefined scenarios, a fund manager can employ active management on the portfolio using a minimax-based strategy, and complement this with out-of-the-money options. With this combination of minimax and options, the fund manager would be able to achieve at least the guaranteed minimum performance via minimax, complemented by the lower premium paid on an out-of-the-money option.

We formulate the combined minimax and option strategy for a portfolio with long positions in stocks. For the minimax strategy, we define upper and lower bounds on the prices of the stocks that constitute the portfolio. Minimax optimization would result in the simultaneous identification of the worst-case scenario and the asset allocation that minimizes the impact of the worst-case scenario on performance. The resulting asset allocation provides the minimum guaranteed performance. Hence, if the worst-case scenario does not materialize and if the future realization of stock prices falls within the predefined bounds, then we can expect an improvement in performance relative to the identified guaranteed minimum performance. In the formulation below, we concentrate on a portfolio of long stock positions only and on put options to provide insurance against a fall in stock prices. The analysis can easily be extended to a portfolio of both long and short stock positions by adding the appropriate options to provide insurance for the short stock positions.

Let $x \in \mathfrak{R}^n$ represent the unknown asset allocation to n stocks, and $C \in \mathfrak{R}^{n \times n}$ be the covariance matrix of returns on the stocks. Let $y_t^S \in \mathfrak{R}^n$, $y_{t+1}^S \in \mathfrak{R}^n$ represent the stock price vectors at times t and $t + 1$, with $y_{t+1}^{S,\text{upper}} \in \mathfrak{R}^n$ and $y_{t+1}^{S,\text{lower}} \in \mathfrak{R}^n$ representing the upper and lower bounds on y_{t+1}^S. Let $\alpha \in [0, 1]$ represent the trade-off between return-maximization and risk-minimization[15]. The minimax formulation is given[16] by

$$\min_{x \in \mathfrak{R}^n} \max_{y_{t+1}^S \in \mathfrak{R}^n} \{-\alpha \langle x, r_{t+1} \rangle + (1 - \alpha) \langle x, Cx \rangle\} \qquad (7.1)$$

subject to

$$y_{t+1}^{S,\text{lower}} \leq y_{t+1}^S \leq y_{t+1}^{S,\text{upper}}$$

$$\langle 1, x \rangle = 1$$

$$x \geq 0$$

$$r_{i,t+1} = \frac{y_{i,t+1}^S}{y_{i,t}^S} - 1, \quad \forall i = 1, \dots, n.$$

[15] The superscript S in this section is an asset identifier (i.e., S for stock) in contrast to s in previous sections which denote scenarios.

[16] This is just one of the many formulations that one can use. For the purpose of clarity, we confine the discussion to this basic formulation.

The solution to (7.1) is a vector x^*, the optimal asset allocation, that minimizes the worst-case performance where the worst-case is given by $y_{t+1}^{S^*}$.

The worst-case, as given by $y_{t+1}^{S^*}$, is a vector of future stock prices that effectively define the lower bound on the price of each stock for which the minimum guaranteed performance applies. We emphasize that if the future realization of stock prices go even further below $y_{t+1}^{S^*}$, then the minimax strategy's minimum guarantee will fail. The use of options may complement the above minimax strategy to cover the situation where the future stock prices do fall below $y_{t+1}^{S^*}$. By buying put options whose strike prices coincide with $y_{t+1}^{S^*}$, the fund manager is paying for out-of-the-money options, where option premiums are relatively low, to provide the insurance cover for price falls below $y_{t+1}^{S^*}$. With a minimax strategy and with put options used in this manner for insurance purposes, the manager has locked in the value of the guaranteed minimum performance.

Three issues are worth mentioning with regard to the use of options to cover the eventuality of prices going further below $y_{t+1}^{S^*}$. The first concerns the availability of options at the strike prices defined by $y_{t+1}^{S^*}$. If options cannot be traded at the levels given by $y_{t+1}^{S^*}$, the fund manager may wish to consider revising the upper and lower bounds used in the minimax optimization. If options exist with strike prices very close to $y_{t+1}^{S^*}$, then the manager may wish to choose the options with the closest strike prices.

The second issue concerns type of options used and the number of options to hold. As discussed extensively in Chapter 8, the manager may wish to fully cover the portfolio, or to partially cover and dynamically hedge. We assume the manager chooses the hedge ratio via some strategy, perhaps by using strategies similar to those suggested in Chapter 8. We shall not consider this aspect further in this section.

The third issue is the availability of options for all the stocks that constitute the portfolio. If options are not available, the fund manager may choose to use instruments in the futures market. In our example, where a stock portfolio is being managed, the fund manager may wish to trade stock index futures options in lieu of individual stock options. As the strike prices as suggested by $y_{t+1}^{S^*}$ are no longer relevant, the manager would necessarily be exposed to basis[17] risk. The choice of strike level for the relevant stock index future requires further modeling to ascertain the strike level most consistent with $y_{t+1}^{S^*}$ such that basis risk is minimal.

The discussion has so far concentrated on a combined minimax and option strategy where the choice of options is decided after the optimization, that is, after knowing the worst-case stock prices given by $y_{t+1}^{S^*}$. This strategy does not take advantage of the potential benefit from an optimal identification of the

---

[17] This is a type of risk attributable to the difference between the assets to be hedged and the derivative's underlying instrument.

option consistent with $y_{t+1}^{S^*}$ and x^*. By including the option within the formulation, the manager benefits from the refinement of $y_{t+1}^{S^*}$ such that the resulting option, with strike prices given by $y_{t+1}^{S^*}$, are simultaneously optimal with the performance of the portfolio, plus the insurance provided by the option.

We use Black and Scholes' option pricing formula[18] to illustrate how a formulation can incorporate the identification of the out-of-the-money options that is simultaneously optimal with both the worst-case scenario, as given by $y_{t+1}^{S^*}$, and the resulting asset allocation, as given by x^*. As before, we assume a fund manager has a stock portfolio with long positions only.

The Black and Scholes formula for a European put option is given below, using the same notation as in Chapter 8. We use this formula to illustrate the generic formulation that can be used without going into the detail of different pricing formulas. While there is no exact analytic formula for the value of an American put option, practitioners can use numerical procedures for its estimation. The Black and Scholes formula for a put option for stock i is

$$B_i = Xe^{-r(T-t)}\Theta(-d_2) - y_{i,t}^S\Theta(-d_1) \tag{7.2}$$

$$d_1 = \frac{\ln(y_{i,t}^S/X) + (r + (\sigma^2/2))(T - t)}{\sigma\sqrt{T - t}} \tag{7.3}$$

$$d_2 = \frac{\ln(y_{i,t}^S/X) + (r - (\sigma/2))(T - t)}{\sigma\sqrt{T - t}} = d_1 - \sigma\sqrt{T - t} \tag{7.4}$$

where $\Theta(d)$ is the cumulative normal distribution function (see CN 8.3), that is,

$$\Theta(d) = \int_{-\infty}^{d} \frac{1}{\sqrt{2\pi}} e^{-z^2/2} \, dz. \tag{7.5}$$

The strike price X is a required parameter input into the above expressions. From our previous discussion of $y_{t+1}^{S^*}$, recall that the worst-case, as given by $y_{t+1}^{S^*}$, should be viewed as the strike price vector to consider when buying relevant options. By using y_{t+1}^S instead of X in (7.2)–(7.5), the strike price can be optimally determined. Let W be the total wealth in dollar value. Consider the revised formulation given by

$$\min_{x\in\mathcal{R}^n} \max_{y_{t+1}^S\in\mathcal{R}^n} \left\{ -\alpha\left(\langle x, r_{t+1}\rangle + \sum_{i=1}^{n} \frac{x_i W}{y_{i,t}^S} B_i \right) + (1 - \alpha)\langle x, Cx\rangle \right\} \tag{7.6}$$

subject to

$$y_{t+1}^{S,\text{lower}} \leq y_{t+1}^S \leq y_{t+1}^{S,\text{upper}}$$

---

[18] This is presented in more detail in Chapter 8.

$$\langle 1, x \rangle = 1$$

$$x \geq 0$$

$$r_{i,t+1} = \frac{y_{i,t+1}^{S}}{y_{i,t}^{S}} - 1$$

$$B_i = y_{i,t+1}^{S} e^{-r(T-t)} \Theta(-d_2) - y_{i,t}^{S} \Theta(-d_1), \quad \forall i = 1, ..., n.$$

In this formulation, the upper and lower bounds on $y_{i,t+1}^{S}$ are determined by the availability of strike prices in the market, that is, the lower bound $y_{i,t+1}^{S,lower}$ is determined by the lowest quoted strike price in the market, and the upper bound $y_{i,t+1}^{S,upper}$ is determined by the highest quoted strike price. In the objective function, the number of stock i, given by the proportion of wealth x_i times the total wealth W divided by the stock price $y_{i,t}^{S}$, is multiplied with the option price B_i to get the option premium to be paid for the insurance.

By bounding the maximizing variable by the actual strike prices available in the market, the solution to the above formulation is an allocation x^* that is simultaneously optimal with actual market options whose strike prices are given by $y_{t+1}^{S^*}$.

8 CONCLUDING REMARKS

In this chapter we have presented potential uses of minimax in portfolio construction and have demonstrated the benefits from using minimax.

Portfolios benefit from minimax portfolio optimization when managers with differing forecasts fail to find a consistently acceptable level of compromise. By using minimax, the compromise is optimally determined at the decision level and not at the forecast level where a compromise may not be politically optimal. We have demonstrated the benefits in terms of the decreased suboptimality across models by showing that when portfolios were created under a specific model and then placed in an environment defined by a different model, these portfolios underperformed markedly against expectations based on the first model, but when portfolios are minimax-optimal, their performance is consistent across models.

Portfolios benefit from dual benchmarking when the manager has flexibility in the type of portfolio can be constructed. In general, the resistance to deviate away from a benchmark in order to perform within the peer group norm prevents the manager from exploiting the power of dual benchmarking. In view of this resistance, the manager can still benefit by tilting the portfolio towards the dual portfolio which is nicely balanced between risk and return. We have tried to show that while one benchmark, being simple and transparent, is good, two are better; but two are better only if managers combine them on the basis of an optimized solution to the problem of meeting multiple objectives; in short, if they use dual benchmarking.

In Section 5 we presented alternative minimax portfolio selection models for investors interested in benchmark tracking, as well as downside risk-averse investors. In Section 6 we considered a multistage framework that can address the problem of finding portfolios that are robust across different scenarios and across different time stages. This formulation would enable a decision maker to form a robust asset allocation in view of a future or a number of future re-allocations. Finally, in Section 7, we presented a formulation where the guaranteed minimum performance of a minimax portfolio is preserved by using options to cover those eventualities outside the predefined minimax scenarios. In the formulations given in this section, the complementary interaction between minimax and the use of options is highlighted where the benefit to the fund manager comes in terms of lower insurance premium.

References

Clarke, R.G. and H. de Silva (1998). "State-Dependent Asset Allocation", *Journal of Portfolio Management*, Winter Edition, 57–64.

Dembo, R.S. (1991). "Scenario Optimization", *Annals of Operations Research*, 30, 63–80.

Don Ezra, D. (1991). "Asset Allocation by Surplus Optimisation", *Financial Analysts Journal*, 51–57.

Duarte, Jr., A.M. (1994). "A Comparative Study of Downside Risk and Volatility in Asset Allocation", *Investigation Operative*, 4, 213–228.

Gulpinar, N., B. Rustem and R. Settergren (2000). "Multi-Stage Stochastic Programming in Computational Finance", Imperial College, DOC Report.

Leibowitz, M.L. and S. Kogelman (1991a). "Return Enhancement from "Foreign" Assets: A New Approach to the Risk/Return Trade-off", *Journal of Portfolio Management*, Summer Edition, 5–13.

Leibowitz, M.L. and S. Kogelman (1991b). "Asset Allocation Under Shortfall Constraints", *Journal of Portfolio Management*, Winter Edition, 18–23.

Levy, H. and H. M. Markowitz (1979). "Approximating Expected Utility by a Function of Mean and Variance", *American Economic Review*, 69 (3), 308–317.

Lipman, R.A. (1990). "Utility, Benchmarks and Investment Objectives", *Transactions of the Institute of Actuaries of Australia*, 842–869.

Maenhout, P.J. (1999). "Robust Portfolio Rules and Asset Pricing", Job Market Paper, Department of Economics, Harvard University.

Markowitz, H. (1952). "Portfolio Selection", *Journal of Finance*, 7 (1), 77–91.

Morgan Guaranty Trust Co. and Reuters Ltd. (1996). "RiskMetrics – Technical Document", New York.

Nielsen, S. (1997). "Mathematical Modeling and Optimisation with Applications to Finance", unpublished lecture notes.

Price, K., B. Price and T.J. Nantell (1982). "Variance and Lower Partial Moment Measures of Systematic Risk: Some Analytical and Empirical Results", *Journal of Finance*, 37, 843–855.

Roll, R. (1992). "A Mean-Variance Analysis of Tracking Error", *Journal of Portfolio Management*, Summer Edition, 13–22.

Rustem, B., Becker, R.G. and W. Marty (2000). "Robust Min-Max Portfolio Strategies For Rival Forecast and Risk Scenarios", *Journal of Economic Dynamics and Control*, 24, 1591–1623.

Sharpe, W.F. (1994). "The Sharpe Ratio", *Journal of Portfolio Management*, Fall Edition, 49–58.

Worzel, K.J., C. Vassiadou-Zeniou and S.A. Zenios (1994), "Integrated Simulation and Optimisation Models for Tracking Indices of Fixed Income Securities", *Operations Research*, 42 (2), 223–233.

COMMENTS AND NOTES

CN 1: Benchmarks and Benchmarking

Benchmarks in portfolio management refer mainly to public indices or portfolios used mainly for the purpose of performance attribution. Benchmarks are broad-based portfolios or indices that reflect such broad-based portfolios. An example of a widely used benchmark is the MCSI EAFE which represents a capitalization-based allocation to countries in Europe, Asia and the Far East. This is a relevant benchmark for US fund managers who have asset and currency exposures to these regions. Another example is LIBOR, the London Inter-Bank Offered Rate, which represents the most liquid sector of the money market. As a fund manager's portfolio is assessed relative to a benchmark, the fund manager would tend to make decisions with a view to outperforming the benchmark. This need for relative attribution leads to a portfolio management technique called benchmark tracking, where deviations from a benchmark's allocation become the focus of decision-making.

CN 2: Downside Risk

Downside risk is a measure of a potential shortfall from a target or threshold return. Several versions of downside risk can be used depending on the nature of the problem to be solved. Below is a generalized formulation, from which variations can be applied to fit the problem's requirements.

$$D = \left(\int_{r=-\infty}^{r=t} (T - r)^N P(r) \partial r \right)^{\frac{1}{N}}$$

where T is the target return, P is the probability of occurrence of r, and N is a parameter that reflects risk aversion. When $N=2$, the formulation takes the form given in Price, et.al (1982).

Chapter 10

Asset/liability management under uncertainty

In this chapter, we present robust methods for solving Asset/Liability Management (ALM) problems. Whereas Chapter 9 concentrates on assets-only optimization, this chapter explores the difficulty of simultaneously optimizing both the asset and the liability sides of a portfolio. Alternative minimax formulations with differing objective functions are presented, depending on the targets or objectives of the ALM portfolios. We illustrate the robustness property of the minimax formulation when the liability structure of an ALM portfolio is sensitive to shifts in yield curves. It can be shown that the minimax solution, as compared to standard immunization, provides the least deterioration in the value of the ALM portfolio.

We also present extended stochastic ALM models that deal with multiperiod objectives and varying performance horizons as well as varying investment, or benchmarking, horizons. The minimax formulations are based on scenarios describing evolutionary paths for both assets and liabilities. These stochastic ALM models are useful for making a comprehensive evaluation of an ALM strategy, whether it is based on minimax, where portfolios are designed to be robust, or on standard ALM techniques, where specific objectives may have dominated the portfolio's construction.

1 INTRODUCTION

In Chapter 9, we present applications of minimax to asset management while in this chapter we consider an application to asset/liability management (ALM). In both areas the objective of the decision maker is asset returns enhancement. However, whereas in asset management the concern is with the management of the volatility of returns, in ALM this is compounded by the management of liabilities to ensure payments are met as they fall due. This implies that the modeler providing a suite of tools to the decision maker has to consider a more complex measure of risk and devise strategies for managing this risk.

The problem of setting up an asset portfolio such that the cashflows from this portfolio are used for managing liabilities is complicated by a number of issues. These are asset return enhancement, volatility of asset returns, full or partial cover of all future liabilities, surplus requirements, the existence of contingent assets and contingent liabilities, and performance monitoring and measurement. Some of these issues may be conflicting in nature. An example is the maximization of asset returns and the minimization of the volatility of those returns. Additionally, asset management objectives may be in conflict with the objective of meeting liabilities as they fall due. The challenge for the modeler is to strike a balance across varying objectives that an ALM decision maker may wish to address.

Some ALM models concentrate on the management of liabilities as this is relatively more difficult than the management of assets. A straightforward way to achieve the objective of meeting all liabilities is to fully match assets and liabilities. This method, referred to as cash-matching, or dedication, is a passive management system that ties up some assets now in order to service liabilities far out in the future. Other ALM models aim to match changes in assets with changes in liabilities. The rationale for these models is that any loss (gain) on the liabilities due to say, an interest rate move, is offset by the gain (loss) on the assets. These models are referred to as immunization models; they match interest rate sensitivities between the asset portfolio and the liability portfolio.

An immunized portfolio is deemed constructed when three basic conditions are satisfied. The first condition is that the present value (usually taken as the market value) of the assets must match the present value of the liabilities. The second is that the assets and liabilities also have the same average life when weighted by the present value of their respective flows. This is the "duration" measure of average life. The third is that the assets are more "convex" than the liabilities[1]. When these three conditions are met, the asset portfolio is said to be dominant over the liability portfolio. Immunization aims to maintain the dominance of the assets over the liabilities at minimum cost. Consequently, the asset and liability flows must be matched at the outset in terms of both initial present value and interest rate sensitivity.

Unlike cash-matching, immunization inherently requires portfolio changes over time. Where cash matching fulfils liabilities through the originally promised flows from coupons and principal payments, immunization generally depends upon offsetting changes in the value of the assets and that of the liabilities. The requirement for continuing changes in immunized portfolios arises from the need to preserve dominance. Such forced rebalancing demonstrate the dynamic nature of the immunization process.

Within the framework of immunization, the portfolio structure can take on many forms as long as the interest rate sensitivity meets the several conditions required to achieve dominance. In particular, the schedule of cashflows from the immunized portfolio need not correspond exactly to the period-by-period

---

[1] This means that the second derivative matrix of the combined assets and liabilities with respect to some underlying variable is positive semi-definite.

payouts of the liability stream. This provides a high level of flexibility in choosing an immunized portfolio. It also means that the ultimate fulfillment of the liabilities depends on a separate strategy that ensures the benefits from immunization are translated into benefits in the context of liability servicing.

In this chapter we consider both immunization and cash-matching, and present formulations of these two types of model in a minimax framework. We first study immunization and its development as a short-term oriented ALM model. We then study dedication and its development as a long-term oriented ALM model. Whereas dedication is long-term oriented, immunization is short term and it is deemed consistent with the short-term nature of asset management where yearly performance reporting is generally required. Because if this, immunization coupled with asset management has gained a lot of attention from various modelers.

Redington's (1952) initial proposal on immunization requires that interest rates be restricted to a flat yield curve, where a unique interest rate applies to all maturities, subject only to parallel movements. He defined duration in the context of a firm's net worth. However, it was Macaulay (1938) who gave the first definition of duration in a financial context (see CN 1). Grove (1974) extended Redington's work by immunizing a nonzero initial net worth. Kaufman (1984) investigated the immunization of the net worth asset ratio. Stock and Simonson (1988) looked at tax-adjusting the duration measure. Vanderhoof (1972) adapted Redington's approach and applied it to insurance companies. Common to all of the above is the assumption of a single interest rate for all discountings of cash flows, that is, a flat yield curve. Shiu (1986) notes that a flat yield curve would provide arbitrage opportunities. This defect has also been pointed out by Boyle (1978).

Fisher and Weil (1971) first extend Redington's model to reflect a nonflat term structure and developed a duration measure that is different from the Macaulay (1938) duration. They formalize the theory of immunization as presented by Redington, within a framework involving parallel yield curve shifts. They develop an immunization strategy based on duration that achieves the desired immunization objectives if, for a single liability problem with a known payout date, the duration of the asset portfolio is equal to the time to the liability payout date. In the general context of immunization of long duration liabilities, long duration assets should be used to achieve immunization. The existence of a demand for single payment notes is then inferred. The conclusion is that the market does not require single-payment notes with a variety of maturities because the duration of a portfolio is a linear function of the durations of its components. All that is needed is a single payment note whose duration is at least as long as the longest liability. It is also noted that such an asset can be combined with assets of short duration to achieve any duration in between.

Fisher and Weil's conclusion may be viewed as a statement of a necessary but not sufficient condition for immunization. In Bierwag et al. (1983a), the authors discuss stochastic process risk and conjecture that an immunization strategy implemented at a high level with as much compression as possible around each of the liability payout dates will tend to minimize stochastic process risk. This implies that, compared to the Fisher and Weil's conclusion, single payment notes with a variety of maturity may be required by the market in order to achieve the desired level of compression around liability payout dates to provide proper immunization. In other words, bullet strategies[2] around the liability payout dates are more effective than barbell strategies[3]. Fong and Vasicek (1984)discuss the effect of a steepening of the yield curve on barbell and bullet portfolios and show that barbell portfolios[4] result in a bigger decline in the value of the portfolio.

Fong and Vasicek discuss that an optimally immunized portfolio that has minimum exposure to interest rate changes is obtained by minimizing a risk measure which the authors refer to as M2, subject to the constraint that the asset portfolio duration equals the liability portfolio duration, where M2 is the weighted variance of time to payments around the horizon date. Shiu (1988) considers the problem of immunizing multiple liabilities and shows that a necessary and sufficient condition for the immunization of multiple liabilities is the separate immunization of each liability.

Cox et al. (1979) propose a duration measure that is different from the one proposed by Macaulay (1938). Their duration measure, which they refer to as a measure of basis risk, is stochastic by construction that depends on the parameters of a specified interest rate process and liquidity preferences of individuals. Boyle (1978) also develops an immunization strategy using stochastic term structure models. Hiller and Shapiro (1989) develop a stochastic programming model for the ALM problem where the term structure is superimposed on a deterministic ALM model to create the stochastic ALM model. This stochastic ALM model is then reformulated back into a deterministic scenario-based framework. We show, in Section 7, how their model can be adapted into a minimax formulation. Consigli and Dempster (1998) also develop a stochastic programming ALM model that uses a multistage recourse framework. We present their formulation and the equivalent minimax formulation in Section 8.

Hiller and Schaack (1990) propose a classification of structured bond portfolio modeling techniques in terms of the asset/liability problem type (deterministic or stochastic) and hedging methodology (dedication and duration

---

[2] In bullet strategies, a liability is offset by an asset or combination of assets of the same, or nearly the same, timing as the liability.

[3] In barbell strategies, a liability is offset by a combination of assets with earlier timing and assets with later timing compared to the timing of the liability.

[4] Barbell portfolios are portfolios of assets and liabilities that conform to barbell strategies. Similarly, bullet portfolios are those that conform to bullet strategies. See previous two footnotes.

matching). They note the differences between Macaulay duration, dollar duration and modified duration and point out that it is dollar duration that is an expression of a change in the value of an asset for a unit change in interest rate, that is, interest rate sensitivity (see CN 1). The unitless measure given by the modified duration is equal to the dollar duration divided by the value of the asset. Other definitions of duration corresponding to other yield curve dynamics may be found in Bierwag (1987).

While modern techniques have enabled immunization to address a wider range of yield curve behaviors, immunization still remains vulnerable to certain market movements. Shiu (1986) shows that immunization against a yield curve shift fails to provide protection against general yield curve shifts. Chambers et al. (1988) utilize a duration vector approach where they defined a vector in which the components reflect moments of adjusted times-to-receipt of the underlying cashflows. A general nonparallel shift approach to duration analysis is given by Reitano (1989, 1990a,b, 1991a,b, 1992), with applications of the nonparallel yield curve approach to measuring potential yield curve risk. Illustrations are provided on price behavior given nonparallel yield curve shifts and how a duration vector is superior to a single duration measure in describing interest rate sensitivity.

Sharpe and Tint (1990) present a different approach to managing assets and liabilities based on the *liability hedging credit* which is a measure of the degree to which a particular asset or asset class can provide utility for an investor with a particular set of liabilities. The emphasis is not on immunization but on asset allocation with a view to maximizing the investor's utility, taking into account the liabilities of the investor. Because of this approach, the method is not designed for immunizing liabilities but for penalizing an expected return criterion within an optimization framework. Several authors have similarly investigated the issue of immunization, including Shiu (1988); Bierwag et al. (1983b–d). An introduction to the topic is given in Fabozzi (1991, 1993).

A method is offered in Reitano (1989, 1990a, 1991a) for describing a multi-dimensional duration measure that represents interest rate sensitivity for a specific yield curve movement. The duration measure overcomes the limitation of earlier duration measures that deal with parallel shifts only. As with the earlier duration measures, multivariate duration measures interest rate sensitivity to one specific yield curve shift, either parallel or nonparallel.

In this chapter we consider the use of discrete and continuous minimax in ALM. In Section 2 we present a multidimensional immunization framework and extend it to a discrete minimax formulation. An illustration is given in Section 3 to highlight the performance of bond portfolios constructed using this framework under different yield curve scenarios. In Section 4, we define a risk measure with emphasis on the liability component of an asset/liability portfolio, and explore the role of this risk measure in identifying the point in

the spectrum of immunization strategies. In Section 5, we give a brief introduction to a continuous minimax formulation of the immunization framework
described in Section 2. In Section 6, we explore other immunization strategies
and present their minimax formulations. In Section 7, we depart from immunization and move close to dedication where liability timings play a critical
role in the formulation of the ALM system. We explore the need to consider a
stochastic ALM system and show how this system can be formulated within a
minimax framework. Similarly, in Section 8 we present a multistage recourse
framework for the stochastic programming ALM model and its minimax
equivalent.

2 THE IMMUNIZATION FRAMEWORK

2.1 Interest Rates

The ALM problems that we discuss cover both cash inflows and cash outflows
whose present values are determined by the timing of the cashflows and the
relevant interest rates for discounting purposes. At time t, different interest
rates, also called spot rates, apply to different cashflows and the full spectrum
of spot rates is referred to as the term structure of interest rates (see CN 2). We
use the notation r^0 to refer to the m-dimensional vector of interest rates with m
key maturities, that is, the term structure

$$r^0 = \begin{bmatrix} r_1^0 \\ \vdots \\ r_m^0 \end{bmatrix}.$$

This notation is used in Sections 2–4 to represent a possible future term
structure of interest rates that would be used as a scenario. The superscript
0 is then replaced by i, and the notation r^i represents a scenario for the term
structure of interest rates.

2.2 The Formulation

Reitano (1990b, 1991b, 1992) demonstrates that classical duration and
convexity analysis (see CN 1) can be readily generalized to include yield
curve shifts that are nonparallel. A model is presented that explicitly recognizes the multivariate nature of yield curve changes. A multivariate price
function $P(r)$ is defined as the price of an asset as a function of interest
rates r, where r is given by a vector of interest rates called the yield curve
vector. Also, the price sensitivity to a change in interest rate is defined as
$P_j(r) = \partial P/\partial r_j$ where j represents the jth element in the yield curve vector. The

model estimates $P(r^0 + \Delta r)$ directly, where r^0 is the initial yield curve vector and $\Delta r = [\Delta r_1, \Delta r_2, ..., \Delta r_m]^T$ is a yield change vector. Both yield and yield change refer to spot values, that is, the spot curve.

Consider the following m-dimensional first and second order expansion of the Price function:

$$P(r^0 + \Delta r) \approx P(r^0) + \sum_j P_j(r^0)\Delta r_j \qquad (2.1)$$

$$P(r^0 + \Delta r) \approx P(r^0) + \sum_j P_j(r^0)\Delta r_j + \frac{1}{2}\sum_j\sum_k P_{jk}(r^0)\Delta r_j\Delta r_k \qquad (2.2)$$

where $P_{jk}(r) = \partial^2 P/\partial r_j\partial r_k$, with j and k representing the jth and kth element of the yield curve vector. Given the multivariate price function $P(r)$, the jth partial duration function, denoted $D_j(r)$, is defined for $P(r) \neq 0$ as follows:

$$D_j(r) \equiv -\frac{P_j(r)}{P(r)} \qquad (2.3)$$

Given the multivariate price function $P(r)$, the (jk)th partial convexity function, denoted $C_{jk}(r)$, is defined for $P(r) \neq 0$ as follows:

$$C_{jk}(r) \equiv \frac{P_{jk}(r)}{P(r)}. \qquad (2.4)$$

Given the above definitions, the total duration vector, denoted $D(r)$, and the total convexity matrix, denoted $C(r)$, are defined as follows:

$$D(r) \equiv [D_1(r), D_2(r), ..., D_m(r)] \qquad (2.5)$$

$$C(r) \equiv \begin{bmatrix} C_{11}(r) & \cdots & C_{1m}(r) \\ \vdots & \ddots & \vdots \\ C_{m1}(r) & \cdots & C_{mm}(r) \end{bmatrix}. \qquad (2.6)$$

Using these definitions, (2.1) and (2.2) can be expressed as

$$P(r^0 + \Delta r) \approx P(r^0)\left(1 - \langle D(r^0), \Delta r\rangle\right) \qquad (2.7)$$

$$P(r^0 + \Delta r) \approx P(r^0)\left(1 - \langle D(r^0), \Delta r\rangle + \frac{1}{2}\langle \Delta r, C(r^0)\Delta r\rangle\right). \qquad (2.8)$$

We state below the conditions for multivariate immunization within an asset/liability framework. In ALM, the surplus, that is, the assets in excess of the liabilities, is a measure of the cushion that the total portfolio has in preventing extreme drawdowns if future liabilities reach high levels. Let $S(r)$ be the value of the surplus given the spot curve r. Let $S(r^0 + \Delta r)$ denote the value of surplus if the spot curve moves from r^0 to $r^0 + \Delta r$. The equivalent formulation for (2.7) and (2.8) for the surplus function is, respectively,

$$S(r^0 + \Delta r) \approx S(r^0)\Big(1 - \langle D_S(r^0), \Delta r\rangle\Big) \tag{2.9}$$

$$S(r^0 + \Delta r) \approx S(r^0)\Big(1 - \langle D_S(r^0), \Delta r\rangle + \frac{1}{2}\langle \Delta r, C_S(r^0)\Delta r\rangle\Big) \tag{2.10}$$

where D_S is the duration of surplus and C_S is its convexity.

From (2.9) and (2.10), in order to have $S(r^0 + \Delta r)$ no smaller than $S(r^0)$, we must have $D_S = 0$. Further, we require C_S to be positive definite to ensure immunization. To implement this surplus immunization, we require relationships to hold between D_S and C_S for both assets and liabilities. For an asset portfolio A, that is, a vector of assets, and a liability portfolio L, that is, a vector of liabilities, these relationships are

$$\langle SD_S(r^0), \Delta r\rangle = \langle AD_A(r^0), \Delta r\rangle - \langle LD_L(r^0), \Delta r\rangle \tag{2.11}$$

$$\langle \Delta r, SC_S(r^0)\Delta r\rangle = \langle \Delta r, AC_A(r^0)\Delta r\rangle - \langle \Delta r, LC_L(r^0)\Delta r\rangle. \tag{2.12}$$

Here, AD_A and AC_A are the asset portfolio's duration vector and convexity matrix. Similarly, LD_L and LC_L are the liability portfolio's duration vector and convexity matrix. These are specified in (2.19)–(2.22) below. From (2.11) and (2.12), the conditions on duration and convexity of surplus can be met if

$$AD_A - LD_L = 0 \tag{2.13}$$

$$\langle x, (AC_A - LC_L)x\rangle > 0 \tag{2.14}$$

for an arbitrary vector x.

In modeling, we can identify conditions under which complete immunization is achieved. Unfortunately, in practice, the conditions on the durational structures of assets and liabilities are very restrictive and potentially difficult to implement. The same holds for the convexity structures. From (2.13), one can attempt to minimize the norm of $(AD_A - LD_L)$. However, a nonzero norm may result in

$$\langle AD_A(r^0), \Delta r\rangle - \langle LD_L(r^0), \Delta r\rangle \neq 0 \tag{2.15}$$

and therefore a failed immunization for unfavorable Δr. This limitation leads to a directional immunization which assumes a specific spot curve shift. Directional immunization requires that, for an assumed shift Δr, we have

$$\langle AD_A(r^0), \Delta r\rangle - \langle LD_L(r^0), \Delta r\rangle = 0 \tag{2.16}$$

$$\langle \Delta r, AC_A(r^0)\Delta r\rangle - \langle \Delta r, LC_L(r^0)\Delta r\rangle > 0. \tag{2.17}$$

Assuming a spot curve shift Δr^1, we can construct an asset portfolio under directional immunization such that $S(r^0 + \Delta r^1) \geq S(r^0)$. Similarly, assuming another spot curve shift Δr^2, we can construct another asset portfolio such that $S(r^0 + \Delta r^2) \geq S(r^0)$. Unfortunately, the asset portfolio that is intended to immunize against Δr^1 may fail to immunize against Δr^2.

Consider, therefore, an immunization framework where we assume a number of scenarios, each scenario being represented by a specific spot curve shift Δr. We wish to find an asset portfolio that gives the best compromise across all the predefined scenarios, that is, an asset portfolio such that the worst case $S(r^0) - S(r^0 + \Delta r)$ drop in surplus is minimized. Within this framework, a solution can be found using minimax optimization.

The minimax directional immunization is

$$\min_{x \in \Re^n} \max_{i \in I} \left\{ \sum_j^n x_j P_j + \left(-\langle AD_A(r^0, x), \Delta r^i \rangle + \langle LD_L(r^0), \Delta r^i \rangle \right)^+ \right\} \quad (2.18)$$

subject to

$$\langle \Delta r^i, AC_A(r^0, x)\Delta r^i \rangle - \langle \Delta r^i, LC_L(r^0)\Delta r^i \rangle \geq \epsilon$$

for all $i \in I$ and

$$x \geq 0$$

$$\langle 1, x \rangle = 1$$

where, as in earlier chapters, $(\cdot)^+ = \max(\cdot, 0)$, 1 indicates the column vector whose elements are all unity, $I = \{1, 2, ..., m^{sce}\}$ is the finite set of spot curve shifts, $\epsilon > 0$ is some predefined convexity restriction, $x \in \Re^n$ is the unknown vector of asset weights, $P \in \Re^n$ is the vector of asset prices,

$$AD_A(r^0, x) = \begin{bmatrix} \sum_j^n x_j P_j D_{A,1}^j(r^0) \\ \vdots \\ \sum_j^n x_j P_j D_{A,m}^j(r^0) \end{bmatrix} \quad (2.19)$$

$$LD_L(r^0) = \begin{bmatrix} \sum_j^{\text{liabilities}} L_j D_{L,1}^j(r^0) \\ \vdots \\ \sum_j^{\text{liabilities}} L_j D_{L,m}^j(r^0) \end{bmatrix} \quad (2.20)$$

$$AC_A(r^0, x) = \begin{bmatrix} \sum_j^n x_j P_j C_{A,11}^j(r^0) & \cdots & \sum_j^n x_j P_j C_{A,1m}^j(r^0) \\ \vdots & \ddots & \vdots \\ \sum_j^n x_j P_j C_{A,m1}^j(r^0) & \cdots & \sum_j^n x_j P_j C_{A,mm}^j(r^0) \end{bmatrix} \qquad (2.21)$$

$$LC_L(r^0) = \begin{bmatrix} \sum_j^{\text{liabilities}} L_j C_{L,11}^j(r^0) & \cdots & \sum_j^{\text{liabilities}} L_j C_{L,1m}^j(r^0) \\ \vdots & \ddots & \vdots \\ \sum_j^{\text{liabilities}} L_j C_{L,m1}^j(r^0) & \cdots & \sum_j^{\text{liabilities}} L_j C_{L,mm}^j(r^0) \end{bmatrix}. \qquad (2.22)$$

We note that previous references to $AD(r^0)$ and $AC(r^0)$ in (2.11) to (2.17) are now parameterized by the decision variable x. Also, from the definitions above, the asset portfolio is impacted by the decision variable x, where $A = \sum_j^n x_j P_j$, whereas the liability portfolio, $L = \sum_j^{\text{liabilities}} L_j$, is not.

Note that when the set I is a singleton, we are immunizing against one specific spot curve shift Δr and the minimax formulation reduces to the Reitano (1989, 1990a,b, 1991a,b, 1992) framework. Additionally, when Δr is a parallel shift, then the minimax formulation reduces further to the classical immunization framework.

3 ILLUSTRATION

In this section, we illustrate the benefits from using the minimax formulation by addressing the problem of a German institution with a stream of bond liabilities denominated in the local currency. The institution wishes to set up an asset portfolio that would immunize against changes in the value of its bond liabilities, presented in Table 3.1. All liabilities and assets are hypothetical. A set of bond assets can be used for immunization; these assets are given in Table 3.2. Portfolios optimized against different yield curve shifts, and a portfolio optimized using minimax are presented in Table 3.3. Finally, Table 3.4 shows the performance of these portfolios if a particular yield curve scenario is realized in the future.

The pricing of all assets and liabilities are based on a hypothetical spot curve given in Figure 3.1 (thick central yield curve). The figure also includes the spot curve scenarios that the institution wishes to consider in the immunization.

Table 3.1   Bond liabilities

Bond	Price	Yearly coupon (%)	Years to maturity	Amount (million DM)
L1	103.8	5.5	3.82	100
L2	110.2	6.75	5.86	100
L3	117.8	7.75	7.86	100
L4	102.5	5.5	9.86	100
L5	114.2	6.5	13.36	100

Table 3.2   Bond assets

Bond	Price	Yearly coupon (%)	Years to maturity	Amount (million DM)
A1	106.9	7	2.82	x_1
A2	115.3	8.5	4.94	x_2
A3	109.8	6.75	6.86	x_3
A4	108.8	6.5	8.86	x_4
A5	125.4	8.5	10.36	x_5
A6	136.6	8.5	15.04	x_6

Table 3.3   Optimal allocations (individually optimized against the 8 SC scenarios, first 8 columns; column 9 for optimization of the expected value; column 10 for minimax)

Asset	SC Scen 1	SC Scen 2	SC Scen 3	SC Scen 4	SC Scen 5	SC Scen 6	SC Scen 7	SC Scen 8	All SC Scens	Minimax
A1	120	122	101	139	149	43	125	98	0	86
A2	9	9	24	11	11	3	18	29	131	12
A3	64	65	135	74	77	131	97	131	96	119
A4	73	74	107	84	89	166	99	112	0	136
A5	50	51	42	58	63	102	51	40	213	73
A6	137	133	63	97	77	27	78	61	11	46

Table 3.4 Portfolio (individually optimized against the 8 SC scenarios, first 8 rows; row 9 for optimization of the expected value; row 10 for minimax) change in value, that is, performance, under each of the 8 SC scenarios.

Portfolio optimized under SC	SC Scen 1 is realized	SC Scen 2 is realized	SC Scen 3 is realized	SC Scen 4 is realized	SC Scen 5 is realized	SC Scen 6 is realized	SC Scen 7 is realized	SC Scen 8 is realized
SC Scen 1 portfolio	0	13	−460	−319	391	237	−643	764
SC Scen 2 portfolio	22	0	−437	−285	363	233	−597	722
SC Scen 3 portfolio	286	−87	0	216	−40	117	159	19
SC Scen 4 portfolio	211	−109	−247	0	128	202	−210	368
SC Scen 5 portfolio	316	−172	−146	155	0	186	−1	176
SC Scen 6 portfolio	−8	162	23	−42	104	0	−60	62
SC Scen 7 portfolio	269	−115	−116	131	23	162	0	170
SC Scen 8 portfolio	291	−88	12	228	−51	112	179	0
All SC Scens portfolio	49	39	−13	6	4	−5	−24	−18
Minimax portfolio	175	0	0	112	11	74	63	36

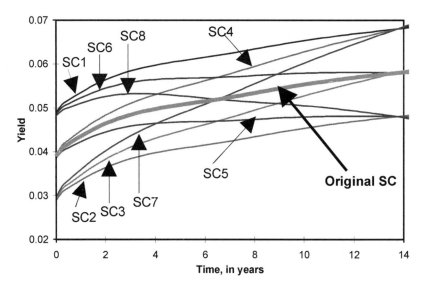

Figure 3.1   The current spot curve and 8 spot curve scenarios.

From Table 3.4, it is clear that a portfolio that has been immunized against a specific spot curve shift has zero change if that anticipated shift is realized. Row 1 shows the performance of a portfolio that has been immunized against SC scenario 1; if SC scenario 1 is realized (Column 1), then this portfolio has zero value change but it has a nonzero value change if other SC scenarios are realized. This portfolio performs well, with a positive value change ($+764$), if SC scenario 8 is realized, but performs badly, with a negative value change (-643), if SC scenario 7 is realized.

Table 3.4 shows that the minimax portfolio (bottom row) has a less volatile performance across the 8 SC scenarios. Clearly, it does not achieve a high performance compared to the other portfolios, but its worst-case performance has been maintained at zero, under SC scenarios 2 and 3. The portfolio that has been optimized across all SC scenarios (penultimate row) performs relatively poorly compared to the minimax portfolio.

4 THE ASSET/LIABILITY (A/L) RISK IN IMMUNIZATION

From the point of view of ALM, the major objective is to meet liabilities as they fall due. Clearly, only exact cash matching can satisfy such an objective. If immunization rather than cash matching is employed, liabilities may not be met with certainty because the objective has shifted from meeting liabilities to minimizing interest rate sensitivities. Fong and Vasicek (1984) suggest that by minimizing the weighted variance around the horizon date for a single liability problem, one obtains an optimally immunized portfolio. This method implies that if the weighted variance is minimized to zero, then one achieves exact cash matching.

In defining the risk measure, we use the definitions presented in Section 2. In that section, the asset portfolio is impacted by the decision variable x, where $A = \sum_j^n x_j P_j$, whereas the liability portfolio, $L = \sum_j^{\text{liabilities}} L_j$, is not. For the aggregate asset portfolio, we calculate its value by aggregating the present value of all cashflows:

$$A = \sum_j x_j \overset{\text{cashflows}}{\underset{t}{\sum}} \frac{A_{j,t}^0}{(1 + r_t^0)} \tag{4.1}$$

where $A_{j,t}^0$ is any cash inflow into the asset portfolio from asset j at time t where the present valuation is based on the current yield curve r^0. We now define a variable $A_{t_c}^0$ that represents a cashflow that has been weighted by the exposure x_j to that asset. Similarly, the present valuation is based on the current yield curve r^0:

$$A_{t_c}^0 \equiv A_{t_c}^0(x_j) = x_j(A_{j,t}^0). \tag{4.2}$$

Cash outflows for liabilities are similarly defined as in (4.1) and (4.2).

In any time period between $t - 1$ and t, there would be times, as indexed by t_c, where cashflows come into the asset portfolio. In general, we would like these asset cashflows to match, or at least not fall too far below, the liability that occurs at the end of the period, that is, at time t. In (4.3) below, the asset cashflow is suitably weighted by the exposure to the asset, where the weighting is given by x_j. In the interest of clarity, we will use $A_{t_c}^0$ to mean this weighted cashflow.

We define the current *A/L risk* as

$$\text{A/L risk} = \sum_{t=1}^{T_H} \left(\underset{\substack{t_c \\ t-1 < t_c \leq t}}{\overset{\text{cashflows}}{\sum}} \left(\frac{A_{t_c}^0}{(1 + r_{t_c}^0)}[t_c] - \frac{L_t^0}{(1 + r_t^0)}[t] \right)^2 \right).$$

It represents the variance of asset timings relative to the timings of the liabilities that those assets are meant to service. In (4.3), T_H is the horizon date. An alternative would be the downside risk:

$$\text{A/L downside risk} = \sum_{t=1}^{T_H} \left(\underset{\substack{t_c \\ t-1 < t_c \leq t}}{\overset{\text{cashflows}}{\sum}} \left(\left(\frac{A_{t_c}^0}{(1 + r_{t_c}^0)}[t_c] - \frac{L_t^0}{(1 + r_t^0)}[t] \right)^- \right)^2 \right)$$

$$\tag{4.3}$$

where $(\cdot)^- = \min(\cdot, 0)$. The A/L risk in (4.3) is calculated in the context of the current environment. For any given scenario i, the A/L risk can be suitably calculated as follows. Here, a superscript is introduced to indicate that asset cashflows, a liability outflow and interest rates are dependent on the nature of the current yield curve.

We define the scenario A/L risk (for scenario i) as

$$\text{A/L risk}_{\text{scenario } i} = \sum_{t=1}^{T_H} \left(\sum_{\substack{t_c \\ t-1 < t_c \leq t}}^{\text{cashflows}} \left(\frac{A_{t_c}^i}{(1 + r_{t_c}^i)}[t_c] - \frac{L_t^i}{(1 + r_t^i)}[t] \right)^2 \right)$$

for all scenarios $i \in I$. Here, the same superscript is introduced to indicate that asset cashflows, a liability outflow and interest rates are dependent on the nature of a scenario i.

The A/L risk measure is always nonnegative. It attains its lowest possible value of zero if and only if the assets are pure discount bonds, each having a maturity equal to that of the liability it is meant to service. This implies exact cash matching, where no interest rate change affects the value of the portfolio. Any other portfolio is to some extent vulnerable to an adverse interest rate movement. The A/L risk in effect measures how much a given portfolio differs from an exact cash-matched portfolio.

Let us investigate the potential of the current A/L risk and the scenario A/L risks in modifying the overall structure of the resultant portfolio. Suppose we consider the component of the current A/L risk relevant to the first liability.

$$\text{Current A/L risk for } L_1 = \left(\sum_{\substack{t_c \\ 0 < t_c \leq t_1}}^{\text{cashflows}} \left(\frac{A_{t_c}^0}{(1 + r_{t_c}^0)}[t_c] - \frac{L_{t_1}^0}{(1 + r_{t_1}^0)}[t_1] \right)^2 \right) \tag{4.4}$$

and the component of all the scenario A/L risks relevant to the first liability

$$\text{Scenario } i \text{ A/L risk for } L_1 = \left(\sum_{\substack{t_c \\ 0 < t_c \leq t_1}}^{\text{cashflows}} \left(\frac{A_{t_c}^i}{(1 + r_{t_c}^i)}[t_c] - \frac{L_{t_1}^i}{(1 + r_{t_1}^i)}[t_1] \right)^2 \right)$$

for all scenarios $i \in I$.

Incorporating all these components into the minimax framework, we increase the likelihood of meeting the first liability. The minimax formulation is

$$\min_{x \in \mathcal{R}^n} \max_{i \in I} \left\{ \begin{array}{l} \left[\sum_{j}^{n} x_j P_j + \left(-\langle AD_A(r^0, x), \Delta r^i \rangle + \langle LD_L(r^0), \Delta r^i \rangle \right)^+ \right. \\ \\ + \left(\sum_{\substack{t_c \\ 0 < t_c \le t_1}}^{\text{cashflows}} \left(\frac{A_{t_c}^0}{(1 + r_{t_c}^0)} [t_c] - \frac{L_{t_1}^0}{(1 + r_{t_1}^0)} [t_1] \right)^2 \right) \\ \\ \left. + \left(\sum_{\substack{t_c \\ 0 < t_c \le t_1}}^{\text{cashflows}} \left(\frac{A_{t_c}^i}{(1 + r_{t_c}^i)} [t_c] - \frac{L_{t_1}^i}{(1 + r_{t_1}^i)} [t_1] \right)^2 \right) \right] \end{array} \right\}$$

subject to

$$\langle \Delta r^i, AC_A(r^0, x) \Delta r^i \rangle - \langle \Delta r^i, LC_L(r^0) \Delta r^i \rangle \ge \epsilon \qquad (4.5)$$

for all scenarios $i \in I$ and

$$x \ge 0$$

$$\langle 1, x \rangle = 1$$

where $AD_A(r^0, x)$, $LD_L(r^0)$, $AC_A(r^0, x)$ and $LC_L(r^0)$ are defined in (2.19)–(2.22).

In the above formulation, $\epsilon > 0$ is some predefined convexity requirements on the combined asset-liability portfolio. A portfolio defined by the solution of the minimax formulation would be more likely able to meet the first liability compared to the portfolio defined by the solution of the original formulation. Suppose we consider the component of the current A/L risk relevant to the first two liabilities.

Current A/L risk for L_1 and L_2

$$= \left\{ \begin{array}{l} \left[\left(\sum_{\substack{t_c \\ 0 < t_c \le t_1}}^{\text{cashflows}} \left(\frac{A_{t_c}^0}{(1 + r_{t_c}^0)} [t_c] - \frac{L_{t_1}^0}{(1 + r_{t_1}^0)} [t_1] \right)^2 \right) \right. \\ \\ \left. + \left(\sum_{\substack{t_c \\ t_1 < t_c \le t_2}}^{\text{cashflows}} \left(\frac{A_{t_c}^0}{(1 + r_{t_c}^0)} [t_c] - \frac{L_{t_2}^0}{(1 + r_{t_2}^0)} [t_2] \right)^2 \right) \right] \end{array} \right\}$$

and the components of all the scenario A/L risks relevant to the first two liabilities

Scenario i A/L risk for L_1 and L_2

$$
= \left\{ \begin{array}{l} \left(\overset{\text{cashflows}}{\underset{\substack{t_c \\ 0 < t_c \le t_1}}{\sum}} \left(\frac{A^i_{t_c}}{(1+r^i_{t_c})}[t_c] - \frac{L^i_{t_1}}{(1+r^i_{t_1})}[t_1] \right)^2 \right) \\[2em] + \left(\overset{\text{cashflows}}{\underset{\substack{t_c \\ t_1 < t_c \le t_2}}{\sum}} \left(\frac{A^i_{t_c}}{(1+r^i_{t_c})}[t_c] - \frac{L^i_{t_2}}{(1+r^i_{t_2})}[t_2] \right)^2 \right) \end{array} \right\}
$$

for all scenarios $i \in I$.

By incorporating all these components into the minimax formulation framework, we increase the likelihood of meeting the first two liabilities. Consider therefore the family of minimax immunization formulations parameterized by the liability timing $T_H = \text{Horizon}$ and for a given $\epsilon > 0$:

$$
\min_{x \in \Re^n} \max_{i \in I} \left\{ \begin{array}{l} \sum_j^n x_j P_j + \left(-\langle AD_A(r^0, x), \Delta r^i \rangle + \langle LD_L(r^0), \Delta r^i \rangle \right)^+ \\[1.5em] + \sum_{t=1}^{T_H} \left(\overset{\text{cashflows}}{\underset{\substack{t_c \\ t-1 < t_c \le t}}{\sum}} \left(\frac{A^0_{t_c}}{(1+r^0_{t_c})}[t_c] - \frac{L^0_t}{(1+r^0_t)}[t] \right)^2 \right) \\[1.5em] + \sum_{t=1}^{T_H} \left(\overset{\text{cashflows}}{\underset{\substack{t_c \\ t-1 < t_c \le t}}{\sum}} \left(\frac{A^i_{t_c}}{(1+r^i_{t_c})}[t_c] - \frac{L^i_t}{(1+r^i_t)}[t] \right)^2 \right) \end{array} \right\}
$$

subject to

$$
\langle \Delta r^i, AC_A(r^0, x) \Delta r^i \rangle - \langle \Delta r^i, LC_L(r^0) \Delta r^i \rangle \ge \epsilon \qquad (4.6)
$$

for all scenarios $i \in I$ and

$$
x \ge 0
$$

$$
\langle 1, x \rangle = 1.
$$

The parameterization by the liability timing $T_H = \text{Horizon}$ can be interpreted as the referred "horizon" under a horizon matching framework. T_H can be viewed as the time before which we are aiming to satisfy cash matching to increase the likelihood of meeting all liabilities prior to this horizon. It can also be viewed as the time after which duration matching is satisfied in order to immunize our portfolio across interest rate scenarios.

5 THE CONTINUOUS MINIMAX DIRECTIONAL IMMUNIZATION

In this section, the vector of spot rates, r^0, is replaced by i^0 in order to distinguish between the discrete variable r^0 and the continuous variable i^0. Whereas in the discrete formulations in Sections 2–4, scenario vectors r^i are used, in this section a continuous variable i^{T_H} is used whose elements vary within uncertainty bounds.

Let $i^{T_H} \in \Re^m$ be a vector of future interest rates at a horizon date T_H, where m represents the number of key maturity rates along the spot curve. In a continuous minimax immunization, we let each element of i^{T_H} vary between upper and lower bounds. For $\epsilon > 0$, we present the continuous version of the discrete minimax immunization:

$$\min_{x \in \Re^n} \max_{i^{T_H} \in \Re^m} \left\{ \sum_j^n x_j P_j + \left(\left\langle AD_A(i^0,x),(i^0 - i^{T_H}) \right\rangle - \left\langle LD_L(i^0),(i^0 - i^{T_H}) \right\rangle \right)^2 \right\}$$

(5.1)

subject to

$$\left\langle (i^0 - i^{T_H}),\left(AC_A(i^0,x) \right)(i^0 - i^{T_H}) \right\rangle - \left\langle (i^0 - i^{T_H}),\left(LC_L(i^0) \right)(i^0 - i^{T_H}) \right\rangle \geq \epsilon$$

$$x \geq 0$$

$$\langle 1,x \rangle = 1$$

$$i^{T_H,\text{lower}} \leq i^{T_H} \leq i^{T_H,\text{upper}}$$

and with $AD_A(i^0,x)$, $LD_L(i^0)$, $AC_A(i^0,x)$, and $LC_L(i^0)$ as described by (2.19) to (2.22) for $i^0 = r^0$ while we consider variations from i^0.

We need to consider the expected spot curve at the horizon and to consider smoothness constraints. To generate the expected spot curve at the horizon, we can use the current spot curve and its corresponding forward curve. We derive the expected spot curve at any future time based on the current forward curve.

A smoothing function (e.g., a spline) may be used in order to define the whole default spot curve at the horizon. In defining the uncertainty scenarios (in the continuous case) on the maximizing variable, we allow i^{T_H} to vary within its bounds

$$i_k^{T_H,\text{lower}} \leq i_k^{T_H} \leq i_k^{T_H,\text{upper}}, \quad k = 1,\dots,m.$$

The constraints on the variation in i^{T_H} depend on its corresponding forward curve. Each element of i^{T_H} has a corresponding forward curve point with the same key maturity. Let

$$f^{T_H} = \begin{bmatrix} f_1^{T_H} \\ f_2^{T_H} \\ \vdots \\ f_m^{T_H} \end{bmatrix}.$$

By ensuring that all the elements of f^{T_H} are within reasonable bounds, we can control the choice of values for the maximizing variable i^{T_H}. Essentially, we are limiting the derivative of the expected spot curve within reasonable boundaries. The expected forward curve at the horizon may be constrained as

$$f_k^{T_H,\text{lower}} \le f_k^{T_H} \le f_k^{T_H,\text{upper}}, \quad k = 1, ..., m.$$

6 OTHER IMMUNIZATION STRATEGIES

We explore other immunization strategies and show how these can be adapted to a minimax framework. In the classical duration and convexity analyses, interest rate sensitivity analysis is done with respect to parallel yield curve moves only. Yield curves provide alternative representations of the term structure of interest rates (see CN 2). As sensitivity analyses to yield curves are widely used as immunization strategies, we consider these and their minimax formulations.

6.1 Univariate Duration Model

Let $\hat{r}$ be the yield and $\Delta\hat{r}$ be its change corresponding to a parallel shift of the yield curve. We refer to a duration model that is dependent on a scalar r only as a *univariate* duration model. Given the univariate price function $P(\hat{r})$, the duration function, denoted $\hat{D}(\hat{r})$, is defined for $P(\hat{r}) \ne 0$ as follows:

$$\hat{D}(\hat{r}) \equiv -\frac{dP(\hat{r})}{P(\hat{r})d\hat{r}}. \tag{6.1}$$

Given the univariate price function $P(\hat{r})$, the convexity function, denoted $\hat{C}(\hat{r})$, is defined for $P(\hat{r}) \ne 0$ as follows:

$$\hat{C}(\hat{r}) \equiv \frac{d^2P(\hat{r})}{P(\hat{r})d\hat{r}^2}. \tag{6.2}$$

Given the above definitions, the dollar duration, denoted $D^d(\hat{r})$, or the dollar sensitivity of the asset to a unit change in $\hat{r}$, is given by $dP(\hat{r})/d\hat{r}$, and the dollar convexity, denoted $C^d(\hat{r})$, is given by $d^2P(\hat{r})/d\hat{r}^2$. For bonds, where the price is defined by the present value of its cashflows, denoted c, discounted using $\hat{r}$, that is,

$$P(\hat{r}) = \sum_{t} c_t(1 + \hat{r})^{-t} \tag{6.3}$$

the dollar duration and dollar convexity are, respectively,

$$D^d(\hat{r}) = -\sum_{t} tc_t(1 + \hat{r})^{-(t+1)} \tag{6.4}$$

$$C^d(\hat{r}) = \sum_{t} t(t + 1)c_t(1 + \hat{r})^{-(t+2)}. \tag{6.5}$$

Classical durational immunization requires that the present value of the asset portfolio A is equal to that of the liability portfolio L. Additionally, it requires that their dollar durations are also equal; this condition is similar to (2.13). If these two conditions are met, then the value of the portfolio is hedged against small movements in $\hat{r}$; it therefore makes sense to construct a portfolio that maximizes its yield subject to meeting these two conditions. Whereas in Section 2 yield refers to a rate on the spot curve, in this section, *yield refers to the internal rate of return or the yield to maturity*. A first-order approximation to the portfolio yield is the dollar duration weighted average yield of the individual securities in the portfolio (Dahl et al., 1993), that is,

$$\hat{r}_P \approx \frac{\sum_{j}^{n} \left(D_j^d(\hat{r}) \right) \hat{r}_j x_j}{\sum_{j}^{n} \left(D_j^d(\hat{r}) \right) x_j}. \tag{6.6}$$

As the portfolio is constrained to have equal asset and liability dollar duration, then the denominator of (6.6) is equal to a fixed value given by the liability dollar duration. The formulation for an immunization strategy with a yield maximization objective is

$$\max_{x \in \mathfrak{R}^n} \frac{\sum_{j}^{n} \left(D_j^d(\hat{r}) \right) \hat{r} x_j}{\sum_{j}^{n} \left(D_j^d(\hat{r}) \right) x_j} \tag{6.7}$$

subject to

$$\sum_{j}^{n} P_j(\hat{r}) x_j = L$$

$$\sum_{j}^{n} D_j^d(\hat{r}) x_j = D_L^d(\hat{r})$$

$$x \geq 0$$

$$\langle 1, x \rangle = 1.$$

In (6.7) the first constraint requires the value of the asset portfolio to be equal to that of the liability portfolio. The second constraint imposes the equivalence of durations. Maximizing the yield of a portfolio is analogous to minimizing its cost. Whereas in Section 2 we concentrate on the cost of the portfolio, in this section we concentrate on its yield. Although the cost-based formulation makes intuitive sense, the yield-based formulation is common practice (e.g., within the insurance industry). It is useful to explore this formulation further. As a drop in the yield of an asset portfolio implies an increase in its value, it makes sense to construct the portfolio such that its future yield is minimized. Thus we have the problem

$$\min_{x \in \mathfrak{R}^n} \frac{\sum_{j}^{n} \left(D_j^d(\hat{r} + \Delta\hat{r}) \right)(\hat{r} + \Delta\hat{r})x_j}{\sum_{j}^{n} \left(D_j^d(\hat{r} + \Delta\hat{r}) \right)x_j} \tag{6.8}$$

subject to

$$\sum_{j}^{n} P_j(\hat{r})x_j = L$$

$$\sum_{j}^{n} D_j^d(\hat{r})x_j = D_L^d(\hat{r})$$

$$x \geq 0$$

$$\langle 1, x \rangle = 1.$$

For an assumed shift in the yield curve, defined by $\Delta\hat{r}$ for all securities under consideration, one can construct a portfolio of minimum future yield, that is, maximum future value. However, similar to the argument in Section 2 with regards to differing assumptions on the spot curve, differing assumptions on the yield curve shift would produce different solutions to (6.8) and one particular solution may not be effective in view of an erroneous assumption of yield curve shift.

Further constraints have to be imposed on (6.8) with regards to allowable yield curve shifts. Because a yield curve has an underlying spot curve, any assumption on the yield curve shift must take into account the future yield curve's consistency with its underlying spot curve. Therefore the choice of $\Delta\hat{r}$ for all securities under consideration is constrained to be consistent with a unique spot curve shift assumption.

For the current yield curve, the following relationship must hold true for all securities under consideration:

$$P(\hat{r}) = \sum_t c_t(1 + \hat{r})^{-t} = \sum_t c_t(1 + r_t)^{-t} \tag{6.9}$$

where $\hat{r}$ is the yield to maturity and r_t is the spot rate that corresponds to cashflow c_t. This relationship must also hold true at a future time, that is, when $\hat{r}$ shifts to $(\hat{r} + \Delta\hat{r})$ for each security then

$$P(\hat{r} + \Delta\hat{r}) = \sum_t c_t(1 + \hat{r} + \Delta\hat{r})^{-t} = \sum_t c_t(1 + r_t + \Delta r_t)^{-t} \tag{6.10}$$

must hold true for all securities, subject to a unique spot curve that defines all r_t, that is, across all maturities t.

We extend the above formulation to a minimax framework. We introduce scenarios under which the future yield of each security determines the future yield on the asset portfolio. Consider the problem

$$\min_{x \in \mathbb{R}^n} \max_{i \in I} \left\{ \frac{\sum_j^n \left(D_j^d(\hat{r} + \Delta\hat{r}^i)\right)(\hat{r} + \Delta\hat{r}^i)x_j}{\sum_j^n \left(D_j^d(\hat{r} + \Delta\hat{r}^i)\right)x_j} \right\} \tag{6.11}$$

subject to

$$\sum_j^n P_j(\hat{r})x_j = L$$

$$\sum_j^n D_j^d(\hat{r})x_j = D_L^d(\hat{r})$$

for all $i \in I$ and

$$x \geq 0$$

$$\langle 1, x \rangle = 1$$

where $I = \{1, 2, ..., m^{sce}\}$ is the finite set of yield curve shifts. Formulation (6.11) is also constrained to satisfy (6.10). This ensures that, subject to the assumed yield curve shifts, the future yield of the portfolio is minimized while maintaining an immunized state at the current time.

6.2 Univariate Convexity Model

Similar to dollar duration, dollar convexity, given by (6.5), is useful to optimize because it affects the value of a bond, convexity being the second order derivative of price with respect to yield. An immunization strategy that minimizes dollar convexity while satisfying the conditions for duration immunization effectively minimizes the change in duration for a specific yield curve shift. As with dollar duration, dollar convexity is additive and the convexity of a portfolio is given by

$$C_P^d(\hat{r}) = \sum C_j^d(\hat{r})x_j. \tag{6.12}$$

Thus, the optimization model becomes

$$\min_{x \in \Re^n} \sum_j^n C_j^d(\hat{r})x_j \tag{6.13}$$

subject to

$$\sum_j^n P_j(\hat{r})x_j = L$$

$$\sum_j^n D_j^d(\hat{r})x_j = D_L^d(\hat{r})$$

$$\sum_j^n C_j^d(\hat{r})x_j \geq C_L^d(\hat{r})$$

$$x \geq 0$$

$$\langle 1, x \rangle = 1$$

where the constraint on convexity ensures that assets continue to dominate liabilities for small yield curve shifts.

The convexity model can be extended to cover the minimization of the future convexity of the portfolio. However, such an extension only makes sense when the future duration of the assets do not deviate substantially from the future duration of the liabilities. In as much as duration in both assets and liabilities cannot be held unchanged, one can attempt to model the absolute difference or the squared difference between their durations. Consider the extended convexity model below:

$$\min_{x \in \Re^n} \sum_j^n C_j^d(\hat{r})x_j + \left(\left(\sum_j^n D_j^d(\hat{r} + \Delta\hat{r})x_j \right) - \left(D_L^d(\hat{r} + \Delta\hat{r}) \right) \right)^2 + \sum_j^n C_j^d(\hat{r} + \Delta\hat{r})x_j \tag{6.14}$$

subject to

$$\sum_{j}^{n} P_j(\hat{r})x_j = L$$

$$\sum_{j}^{n} D_j^d(\hat{r})x_j = D_L^d(\hat{r})$$

$$\sum_{j}^{n} C_j^d(\hat{r})x_j \geq C_L^d(\hat{r})$$

$$\sum_{j}^{n} C_j^d(\hat{r} + \Delta\hat{r})x_j \geq C_L^d(\hat{r} + \Delta\hat{r})$$

$$x \geq 0$$

$$\langle 1, x \rangle = 1.$$

Similar to the constraint that current asset convexity be greater or equal to current liability convexity, the future asset convexity should also be greater or equal to the future liability convexity.

We present the above formulation within a minimax framework. Consider the following formulation:

$$\min_{x \in \mathfrak{R}^n} \max_{i \in I} \left\{ \sum_{j}^{n} C_j^d(\hat{r})x_j + \left(\left(\sum_{j}^{n} D_j^d(\hat{r} + \Delta\hat{r}^i)x_j \right) \right. \right.$$

$$\left. \left. - \left(D_L^d(\hat{r} + \Delta\hat{r}^i) \right) \right)^2 + \sum_{j}^{n} C_j^d(\hat{r} + \Delta\hat{r}^i)x_j \right\} \qquad (6.15)$$

subject to

$$\sum_{j}^{n} P_j(\hat{r})x_j = L$$

$$\sum_{j}^{n} D_j^d(\hat{r})x_j = D_L^d(\hat{r})$$

$$\sum_{j}^{n} C_j^d(\hat{r})x_j \geq C_L^d(\hat{r})$$

$$\sum_{j}^{n} C_j^d(\hat{r} + \Delta\hat{r}^i)x_j \geq C_L^d(\hat{r} + \Delta\hat{r}^i)$$

for all $i \in I$ and

$$x \geq 0$$

$$\langle 1, x \rangle = 1$$

where $I = \{1, 2, ..., m^{sce}\}$ is the finite set of yield curve shifts. Formulation (6.15) is also constrained to satisfy (6.10).

7 THE STOCHASTIC ALM MODEL 1

Previous sections of this chapter focus on immunization models. In this section, we present a long-term oriented stochastic dedication-based model. In the stochastic ALM model, a term structure (TS) model (see CN 2) is required to define the stochastic nature of the system. In principle, any TS model would be consistent with the framework developed below. However, we initially restrict the discussion to a TS model utilizing a binomial tree. This simplification does not impair the exposition on the ALM model. It keeps the discussion clear of generalized tree structures (see CN 3). We describe a TS model that is fully determined by the short rate, described as a stochastic process, and by the current term structure of interest rates. The short rate is modeled by

$$r_{t+1} = r_t e^{(\mu_t + \sigma)} \tag{7.1}$$

where r_t is the level of short term interest rate at time t, r_{t+1} is the level of short term interest rate at time $t + 1$, μ_t is the drift parameter, and σ is the volatility parameter.

In the binomial model of the term structure tree, we refer to any point on the tree as either a state or a node, interchangeably. A state or node defines a particular term structure of interest rates, that is, a particular set of interest rates of varying maturity. The term *node* refers to a previous state, or the state from which the current states have emanated from. The term *state* refers to one of a number of possible realizations of the system at a particular time. A state evolves from a node which is itself a state. The realization of states over time is given in Figure 7.1. In the setting below of the binomial tree, at each point in time, the set of possible states has a cardinality of 2^t. At time 0, the current time, there is only one state which is nothing more than the current term structure. At time 1, two states emanate from the previous (unique) node. Arrival at these two states is based on a predefined probability of an "up state" and the probability of a "down state". At time 2, four states emanate from the two previous nodes; these four states are again generated based on the "up state" and "down state" for the first node and the "up state" and "down state" for the second node. For each time t, the number of states is twice the number of nodes at time $t - 1$. We note that, for general trees, the number of states that emanate from a node is arbitrary but finite.

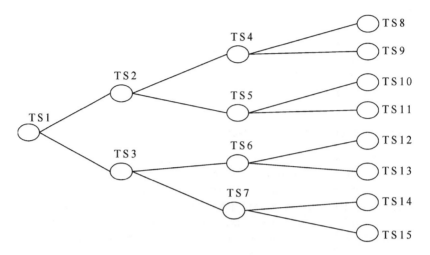

Figure 7.1   The evolution of the term structure in a binomial model with $t = 4$.

Each state or node defines the short-term interest rate as it has evolved, plus the corresponding term structure of future interest rates. The stochastic process that defines the evolution of the short rate requires a drift parameter as well as a volatility parameter. The drift may be defined for every period such that the expected level of short rates is consistent with market derived forward rates. Short-term volatility may be derived using the market price of interest rate options.

In the development of the ALM, we need to establish time frames and reference points that would describe the temporal structure of the ALM problem. The *terminal date* is defined as the date at which all assets and liabilities are measured in a common way for accounting purposes; we assume that both assets and liabilities are turned into cash at the terminal date. For a corporate wishing to structure its balance sheet, the terminal date may be *one* year from time 0. For a pension fund, the terminal date may be 30 years from time 0. The *payout date* is defined as the date at which a liability needs to be serviced. For a single liability problem, the payout date corresponds to the terminal date. For a multiple liability problem, the payout dates may be scattered unevenly between time 0 and the terminal date. The *horizon date* is defined as the time at which we wish to terminate a particular calculation, either for performance measurement purposes or for a more specific management purpose.

In the context of the binomial tree that has been described above, we need to reconcile the time frame in the binomial tree and that required by the ALM system. We use the following assumptions:

- The *terminal or horizon date* T_H of the ALM system coincides with the terminal date of the binomial tree.

- The *payout dates t* of the liabilities coincide with node times in the binomial tree.

- The *date of strategy shift* T_{cm} coincides with a node time in the binomial tree, where the strategy shift refers to the cross-over from stochastic cash-matching to stochastic immunization, to be discussed in detail below.

The above assumptions map the binomial temporal framework onto the ALM temporal framework; this results in a consistent way of reckoning times and time-related calculations. These assumptions also help overcome the problem of having a liability payout date that falls within two node times where it is not possible to value a liability. In practice, known liability payout dates may not be uniformly spaced; in this case, the choice of node times have to be determined heuristically to best fit the timings of the known payout dates. For unknown liability payout dates, for example, liabilities that are contingent on interest rate levels, we use the predefined node times to represent potential liability payout dates, depending on the states of the world at that node time.

Based on the terminal date and the number of node times, we are able to calculate the cardinality of the set of potential states of the world at the terminal date. Each state at the terminal date, hereafter *terminal state*, is arrived at via a series of states starting from state 0, at node 0. The evolution of the term structure from time 0 to the terminal date is given by a particular series of states. This evolution defines a particular scenario in the ALM framework: *a scenario is a path on the term structure tree that fully describes directly the evolution of the term structure and indirectly the valuation of the assets and liabilities in time.* In Table 7.1, we give all the paths of evolution, and therefore the scenarios, based on the binomial tree given in Figure

Table 7.1 The scenarios generated from the binomial model

Scenario	Composition
Scenario 1	TS1, TS2, TS4, TS8
Scenario 2	TS1, TS2, TS4, TS9
Scenario 3	TS1, TS2, TS5, TS10
Scenario 4	TS1, TS2, TS5, TS11
Scenario 5	TS1, TS3, TS6, TS12
Scenario 6	TS1, TS3, TS6, TS13
Scenario 7	TS1, TS3, TS7, TS14
Scenario 8	TS1, TS3, TS7, TS15

7.1. We observe that a scenario is a series of states that starts from state 0 to any one of the terminal states. From Table 7.1, scenario 1 is given by the state series {TS1, TS2, TS4, TS8}. We define the set I as the discrete set of scenarios i. Each scenario i is defined[5] by a unique series of states on the binomial tree that fully describe a unique evolution of the term structure of interest rates.

The smallest time interval in the ALM system is the minimum of the interval between two known liability payout dates and the interval between two node times. The design of the ALM includes a decision on the granularity of the temporal framework. As the time interval becomes smaller, the level of detail of the valuation process becomes better. Additionally, as the time interval becomes smaller, the scenarios increase since the number of scenarios is equivalent to the cardinality of the set of terminal states.

Consider the ALM problem for an institution that wishes to restructure its balance sheet. The latter is a statement of the assets held, the cash at hand, the known liabilities with known timings, and the contingent assets and liabilities valued at the current time. The contingent nature of the contingent assets is based on some underlying asset that may have embedded optionality that is interest rate dependent. The contingent nature of the contingent liabilities may be viewed similarly. The difference between a contingent asset and a contingent liability is that the contingent asset is a potential inflow to the balance sheet that may improve the health of the balance sheet whereas the contingent liability is a potential outflow that may cause a deterioration of the health of the balance sheet.

The assumption that the institution has a stream of known liabilities with known timings is not unreasonable. A reserve bank may have a well-defined liability stream that represents its country's import payments. A pension fund may have a well-defined liability stream that has been determined using actuarial calculations. An insurance company may have a well-defined liability stream given by the average liability amounts and timings that have been estimated over a long period of historical data.

The uncertainty caused by the presence of contingent assets and contingent liabilities is handled by a term structure (TS) model which in this section is a binomial tree[6]. The valuation of a contingent asset or liability depends on the term structure of interest rates at a specific node on the tree. The potential payout of a contingent liability is evaluated at a node. As liabilities are valued at the nodes, we can ascertain whether a liability matures, and therefore

---

[5] We use the same indexing as in Section 2; here we give details of how individual scenarios are defined. Whereas in Section 2 we defined scenarios in terms of yield curve shifts, here we define them in terms of yield curve evolutions.

[6] We remind the reader that other tree structures may be used but in this section, we use a binomial model for a clearer discussion.

requires payment, at a specific state under a specific node. The potential cash inflow of a contingent asset is similarly evaluated.

The types of ALM problems considered in this section only include liabilities with known payout dates that can be formulated within the above temporal framework. The same applies to contingent assets and liabilities. In general, contingent assets and liabilities that are not highly correlated with the term structure of interest rates would not fall in the category of assets and liabilities that the proposed system can handle with high precision. An example of such an asset would be a mortgage-backed security (MBS) whose prepayment model fails to show a high correlation with the term structure compared to other securities due to the irrational component of the optionality in the MBS. Because a prepayment model associated with MBS may attempt to capture both the rational part and the irrational part of the option embedded in the MBS, there may be a vague definition of the relationship between the valuation of the MBS and the prevailing term structure of interest rates at any given node in the binomial tree. As the prepayment model becomes better in defining this relationship, the degree of precision improves.

The institution has a pool of assets and cash reserves that it can manipulate to restructure its balance sheet. It also has access to a bigger universe of assets available in the markets that it can choose to buy by drawing on its cash reserves. If it can finance the purchase of new assets from the sale of current assets, then it does not have to draw on the cash reserves and it is therefore self-financing. However, self-financing is rarely easily achieved. In general, a restructuring may result in either a better short-term position at the expense of the long-term position, or vice versa.

We use the following notation:

n	universe of all assets at time 0
m	number of scenarios
a	the number of assets in the current portfolio
$\hat{x}$	vector of previous allocations to all assets
$\hat{x}_j$	previous nominal amount of asset j that exists in the current portfolio, $j \in [1, a]$
x_j	decision variable nominal amount of asset j that exists in the current portfolio, $j \in [1, a]$
x_k	decision variable nominal amount of asset k that does not exist in the current portfolio, $k \in [a + 1, n]$
x	vector of current allocations to all assets, that is, the decision variable with elements x_j and x_k
P_j	value of asset j, $j \in [1, a]$

P_k value of asset k, $k \in [a+1,n]$

$W^i_{T_H}$ terminal wealth under scenario i

W_D desired terminal wealth, based on an exogenously deter-
                    mined benchmark return

W^i_t wealth at time t under scenario i

$F(f,t_a,t_b)$ future valuing factor from time t_a to t_b given cashflow f[7]

L^i_t liability at time t under scenario i

$c^i_{d,e}$ the eth cashflow from asset d under scenario i

$t_{d,e}$ the payment time of cashflow $c^i_{d,e}$

I set of scenarios $i = 1,...,m$.

The ALM problem we address involves the rebalancing of a balance sheet of a corporate where holdings in existing assets may have to be changed, with the possibility of completely dropping some assets and introducing new assets. The universe of assets will cover both assets in the current portfolio prior to restructuring and assets not in the current portfolio. Some assets and liabilities may be contingent on the term structure of interest rates in the future. The main objective of the corporate is to be able to meet its liabilities as they fall due and at the same time ensure that there is a positive return at the terminal date.

The formulation, given below, minimizes the worst-cash position possible based on the identified scenarios. This objective is constrained mainly by a cash balance equation. Essentially, any cash balance at the start of a period, that is, the first term in the constraint below, plus any cashflow during the period, that is, the second term in the constraint below, has to account for the cash balance at the end of the period.

The minimax formulation is

$$\min_{x \in \Re^n} \max_{i \in I} \left\{ \sum_{j=1}^{a} P_j(x_j - \hat{x}_j) + \sum_{k=a+1}^{n} P_k x_k + \left(W_D - W^i_{T_H} \right)^+ \right\} \quad (7.2)$$

subject to liability constraints at each node on the term structure

$$W^i_{t-1} F(W^i_{t-1}, t-1, t) + \sum_{d=1}^{n} \sum_{\substack{e \text{ such that} \\ t-1 \leq t_{d,e} \leq t}} x_d(c^i_{d,e}) F(x_{d,0}(c^i_{d,e}), t_{d,e}, t) - L^i_t = W^i_t$$

for all $t \in [1,...,T_H]$,

$$x \geq 0$$

---

[7] Cashflow f is a generic variable that represents either a cash inflow or a cash outflow.

$$\langle 1, x \rangle = 1$$

$$W_0^i \geq 0$$

for all scenarios i where a scenario is defined by a path on the term structure tree.

The first constraint is a set of linear equations $W_1^i, \dots, W_t^i, \dots, W_{T_H}^i$ for each scenario i that forms the constraints of the optimization. As $W_1^i, \dots, W_t^i, \dots, W_{T_H}^i$ represent the surpluses or deficits at dates $1, \dots, t, \dots, T_H$ after liabilities $L_1^i, \dots, L_t^i, \dots, L_{T_H}^i$ have been paid, then in order to meet the liabilities on their payment date, without borrowing, we require $W_1^i, \dots, W_t^i, \dots, W_{T_H}^i$ to be nonnegative for all scenarios. The first term in this constraint refers to the future-valued wealth generated at time $t - 1$, while the second term refers to the future-valued cashflows that occur in between times $t - 1$ and t, as indicated by the index e that has been constrained to identify these cashflows.

The problem with ensuring nonnegativity is that all the constraint sets for all scenarios restrict the solution space severely because the formulation is essentially a stochastic cashmatching framework. If we relax the constraints such that only $W_{T_H}^i \geq 0$, then the stream of liabilities $L_1^i, \dots, L_t^i, \dots, L_{T_H}^i$ will be met by the portfolio, but only by the terminal date T_H. This relaxation of constraints implies that at dates $t \in [1, \dots, T_H - 1]$, the liabilities L_t^i are not paid with certainty, but that the sum of all liabilities $L_t^i, t \in [1, \dots, T_H]$ is met in full at date T_H. Because $W_t^i, t \in [1, \dots, T_H - 1]$, are allowed to be negative, the cashflows between dates $T_H - 1$ and T_H are used to cover these shortfalls. The constraint $W_{T_H}^i \geq 0$ ensures that the portfolio will meet all the liabilities by the time the last liability $L_{T_H}^i$ needs to be paid.

The variable $F(f, t_a, t_b)$, represents the factor that computes the value of the cashflows in the future based on current interest rates. The cashflow is either an asset cashflow or a W_t^i. The interest rate to be used within $F(f, t_a, t_b)$ depends on whether the cashflow is a surplus or a deficit. In the case of the future-valuing factor $F(W_{t-1}^i, t - 1, t)$, if W_{t-1}^i is a surplus ($W_{t-1}^i \geq 0$), then the interest rate to use is the lending rate. If it is a deficit ($W_{t-1}^i < 0$), then the interest rate to use is the borrowing rate. In the latter case, the institution essentially borrows to service a liability that falls due. This distinction is important as lending rates are generally much lower than borrowing rates, and the discrepancy between these rates could have a large impact on the cash balance at the end of a period.

The formulation above ensures that the terminal wealth of the corporate is well considered in the decision-making process. However, the servicing of liabilities is no longer certain when W_t^i are unconstrained and only $W_{T_H}^i$ is required to be greater than zero.

Adding the following constraint ensures that the first liability is paid:

$$W_0^i F(W_0^i, 0, 1) + \sum_{d=1}^{n} \sum_{\substack{e \text{ such that} \\ 0 \le t_{d,e} \le 1}} x_d(c_{d,e}^i) F(x_d(c_{d,e}^i), t_{d,e}, 1) - L_1^i = W_1^i$$

$W_1^i \ge 0$ for all scenarios $i \in I$. The constraint ensures a surplus at the first payout date. Thus, the solution will meet the first liability. In addition, by *adding* the following constraint, we ensure further that the first and second liabilities are paid out.

$$W_1^i F(W_1^i, 1, 2) + \sum_{d=1}^{n} \sum_{\substack{e \text{ such that} \\ 1 \le t_{d,e} \le 2}} x_d(c_{d,e}^i) F(x_d(c_{d,e}^i), t_{d,e}, 2) - L_2^i = W_2^i.$$

$W_2^i \ge 0$ for all scenarios $i \in I$.

We have therefore created flexibility for addressing liability servicing. As more and more W_t^i are required to be nonnegative, the system becomes increasingly closer to full stochastic cashmatching.

A formulation where only $W_1^i \ge 0$ and $W_{T_H}^i \ge 0$ ensures a surplus at time 1 and T_H. Because liabilities L_t^i, $t \in [2, ..., T_H - 1]$, may not be paid with certainty, we wish to provide immunization for these liabilities. To achieve this, we consider the asset/liability risk (A/L risk) defined for scenario i as

$$\text{A/L risk}_i = \sum_{t=1}^{T_H} \left(\sum_{d=1}^{n} \sum_{\substack{e \text{ such that} \\ t-1 \le t_{d,e} \le t}} x_d(c_{d,e}^i) F(x_d(c_{d,e}^i), t_{d,e}, t)(t - t_{d,e}) \right)^2.$$

The family of minimax ALM formulations given below is parameterized by T_{cm}, the date of strategy shift, where the system prior to T_{cm} is stochastic cashmatching and the system after T_{cm} is stochastic immunization.

$$\min_{x \in \Re^n} \max_{i \in I} \left\{ \begin{array}{l} \sum_{j=1}^{a} P_j(x_j - \hat{x}_j) + \sum_{k=a+1}^{n} P_k x_k + \left(W_D - W_{T_H}^i \right)^+ \\ \\ + \left(\sum_{t=1}^{T_H} \left(\sum_{d=1}^{n} \sum_{\substack{e \text{ such that} \\ t-1 \le t_{d,e} \le t}} x_d(c_{d,e}^i) F(x_d(c_{d,e}^i), t_{d,e}, t)(t - t_{d,e}) \right)^2 \right) \end{array} \right\}$$

(7.3)

subject to liability constraints at each node on the term structure:

$$W_{t-1}^i F(W_{t-1}^i, t-1, t) + \sum_{d=1}^{n} \sum_{\substack{e \text{ such that} \\ t-1 \le t_{d,e} \le t}} x_d(c_{d,e}^i) F(x_d(c_{d,e}^i), t_{d,e}, t) - L_t^i = W_t^i$$

(7.5)

for all $t \in [1, ..., T_H]$,

$$W_1^i, ..., W_{T_{cm}}^i \geq 0, \quad \text{stochastic cashmatching up to time } T_{cm}$$

$$W_{T_H}^i \geq 0, \quad \text{terminal wealth constraint}$$

$$x \geq 0$$

$$\langle 1, x \rangle = 1$$

$$W_0^i \geq 0$$

for all scenarios where a scenario is defined by a path on the term structure tree.

The formulation emphasizes the minimization of the effect of the worst-case scenario. This effect is expressed in terms of trading volume, downside deviation from the target terminal wealth and cumulative deviations of assets from liabilities. Minimization of the worst trading volume is effectively a minimization of the worst transaction costs. Minimization of the worst downside deviation from the target terminal wealth is effectively a minimization of the worst terminal wealth. Minimization of the cumulative deviations of assets from liabilities is effectively a minimization of the worst duration mismatch. We note that the objective function in (7.5) does not contain any term that represents portfolio return. Thus, it does not admit any trade-off specification between risk and return. This is considered in (7.7) below. The minimax return framework is given by

$$\min_{x \in \Re^n} \max_{i \in I} \left\{ -W_{T_H}^i \right\} \tag{7.6}$$

subject to liability constraints at each node on the term structure

$$W_{t-1}^i F(W_{t-1}^i, t-1, t) + \sum_{d=1}^{n} \sum_{\substack{e \text{ such that} \\ t-1 \leq t_{d,e} \leq t}} x_d(c_{d,e}^i) F(x_d(c_{d,e}^i), t_{d,e}, t) - L_t^i = W_t^i$$

for all $t \in [1, ..., T_H]$,

$$W_1^i, ..., W_{T_{cm}}^i \geq 0, \quad \text{stochastic cashmatching up to time } T_{cm}$$

$$W_{T_H}^i \geq 0, \quad \text{terminal wealth constraint}$$

$$x \geq 0$$

$$\langle 1, x \rangle = 1$$

$$W_0^i \geq 0$$

for all scenarios where a scenario is defined by a path on the term structure tree.

In (7.6) T_{cm} defines the demarcation between stochastic cashmatching and stochastic immunization. We can thus have a uniform framework for dealing with risk and return. This is the minimax formulation below parameterized by α where $\alpha \in [0, 1]$ represents the trade-off between the minimization of the worst case risk and the maximization of the worst case return. The worst-case scenario associated with the worst case risk may not be the same as the worst-case scenario associated with the worst case return. Hence, consider the problem

$$
\min_{x \in \mathfrak{R}^n} \max_{i \in I} \left\{ (-\alpha W_{T_H}^i) + (1 - \alpha) \right.
$$

$$
\times \left\{ \begin{array}{l} \sum_{j=1}^{a} P_j(x_j - \hat{x}_j) + \sum_{k=a+1}^{n} P_k x_k + \left(W_D - W_{T_H}^i \right)^+ \\[3mm] + \left(\sum_{t=1}^{T_H} \left(\sum_{d=1}^{n} \sum_{\substack{e \text{ such that} \\ t-1 \le t_{d,e} \le t}} x_d(c_{d,e}^i) F(x_d(c_{d,e}^i), t_{d,e}, t)(t - t_{d,e}) \right)^2 \right) \end{array} \right\} \right\}
$$

subject to liability constraints at each node on the term structure

$$
W_{t-1}^i F(W_{t-1}^i, t-1, t) + \sum_{d=1}^{n} \sum_{\substack{e \text{ such that} \\ t-1 \le t_{d,e} \le t}} x_d(c_{d,e}^i) F(x_d(c_{d,e}^i), t_{d,e}, t) - L_t^i = W_t^i
$$

(7.7)

for all $t \in [1, ..., T_H]$,

$$
W_1^i, .., W_{T_{cm}}^i \ge 0, \quad \text{stochastic cashmatching up to time } T_{cm}
$$

$$
W_{T_H}^i \ge 0, \quad \text{terminal wealth constraint}
$$

$$
x \ge 0
$$

$$
\langle 1, x \rangle = 1
$$

$$
W_0^i \ge 0
$$

for all scenarios where a scenario is defined by a path on the term structure tree. We also require that $0 \le \alpha \le 1$.

When $\alpha = 1$, then the formulation reduces to the maximization of the worst case return and when $\alpha = 0$, the formulation reduces to the minimization of the worst case risk. Varying α between 0 and 1 will result in a set of solutions that defines an efficient frontier. The horizontal axis indicates the risk in terms of (7.5) and the vertical axis gives the return in terms of (7.6).

The risk in (7.5) and (7.7) is a combination of three components: transaction cost-related risks, downside risk given by a possible deficit terminal wealth, and duration-mismatch risk given by the deviations of assets from liabilities. It is important to note that any of these risks can be used on their own in the minimax formulation. These three risk measures, whether used individually or in combination, would result in different efficient frontiers. The formulation above is in a general form and the associated efficient frontier is unique for a given value of T_{cm}.

8 THE STOCHASTIC ALM MODEL 2

In Section 7 we present a stochastic ALM model where the source of uncertainty is represented by a finite number of scenarios. The decision variables are solved by formulating the stochastic problem into its deterministic equivalent. Thus, the expectations operator in the stochastic framework is replaced by a finite number of scenarios weighted by their probability. In Section 7 we also present the minimax formulation where the effect of the worst-case scenario is minimized. One can infer that the minimax solution may be conservative as it provides a buffer against the worst case evaluated across all scenarios. As the decision horizon gets longer, the worst case may become increasingly pessimistic. The formulations in Section 7 do not allow for any dynamic revision of the decision variables in response to new information. In this section we present a dynamic multistage recourse stochastic ALM model together with its multiperiod minimax representation. Finally, we present a one-period minimax formulation where the worst case is evaluated only over the next period with expectations used to evaluate the remaining period until T_H.

8.1 A Dynamic Multistage Recourse Stochastic ALM Model

Consigli and Dempster (1998) consider two ways of representing the terminal wealth in the objective function: first, in terms of maximizing the terminal wealth, and second, in terms of minimizing its deviation from a target terminal wealth. In the first case, the optimization becomes

$$\max_{x_A} \mathrm{E}\left(W_{T_H}\right)$$

where $\mathrm{E}(\cdot)$ is the expectations operator and W_{T_H} is the terminal wealth, to be defined below. In the second case, the optimization is given by

$$\min_{x_A} \mathrm{E}\left(W_{T_H} - W^d\right)^2$$

where W^d is the desired terminal wealth, that is, its target value.

We adopt the following notation[8] in this section:

T_H	horizon
$t_1, t_2 = 1, \dots, T_H + 1$	time periods
$i = 1, \dots, I$	asset type
$j = 1, \dots, J$	liability type
$k = 1, \dots, K$	riskless instrument type
$x_{A,it}^+$	amount purchased of asset i in period t
$x_{A,it_1 t_2}$	amount held of asset i in period t_2 which was purchased in period $t_1 \leq t_2$
$x_{A,it_1 t_2}^-$	amount sold of asset i in period t_2 which was purchased in period $t_1 < t_2$
$y_{L,jt}^+$	amount incurred of liability j in period t
$y_{L,jt_1 t_2}$	amount held of liability j in period t_2 which was incurred in period $t_1 \leq t_2$
$y_{L,jt_1 t_2}^-$	amount discharged of liability j in period t_2 which was incurred in period $t_1 < t_2$
z_{kt}^+	amount held of riskless asset k in period t
z_{kt}^-	amount owed of riskless asset k in period t
$\delta_{x_{A,it}^+}$	binary action of buying asset i in period t
$\delta_{x_{A,it_1 t_2}^-}$	binary action of selling asset i in period t_2 $(t_1 < t_2)$
$\delta_{y_{L,jt}^+}$	binary action of incurring liability j in period t
$\delta_{y_{L,jt_1 t_2}^-}$	binary action of discharging liability j in period t_2 $(t_1 < t_2)$

The following additional parameters are conditioned by their histories:

$r_{A,it_1 t}$	cash return in period t on asset i purchased in period $t_1 < t$
$r_{L,jt_1 t}$	cash return in period t on liability j incurred in period $t_1 < t$
r_{kt}^+	return in period t on riskless asset k held in period $t - 1$
r_{kt}^-	unit cost of borrowing riskless asset k in period $t - 1$
e_{it}	lump sum transaction cost of purchasing asset i in period t
e_{jt}	lump sum transaction cost of incurring liability j in period t

[8] This follows the notation in Consigli and Dempster (1998) for ease of cross-referencing.

f_{it} unit cash outflow upon purchasing asset i in period t

f_{jt} unit cash inflow upon incurring liability j in period t

$g_{A,it_1 t}$ unit cash inflow upon selling in period t asset i purchased in period $t_1 < t$

$g_{L,jt_1 t}$ unit cash outflow upon discharging in period t liability j incurred in period $t_1 < t$

$h_{A,it_1 t}$ lump sum transaction cost of selling in period t asset i purchased in period $t_1 < t$

$h_{L,jt_1 t}$ lump sum transaction cost of discharging in period t liability j incurred in period $t_1 < t$

ρ_{it} exchange rate appropriate to asset i in period t

ρ_{jt} exchange rate appropriate to liability j in period t

ρ_{kt} exchange rate appropriate to riskless asset k in period t

P_{A,it_1} market value at the horizon of asset i purchased in period $t_1 \leq T_H$

P_{L,jt_1} market value at the horizon of liability j incurred in period $t_1 \leq T_H$

The terminal wealth function is given by

$$W_{T_H} = \sum_{t_1=1}^{T_H} \left[\sum_{i=1}^{I} P_{A,it_1} x_{A,it_1 T_H} - \sum_{j=1}^{J} P_{L,jt_1} x_{L,jt_1 T_H} \right]$$

$$+ \sum_{k=1}^{K} \left[(1 + r^+_{k(T_H+1)}) z^+_{kT_H} - (1 + r^-_{k(T_H+1)}) z^-_{kT_H} \right]. \quad (8.1)$$

The constraints on the variables are

Cash balance constraints, for $t = 1, ..., T_H + 1$:

$$\sum_{i=1}^{I} \rho_{it} \left[-e_{it}\delta_{x^+_{A,it}} - f_{it}x^+_{A,it} + \sum_{t_1=1}^{t-1} \left(r_{A,it_1 t}x_{A,it_1(t-1)} + g_{A,it_1 t}x^-_{A,it_1 t} - h_{A,it_1 t}\delta_{x^-_{A,it_1 t}} \right) \right]$$

$$- \sum_{j=1}^{J} \rho_{jt} \left[-e_{jt}\delta_{x^+_{L,jt}} - f_{jt}x^+_{L,jt} + \sum_{t_1=1}^{t-1} \left(r_{L,jt_1 t}x_{L,jt_1(t-1)} + g_{L,jt_1 t}x^-_{L,jt_1 t} - h_{L,jt_1 t}\delta_{x^-_{L,jt_1 t}} \right) \right]$$

$$+ \sum_{k=1}^{K} \rho_{kt} [(1 + r^+_{kt})z^+_{k(t-1)} - (1 + r^-_{kt})z^-_{k(t-1)} - z^+_{kt} + z^-_{kt}] = 0$$

Asset purchase inventory balance, $i = 1, ..., I$ and $t = 1, ..., T_H$:

$$x_{A,itt} - x_{A,it}^+ = 0$$

Liability discharge inventory balance, $j = 1, ..., J$ and $t = 1, ..., T_H$:

$$x_{L,jtt} - x_{L,jt}^+ = 0$$

Asset sale inventory balance, $i = 1, ..., I$, $t_1 = 1, ..., t - 1$ and $t = 1, ..., T_H + 1$:

$$x_{A,it_1t} - x_{A,it_1(t-1)} + x_{A,it_1t}^- = 0$$

Liability discharge inventory balance, $j = 1, ..., J$, $t_1 = 1, ..., t - 1$ and $t = 1, ..., T_H + 1$:

$$x_{L,jt_1t} - x_{L,jt_1(t-1)} + x_{L,jt_1t}^- = 0$$

No decisions past the horizon, $i = 1, ..., I, j = 1, ..., J$ and $t_1 = 1, ..., T_H$:

$$x_{A,i(T_H+1)}^+ = 0$$

$$x_{A,it_1(T_H+1)}^- = 0$$

$$y_{L,j(T_H+1)}^+ = 0$$

$$y_{L,jt_1(T_H+1)}^- = 0$$

Investment limits by type, $i = 1, ..., I$ and $t_1 = 1, ..., T_H$:

$$X_{A,it}^L \delta_{x_{A,it}^+} \le x_{A,it}^+ \le X_{A,it}^U \delta_{x_{A,it}^+}$$

Liability limits by type, $j = 1, ..., J$ and $t_1 = 1, ..., T_H$:

$$X_{L,jt}^L \delta_{x_{L,jt}^+} \le x_{L,jt}^+ \le X_{L,jt}^U \delta_{x_{L,jt}^+}$$

Short position limit by type, $k = 1, ..., K$ and $t_1 = 1, ..., T_H$

$$0 \le z_{kt}^+$$

$$0 \le z_{kt}^- \le Z_{kt}^{-,U}$$

Maximum new investments, $t_1 = 1, ..., T_H$:

$$\sum_{i=1}^{I} x_{A,it}^+ \le X_{A,t}^{+,U}$$

Maximum new liabilities, $t_1 = 1, ..., T_H$:

$$\sum_{j=1}^{J} y_{L,jt}^+ \le Y_{L,t}^{+,U}$$

Maximum liability per period, $t_1 = 1, \ldots, T_H$:

$$\sum_{j=1}^{J} \sum_{t_1=1}^{T_H} y_{L,jt_1t} \leq Y_{L,t}^{U}$$

Consigli and Dempster (1998) suggest the following stochastic programming multistage recourse framework for solving the ALM formulation given above. In (8.2), $E_{\omega_{T_H}|\omega_{T_H-1}}(\cdot)$ is the conditional expectation based on previous information at $T_H - 1$.

$$\min_{x_1 \in \mathfrak{R}^{n_1}} \left\{ f_1(x_1) + E_{\omega_2} \left[\min_{x_2 \in \mathfrak{R}^{n_2}} \left\{ f_2(x_2) + \cdots + E_{\omega_{T_H}|\omega_{T_H-1}} \left[\min_{x_{T_H} \in \mathfrak{R}^{n_{T_H}}} \left\{ f_{T_H}(x_{T_H}) \right\} \right] \right\} \right] \right\}$$

(8.2)

such that

$$A_1 x_1 = b_1$$

$$B_2 x_1 + A_2 x_2 = b_2, \quad \text{a.s.}^9$$

$$B_3 x_2 + A_3 x_3 = b_3, \quad \text{a.s.}$$

$$B_{T_H} x_{T_H-1} + A_{T_H} x_{T_H} = b_{T_H}, \quad \text{a.s.}$$

$$l_1 \leq x_1 \leq u_1$$

$$l_t \leq x_t \leq u_t, \quad \text{a.s. } t = 2, \ldots, T_H$$

Problem (8.2) states that the first period is deterministic and the remaining periods are stochastic. We shall adhere to this formulation in the minimax framework in the next section.

In this section, all formulations, both the original Consigli and Dempster (1998) and the minimax formulation below, cover a period up to the horizon T_H. We note that formulations from (8.2) onwards use the same variables as (8.1), and the optimization in all formulations end at time T_H. We also note that (8.1), as in the original Consigli and Dempster (1998), uses the riskfree interest rate variables r_{kt}^{+}, the return in period t on riskless asset k held in period $t - 1$, and r_{kt}^{-}, the unit cost of borrowing riskless asset k in period $t - 1$. Due to the definitions of the riskless interest rates, the applicable rates at T_H are $r_{kT_H+1}^{+}$ and $r_{kT_H+1}^{-}$. The use of the time $T_H + 1$ is simply for consistency with these definitions and does not alter the time frame of the formulations.

---

[9] a.s. means almost surely.

8.2 The Minimax Formulation of the Stochastic ALM Model 2

The formulation above shows that the decision variable x_t is solved depending on the path taken through the scenario tree. The solution to

$$E_{\omega_{T_H}|\omega_{T_H-1}}\left[\min_{x_{T_H}\in\Re^{n_{T_H}}}\left\{f_{T_H}(x_{T_H})\right\}\right] \tag{8.3}$$

is dependent on the state ω_{T_H-1}, that is, the path on the particular scenario. If we take the current state ω_{T_H} as one of the terminal states from Table 7.1, say TS8, then the solution is dependent on the previous state given by TS4. For a different terminal state ω_{T_H}, say TS15, the solution is dependent on another previous state, TS7.

Nielsen (1997) suggests a method of solving the above stochastic formulation by transforming it into a deterministic problem, referred to as the split-variable formulation. In order to explain this formulation, we need to discuss scenarios and states in a scenario tree and how they provide the pattern for the transformation from a stochastic to a deterministic framework. From Figure 7.1 and Table 7.1, a scenario, say S1, is a series of states[10] given by TS1, TS2, TS4 and TS8. Similarly, scenario S8 is represented by the series of states TS1, TS3, TS7 and TS15.

We note that the first period in (8.2) is deterministic and wholly dependent on the outcome of known parameters. To illustrate, if the source of uncertainty is the movement of the yield curve, the Stage 1 is wholly dependent on the current yield curve. In contrast, the uncertainty formulated in Figure 7.1 affects the initial and all subsequent stages.

In the split-variable formulation, a set of decision variables $x_{a,t}^s \in \Re^{n_t}$ are used at each time t, for a particular scenario s; these decision variables represent the deterministic equivalent of the original stochastic variables $x_t \in \Re^{n_t}$. For each scenario s, let the formulation below represent the deterministic scenario subproblem:

$$\min_{x_1\in\Re^{n_1},x_{a,t}^s\in\Re^{n_t},t=2,\ldots,T_H}\left\{f_1(x_1)+f_2^s(x_{a,2}^s)+\cdots+f_{T_H}^s(x_{a,T_H}^s)\right\} \tag{8.4}$$

such that

$$A_1x_1 = b_1$$

$$B_{a,2}^s x_{a,1}^s + A_{a,2}^s x_{a,2}^s = b_{a,2}^s$$

$$B_{a,3}^s x_{a,2}^s + A_{a,3}^s x_{a,3}^s = b_{a,3}^s$$

$$B_{a,T_H}^s x_{a,T_H-1}^s + A_{a,T_H}^s x_{a,T_H}^s = b_{a,T_H}^s$$

---

[10] We remind the reader that a binomial tree is used in Section 7 for illustration and simplification. Other more general tree structures may be used.

$$l_1 \le x_1 \le u_1$$

$$l_t \le x_{a,t}^s \le u_t, \quad t = 2, ..., T_H$$

where the superscript s on $A_{a,t}^s$, $B_{a,t}^s$, $b_{a,t}^s$ and $f_t^s(x_{a,t}^s)$ denotes the unique realization of ω_t associated with scenario s. Additionally, the subscript a on these variables denotes the ancestor state of that scenario. Let $a_{t-1}(s)$ represent the ancestor state, that is, the state from which the current state at time t emanates. For scenario S8, the ancestor state of TS15 is TS7. The ancestor state $a_{t-1}(s)$ is used below to define nonanticipativity constraints, that is, constraints that define the association between future states that have been conditioned by the same ancestor state. These nonanticipativity constraints ensure that any decision taken at a previous state is passed on to future states conditioned by that previous state. Further, let $N_{a_{t-1}(s)}$ represent the number of states from the root state to the ancestor state. If the current state is TS15, $N_{a_{t-1}(s)} = 3$. We use this number in the minimax formulation below.

The nonanticipativity constraints are written as

$$x_{a,t}^s = x_{a,t}^{s+1}, \quad \text{for } t = 2, ..., T_H \text{ and } a_{t-1}(s) = a_{t-1}(s + 1). \tag{8.5}$$

In the stochastic formulation above, the expectation is implemented over a finite set of scenarios with a probability of occurrence, θ^s associated with each scenario. Using the deterministic scenario formulation together with nonanticipativity constraints, the complete split-variable formulation of the stochastic problem becomes

$$\min_{x_1 \in \Re^{n_1}, x_{a,t}^s \in \Re^{n_t}, t=2,...,T_H} \left\{ f_1(x_1) + \sum_{s=1}^{m^{sce}} \theta^s \left(f_2^s(x_{a,2}^s) + \cdots + f_{T_H}^s(x_{a,T_H}^s) \right) \right\} \tag{8.6}$$

subject to

$$A_1 x_1 = b_1$$

$$B_{a,2}^s x_{a,1}^s + A_{a,2}^s x_{a,2}^s = b_{a,2}^s$$

$$B_{a,3}^s x_{a,2}^s + A_{a,3}^s x_{a,3}^s = b_{a,3}^s$$

$$B_{a,T_H}^s x_{a,T_H-1}^s + A_{a,T_H}^s x_{a,T_H}^s = b_{a,T_H}^s$$

$$l_1 \le x_1 \le u_1$$

$$l_t \le x_{a,t}^s \le u_t, \quad t = 2, ..., T_H$$

and

$$x_{a,t}^s = x_{a,t}^{s+1}, \quad \text{for } t = 2, ..., T_H \text{ and } a_{t-1}(s) = a_{t-1}(s + 1).$$

The multistage recourse framework is now re-cast into a single-stage minimax framework, transforming the treatment of uncertainties from the use of expectations to the use of worst-case scenarios.

$$\min_{x_1 \in \mathfrak{R}^{n_1}, x^s_{a,t} \in \mathfrak{R}^{n_t}, t=2,\dots,T_H} \ \max_{s \in \Omega} \left\{ f_1(x_1) + \left(f^s_2(x^s_{a,2}) + \cdots + f^s_{T_H}(x^s_{T_H}) \right) \right\} \qquad (8.7)$$

subject to

$$A_1 x_1 = b_1$$

$$B^s_{a,2} x^s_{a,1} + A^s_{a,2} x^s_{a,2} = b^s_{a,2}$$

$$B^s_{a,3} x^s_{a,2} + A^s_{a,3} x^s_{a,3} = b^s_{a,3}$$

$$B^s_{a,T_H} x^s_{a,T_H-1} + A^s_{a,T_H} x^s_{a,T_H} = b^s_{a,T_H}$$

$$l_1 \leq x_1 \leq u_1$$

$$l_t \leq x^s_{a,t} \leq u_t, \quad t = 2, \dots, T_H$$

and

$$x^s_{a,t} = x^{s+1}_{a,t}, \quad t = 2, \dots, T_H \text{ and } a_{t-1}(s) = a_{t-1}(s+1).$$

Problem (8.7) can be represented by the following nonlinear programming formulation:

$$\min_{\Phi \in \mathfrak{R}^1, x_1 \in \mathfrak{R}^{n_1}, x^s_{a,t} \in \mathfrak{R}^{n_t}, t=2,\dots,T_H} \{\Phi\} \qquad (8.8)$$

subject to

$$\left(f_1(x_1) + f^s_2(x^s_{a,2}) + \cdots + f^s_{a,T_H}(x^s_{a,T_H}) \right) \leq \Phi, \quad s \in \Omega$$

$$A_1 x_1 = b_1$$

$$B^s_{a,2} x^s_{a,1} + A^s_{a,2} x^s_{a,2} = b^s_{a,2}$$

$$B^s_{a,3} x^s_{a,2} + A^s_{a,3} x^s_{a,3} = b^s_{a,3}$$

$$B^s_{a,T_H} x^s_{a,T_H-1} + A^s_{a,T_H} x^s_{a,T_H} = b^s_{a,T_H}$$

$$l_1 \leq x_1 \leq u_1$$

$$l_t \leq x^s_{a,t} \leq u_t, \quad t = 2, \dots, T_H$$

and

$$x_{a,t}^s = x_{a,t}^{s+1}, \quad \text{for } t = 2, ..., T_H \quad \text{and} \quad a_{t-1}(s) = a_{t-1}(s+1).$$

The above strategy guards against the worst-case scenario across all scenarios identified at the starting time. Clearly, this represents a very pessimistic view of the worst case, and perhaps may be useful only for short horizon problems.

8.3 A Practical Single-stage Minimax Formulation

The formulation in Section 8.2 looks at long-term solutions to the ALM problem. Although ALM managers may have long-term horizons, it is generally acknowledged that taking long-term views rarely provide a practical insight into the construction of an ALM portfolio. A major reason for this is the lack of forecasting power to project and estimate a portfolio's performance in the far future. In practice, economists who provide supportive research for ALM construction have forecasting horizons from a few months to 1 or 2 years. Any forecast over a 1 year horizon is regarded as speculative and therefore not reliable for decision making. However, they are useful for forming a general view of global markets and how these markets may affect ALM portfolios in broad terms.

A short-term framework for optimization would be consistent with the limitation on the length of the forecasting horizon. Such a short-term framework is not necessarily inferior to the long-term framework. It is the close overlap between short-term optimization and short-term forecasting that provides a good argument for analyzing ALM portfolios with 1 or 2 year horizons. In this section, we present a single-stage minimax formulation where the worst case is identified over the first stage only. The scenarios are defined by the uncertainties until T_H. These are expectations of uncertainties evaluated from the beginning of the second stage onwards. Consider this scenario tree given in Figure 8.1.

In Figure 8.1, four scenarios have been identified as important at Stage 1. This number of scenarios may reflect four outcomes of the economic environment that forecasters are interested in. For example, Scenario 1 may refer to a rise in interest rates that would result in an upward shift in the yield curve. Scenario 2 may also refer to a rise in interest rates, but forecasts a flattening of the yield curve due to a projected upward shift at the short-end of the curve. Scenario 3 may refer to a hold in interest rates that would result in an upward shift at the long-end of the yield curve. Lastly, Scenario 4 may also refer to a hold in interest rates that would result in no significant shift in the shape of the yield curve. Each of these scenarios reflects a projection of the yield curve in the future, and can be used to provide expectations on the value of unknown variables.

Let $E^s(\cdot)$ be the expectation evaluated for scenario s. Consider the minimax formulation below:

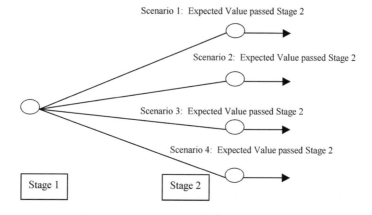

Figure 8.1 A single stage multiple scenario tree.

$$\min_{x_1 \in \Re^{n_1}} \max_{s \in \Re^{m_{sce}}} \left\{ f_1(x_1) + E^s\left[\left\{f_2^{i_2}(x_2) + \cdots + f_{T_H}^{i_{T_H}}(x_{T_H})\right\}\right]\right\} \qquad (8.9)$$

such that

$$A_1 x_1 = b_1$$

$$B_2^s x_1 + A_2^s x_2^s = b_2^s, \quad \text{for all scenario } s$$

$$B_3^s x_2^s + A_3^s x_3^s = b_3^s, \quad \text{for all scenario } s$$

$$B_{T_H}^s x_{T_H-1}^s + A_{T_H}^s x_{T_H}^s = b_{T_H}^s, \quad \text{for all scenario } s$$

$$l_1 \le x_1 \le u_1$$

$$l_t \le x_t^s \le u_t, \quad \text{for all scenario } s, \; t = 2, ..., T_H$$

The above formulation is subject to the deterministic constraints in the first stage, keeping in parallel with Consigli and Dempster (1998), as well as with the stochastic constraints from the second stage onwards. The number of scenarios is given by $m^{sce} = 4$, as defined by Figure 8.1.

Problem (8.9) can be implemented, and is practical and conforms to the well-developed format of economic forecasting. By using a scenario-based optimization as given by (8.9), decision makers are then able to implement their economic forecasts; they also have a better feel for the solution being offered by the optimizer. The formulation acknowledges the importance of worst-case analysis in the short run but admits the use of scenarios based on expectations in the medium term.

9 CONCLUDING REMARKS

In this chapter we have presented potential uses of minimax in asset liability management systems and have demonstrated the benefits from using minimax.

Asset allocations that attempt simultaneously to maximize terminal wealth and minimize default or the probability of default on liabilities as they fall due, generally operate within the expectations framework. This may lead a decision maker to make asset allocation decisions resulting in the nonservicing of liabilities. This becomes critical in certain areas of finance, in particular the pensions industry, where the servicing of long-term liabilities is an integral part of the pensions' objectives and of the allocation decision. ALM systems benefit from minimax optimization when different environmental scenarios imply different liability scenarios. The realization of any one particular liability scenario is critical compared to the realization of some average or expected value liability. ALM decisions with differing liability forecasts require that the asset allocation decision is robust in view of multiple liability scenarios. Such decisions, if based on minimax, would result in the least negative impact in terms of liability servicing. This is clearly an important property of minimax that makes it useful in analyzing and optimizing ALM systems.

In this chapter we have also presented alternative minimax ALM systems that would suit a variety of ALM users. An ALM formulation that would be useful for a particular user may not be useful for others. The choice depends on the nature of the ALM problem and the constraints imposed on the management of their assets and liabilities. Whatever the particular formulation, it is possible to adapt this to a minimax framework and construct robust solutions that address the issue of worst-case environmental, and therefore liability, scenarios.

References

Bierwag, G.O. (1987). *Duration Analysis: Managing Interest Rate Risk*, Ballinger, Cambridge, MA.

Bierwag, G.O., G.C. Kaufman and A. Toevs (1983a). "Bond Portfolio Immunization and Stochastic Process Risk", *Journal of Bank Research*, 13 (4), 282–291.

Bierwag, G.O., G.C. Kaufman and A. Toevs (1983b). "Duration: Its Development and Use in Bond Portfolio Management", *Financial Analysts Journal*, 39 (4), 15–35.

Bierwag, G.O., G.C. Kaufman and A. Toevs (1983c). "Immunization Strategies for Funding Multiple Liabilities", *Journal of Financial and Quantitative Analysis*, 18 (1), 113–123.

Bierwag, G.O., G.C. Kaufman and A. Toevs (1983d). "Recent Developments in Bond Portfolio Immunization Strategies", in: G.C. Kaufman, G.O. Bierwag and A. Toevs (editors), *Innovations in Bond Portfolio Management: Duration Analysis and Immunization*, JAI Press, Greenwich, CT.

Boyle, P.P. (1978). "Immunization under Stochastic Models of the Term Structure", *Journal of the Institute of Actuaries*, 105, 177–187.

Chambers, D.R., W.T. Carleton and R.W. McEnally (1988). "Immunizing Default-free Bond Portfolios with a Duration Vector", *Journal of Financial and Quantitative Analysis*, 23, 89–104.

Consigli, G. and M.A.H. Dempster (1998). "Dynamic Stochastic Programming for Asset Liability Management", Proceedings of the APMOD95 Conference, Brunel University, *Annals of Operations Research*, 81, 131–161.

Cox, J.C., J.E. Ingersoll and S.A. Ross (1979). "Duration and the Measurement of Basis Risk", *Journal of Business*, 52 (1), 51–61.

Dahl, H., A. Meeraus and S. Zenios (1993). "Some Financial Optimization Models: I Risk Management", in *Financial Optimisation*, Cambridge University Press, Cambridge.

Fabozzi, F. (editor) (1991). *The Handbook of Fixed Income Securities*, 3rd edition, Business One-Irwin, Homewood.

Fabozzi, F. (editor) (1993). *Bond Markets, Analysis and Strategies*, 2nd edition, Prentice Hall, Engelwood Cliffs, NJ.

Fisher, L. and R.L. Weil (1971). "Coping with the Risk of Interest Rate Fluctuations: Returns to Bondholders from Naive and Optimal Strategies", *Journal of Business*, 44, 408–431.

Fong, G. and O. Vasicek (1984). "A Risk Minimizing Strategy for Portfolio Immunization", *Journal of Finance*, 39, 1541–1546.

Grove, M.A. (1974). "On 'Duration' and the Optimal Maturity Structure of the Balance Sheet", *Bell Journal Of Economics and Management Science*, 5, 696–709.

Hiller, R.S. and C. Schaack (1990). "A Classification of Structured Bond Portfolio Modeling Techniques", *Journal of Portfolio Management*, 17 (1), 37–48.

Hiller, R.S. and J.F. Shapiro (1989). "Stochastic Programming Models for Asset/Liability Management Problems", International Financial Services Research Center Discussion Paper no. 105-89, Sloan School of Management, MIT.

Kaufman, G.C. (1984). "Measuring and Managing Interest Rate Risk: A Primer", *Economic Perspective, Federal Reserve Bank of Chicago*, 16–19.

Macaulay, F.R. (1938). "Some Theoretical Problems Suggested by the Movements of Interest Rates, Bond Yields and Stock Prices in the U.S. Since 1856", National Bureau of Economic Research, New York.

Nielsen, S. (1997). "Mathematical Modeling and Optimisation with Applications to Finance", unpublished lecture notes.

Redington, F.M. (1952). "Review of the Principle of Life Office Valuations", *Journal of the Institute of Actuaries*, 18, 286–340.

Reitano, R.R. (1989). "Multivariate Approach to Duration Analysis", *ARCH*, 97–181.

Reitano, R.R. (1990a). "A Multivariate Approach to Immunization Theory", *ARCH*, 261–312.

Reitano, R.R. (1990b). "Non-Parallel Yield Curve Shifts and Durational Leverage", *Journal of Portfolio Management*, 16, 62–67.

Reitano, R.R. (1991a). "Multivariate Duration Analysis", *TSA*, XLIII, 335–376.

Reitano, R.R. (1991b). "Non-Parallel Yield Curve Shifts and Spread Leverage", *Journal of Portfolio Management*, 17, 82–87.

Reitano, R.R. (1992). "Non-Parallel Yield Curve Shifts and Immunization", *Journal of Portfolio Management*, 18, 36–43.

Sharpe, W. and L.G. Tint (1990). "Liabilities – A New Approach", *Journal of Portfolio Management*, Winter, 5–10.

Shiu, E.S.W. (1986). "A Generalization of Redington's Theory of Immunization", *ARCH*, 69–81.

Shiu, E.S.W. (1988). "Immunization of Multiple Liabilities", *Insurance: Mathematics and Economics*, 7, 219–224.

Stock, D. and D.G. Simonson (1988). "Tax-Adjusted Duration for Amortizing Debt Instrument", *Journal of Financial and Quantitative Analysis*, 23, 313–327.

Vanderhoof, I.T. (1972). "The Interest Rate Assumptions and the Maturity Structure of the Assets of a Life Insurance Company", *TSA*, XXIV, 157–192.

COMMENTS AND NOTES

CN 1: Duration and Convexity

A measure of the speed of payment on a bond is the average maturity of the stream of cashflows. This is referred to as Macaulay Duration. Dollar Duration is the interest rate sensitivity of a bond, that is, the change in value of the bond for a unit change in yield. Modified Duration represents the percentage change in the value of the bond for a small change in yield. Using the following notation:

P	the full price of a bond
C	annual coupon payment
R	redemption payment
t_m	maturity of the bond in years
$r(t_i)$	spot rate, in decimal, applicable to payment i due at time t_i
y	yield to maturity or redemption yield, in decimal
f	frequency of coupon payments
$d(t_i) = \dfrac{1}{\left(1 + (r(t_i)/f)\right)^{ft_i}}$	discount factor applicable to payment i due at time t_i

The full price of a bond, using spot rates, is given by

$$P = \frac{C}{f} \sum_{i=1}^{m} d(t_i) + Rd(t_m).$$

The full price of a bond, using yield, is given by

$$P = \frac{C}{f} \sum_{i=1}^{m} \frac{1}{(1 + (y/f))^{ft_i}} + R \frac{1}{(1 + (y/f))^{ft_{mi}}}.$$

The Macaulay Duration, using discount rates, is given by

$$\text{Macaulay Duration} = \frac{\sum_{i=1}^{m} t_i d(t_i)(C_i/f) + t_m d(t_m)R}{\sum_{i=1}^{m} d(t_i)(C_i/f) + d(t_m)R}.$$

The Modified Duration is given by

$$\text{Modified Duration} = \frac{1}{P} \frac{\partial P}{\partial y} = -\frac{\text{Macaulay Duration}}{1 + (y/f)}$$

Modified Duration is the measure most commonly used by market practitioners.

Convexity is the second derivative of the price function with respect to yield, that is,

$$\text{Convexity} = \frac{1}{(1 + (y/f))^2 f^2} \frac{\sum_{i=1}^{m} t_i t_{i+1} d(t_i)(C_i/f) + t_m t_{m+1} d(t_m)R}{\sum_{i=1}^{m} d(t_i)(C_i/f) + d(t_m)R}.$$

CN 2: Term Structure of Interest Rates

The term structure of interest rates refers to the set of interest rates for lending or borrowing with different maturities. In the case of government borrowings, normally the most liquid of fixed income instruments, the term structure refers to the set of interest rates that investors demand for holding government debt. The term structure of interest rates is not directly observable. What can be observed is the yield or the internal rate of return on a bond that represents some form of average return over the life of the bond. For all bonds issued by the government, there is a corresponding set of yields called the yield curve. The term structure is implied in the yield curve. A technique called bootstrapping can be used to extract the term structure of interest rates from the yield curve.

CN 3: Generalized Tree Structures

The structure of generalized trees uses nodes and branches of an unspecified number. Trees are generally used as decision tools that aid in the itemization of possibilities that face a decision maker. In other words, trees are explicit versions of permutations of events. A node is any part of a tree where an intermediate decision can be taken with regards to which branch to take. A node normally connotes a state or an environment that has been reached while traversing the tree. A branch is the transition from one node to another. Generalized trees may have varying numbers of branches emanating from a node representing varying sets of possibilities for the decision maker.

At the tips of the tree are leaves that represent the terminal state or environment. The path that a decision maker takes from the first node to a leaf is referred to as a scenario. Therefore, the number of leaves is the same as the number of scenarios.

Chapter 11

Robust currency management

In this chapter we discuss currency management. We consider the strategic point of view where a currency benchmark is identified. We also consider the tactical point of view where the frequent re-balancing of a portfolio's currency hedge is managed in order for the portfolio to benefit from short- and medium-term currency fluctuations. Currency benchmark identification is important in finding the long-term optimal hedge ratio that a portfolio should adopt in order to minimize the negative impact of any currency depreciation in the medium and long term. We discuss benchmark identification, first, for a pure currency portfolio as this simplifies and clarifies the role of strategic currency management, and second, for an international asset portfolio as this clarifies the use of currency hedging as a tool for managing the currency exposure of such a portfolio. The tactical management of a currency portfolio involves currency trades that cause fluctuations about the strategic benchmark. These trades represent short- to medium-term adjustments to the optimal hedge ratio in order to maximize returns and accumulate short-term gains. We consider both the strategic and tactical formulations.

1 INTRODUCTION

The financial modeling of the behavior of currencies has gained attention as portfolios become increasingly international. Currency modeling has been a major focus of research as more and more market participants realize the impact of currencies on the value of their portfolios. By investing in the equity market of a foreign country, a portfolio manager not only receives local market returns but also the currency return associated with a foreign currency-denominated equity market. The manager needs to appraise the local equity market to judge whether the investment is worth the risk. Additionally, she needs to appraise the potential performance of the foreign currency to judge whether the exposure to that foreign currency would yield a desirable return. A positive appraisal of the currency may motivate her to maintain the exposure to that foreign currency whereas a negative appraisal may cause her to hedge the currency risk (see CN 1). Local market forecasting and currency forecasting become essential in this appraisal process. The

manager may find it increasingly difficult to achieve a suitable conjoint modeling of local markets and currencies as more and more foreign markets are
included in the manager's portfolio. Because the skill for appraising equity
markets in general is different from the skill required to appraise currencies,
the portfolio manager, perhaps originally hired for her expertise in stock
picking, may not be able to function as expertly in the currency domain.
There is thus a very strong argument for employing a specialist currency
manager whose role is to develop a currency overlay that would manage the
currency risks for the portfolio manager.

Currency forecasting is also important for borrowers who can issue foreign
currency-denominated debt. Apart from the skill required to appraise the
potential demand for a foreign-denominated debt, the borrower would also
need to have the skill to appraise a potential currency depreciation. Being able
to forecast the depreciation of a currency may cause the borrower to issue debt
in that currency because the depreciation would result in less (debt) repayment
than would have been the case if the currency is stable or if it appreciates.
While a potential currency appreciation is desirable for an investing portfolio
manager, a potential currency depreciation is desirable for a borrower.

Market participants involved in international trade also find that currency
forecasting is an essential part of their planning. Foreign and domestic companies competing for market share know that their competitiveness is a direct
function of exchange rate behavior. Similarly, policy makers may find that the
currency markets influence their decision on the optimal mix of monetary and
fiscal policies.

The increasing proportion of foreign assets that enter asset portfolios gives
increasing importance to currency management. Several authors have investigated the impact of currency hedging decisions on the performance of diversified international portfolios. Perold and Schulman (1988) argue that it is
better to formulate long-run investment policy in terms of hedged portfolios
than unhedged portfolios. The key argument is that, from the point of view of
long-run policy, investors should think of currency hedging as having zero
expected return, and that on average, currency hedging gives substantial risk
reduction at no loss of expected return. Thus, hedging should be the policy,
and lifting the hedge an active investment decision.

Perold and Schulman (1988) also argue that there are good grounds for
believing that, over relatively short periods, the expected return from
currency hedging (equivalently, currency exposure) will be nonzero. Further,
there is no indication that these premia will be consistently positive or
consistently negative, or that pursuit of such returns (by being longer or
shorter currency at the margin) will be worth the added risk. The marginal
decisions with respect to currency exposure should therefore be made by
managers who are in close touch with the changing nature of risk and return
in the currency markets.

One can argue, however, that the long-run view where currency hedging produces a zero expected return is not consistent with real-world portfolio management with relatively shorter time frames. Fund managers are assessed on regular intervals, mainly on a yearly basis. We are therefore forced to look at more realistic investment policy horizons that would not satisfy the zero-expected-return assumption on currency. This horizon constraint also forces us to look at currency risks and returns in conjunction with local risks and returns of assets. With this horizon constraint, we deviate from the Perold and Schulman's (1988) prescription of full hedging as the default investment strategy and of selectively lifting the hedge as an active investment decision.

Jorion (1989) considers the relative merits of hedging and of not hedging for portfolios of foreign stocks and bonds, from the point of view of a usd-based[1] investor. Studying 11 years of data, subdivided into periods of dollar weakness and strength, Jorion (1989) observed: the first 3-year period, characterized by dollar weakness; then the next 4-year period, characterized by dollar strength; and then the last 4-year period, characterized by dollar weakness. It is shown that hedging only makes sense during the period of dollar strength. This study illustrates the point that the long-run investment horizon consistent with zero expected currency return addressed in Perold and Schulman (1988) could be longer than 4 years.

Levy (1981) presents a multicurrency portfolio optimization framework as an extension to a single currency mean-variance system. The extension comes in the form of calculating returns as total returns, that is, the returns on a foreign asset calculated as the proportionate change in base currency denominated price of the asset. Total return calculations are used on a monthly interval for 9 years as input for covariance calculations. Both calculations are then used as input to a mean-variance optimizer. The suggested framework is not sufficient to capture the full range of possibilities for risk and return because it does not allow investors to hedge their currency exposure. The underlying assumption is that investors using the system would invest on an unhedged basis only. An exact framework for incorporating currency risk and return into local assets risk and return as a combined input for a mean-variance optimization system is given by Rustem (1995). This is based on the exact evaluation of the multiplicative currency and local risk expressions.

Eun and Resnick (1985) report on the benefits from international diversification and on the optimal international portfolios from 15 currency perspectives. Similar to Levy (1981), this paper does not capture the full range of diversification benefits and the full range of risk and return possibilities available to international investors because currency hedging is ignored. In particular, the benefits from diversification as a result of asset selection can only be ascertained on a full currency hedging basis. Eun and Resnick (1988) study

---

[1] The term "usd-based" means the base currency of the manager is the US dollar.

risk and return profiles of particular portfolios under no currency hedging and full currency hedging, in order to consider the benefits from international diversification. The framework is an improvement from that used in Eun and Resnick (1985), but it still does not capture the full range of risk and return possibilities, as can be provided with selective hedging.

Hauser and Levy (1991) evaluate three currency hedging strategies for fixed income portfolios: *no hedging*, *full hedging* and *partial hedging*. This study shows that returns on foreign bonds are positively correlated with currency returns, and that long-term foreign bonds are significantly less correlated with exchange rates than short-term bonds. The gains from hedging are higher, the shorter the maturity of the bonds in the portfolio. For low-risk portfolios, a fully hedged (or in some cases over-hedged) strategy is optimal. An investor should refrain from hedging only when the expected return on exchange rates is higher than the forward premium. Given that the expected return on foreign exchange is higher than the forward premium, the optimally hedged efficient set converges to the nonhedged efficient set as the expected return is increased. Hauser and Levy (1991) discuss the results for these three currency hedging strategies, but do not consider in detail how partial hedging is determined. Investors are allowed to simultaneously determine the investment proportion in each asset and the optimal amount of forward hedging. However, the framework that allows this capability is not discussed. The expansion of the analysis to include partial hedging, including cross hedging (see CN 2), may provide a richer set of risk and return possibilities.

Eaker and Grant (1990) discuss the benefits from international diversification using three hedging strategies: *no hedging*, *full hedging* and *selective hedging*. The selective hedging framework is based on a constant partial hedge that does not change during the determination of the efficient frontier. The result of this restriction is that *selective hedging* is not flexible to cover *full hedging* as a possibility for the minimum-risk portfolio and it does not cover *no hedging* as a possibility for the maximum-return portfolio.

In addressing the issue of currency hedging, currency managers need to strike a balance between the optimal long-term currency hedge in view of the long-term direction of particular currencies and the optimal short-term currency hedge in view of short-term fluctuations in these currencies. This balance is generally seen as a trade-off between a strategic currency hedge and a tactical currency hedge. The long-term currency hedge can be identified within a strategic currency management system, where the long-term direction of particular currencies are modeled on fundamental economic factors. On a short-term basis, a tactical currency hedge essentially augments the strategic hedge in order to benefit from short-term fluctuations in the value of the currencies.

In Section 2, we consider a strategic currency management system that is useful in determining optimal currency benchmarks for long-term planning.

The identification of the strategic currency portfolio is dependent on a long-term model of currency movement.

In Section 3, we distinguish between a strategic pure currency portfolio and a strategic currency benchmark designed for the purpose of hedging currency exposures of international portfolios. While Section 2 concentrates on the relatively simpler problem of finding the optimal currency benchmark for a pure currency portfolio, Section 3 focuses on finding the optimal currency benchmark that hedges the risk of different currency exposures. We do not consider pure currency portfolios but study general asset portfolios whose risk and return profiles demand the solution to the currency benchmark identification problem.

In Section 4, we discuss a generic currency model generating signals whose value and direction determine the tactical currency bet that the currency manager would implement. We assume there exists such a framework that enables the currency manager to read signals indicating whether she should be long or short a particular currency (see CN 3).

In Section 5, we generalize the framework developed in Section 4 to address the issues of mis-forecasting. We study this in terms of the downside risk due to mis-forecasting and also in terms of the ability of the modeling framework to provide cushions against worst-case events, such as following the wrong forecasting model, that could potentially harm an investor.

In Section 6, we discuss the interplay between the strategic currency benchmark for an international portfolio and the tactical currency management for such a portfolio. This section integrates for the reader the concept of currency overlay by explaining the dominance of the strategic currency benchmark and the constraints it imposes on the tactical system.

In Section 7, we briefly discuss the use of currency options in the management of currency exposures. While options in general provide insurance at a price, they also serve as return enhancers within a tactical currency system. The use of options as very short-term tools for generating excess returns in currency overlay systems is discussed.

An appendix is provided to consider how fundamental and technical approaches to currency forecasting can be integrated into a single forecasting framework. It is noted that a reliance on fundamental or technical analysis alone can lead to currency management problems and that these can be avoided by adopting an integrated approach to currency forecasting.

2 STRATEGIC CURRENCY MANAGEMENT 1: PURE CURRENCY PORTFOLIOS

We illustrate the importance of a strategic currency benchmark by looking at a time series for yen/usd[2]. Figure 2.1 shows the yen/usd exchange rate from

---

[2] Yen/usd is the exchange rate between the Japanese yen and the US dollar.

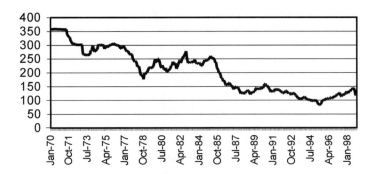

Figure 2.1   Yen/usd rate for the period 1970–98.

1970 to 1998. The figure shows that over 30 years, the yen has been appreciating against the US dollar. For a usd-based currency manager with a 30-year horizon, it would have made sense to keep yen exposure unhedged, thereby benefiting from yen appreciation. However, for a yen-based manager with the same horizon, it would have made sense to hedge all US dollar exposure, thereby avoiding the loss from US dollar depreciation.

Under usual circumstances, the horizon of an international portfolio being managed by a currency manager may span several years, although less than the 30-year period in our illustration. This raises the question as to whether, for a usd-based investor, keeping the yen exposure unhedged is a sensible strategy. Figure 2.2 shows the exchange rate between 1991 and 1994, showing a period of yen appreciation relative to the US dollar, while Figure 2.3 shows the rate between 1995 and 1998, showing a period of yen depreciation. For a usd-based currency manager with a 4-year horizon, the strategic currency hedge at the start of 1991 would have been an unhedged yen exposure. For the same

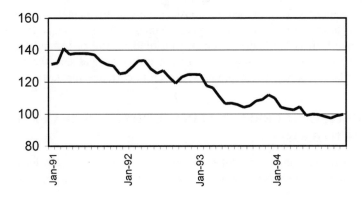

Figure 2.2   Yen/usd rate for the period 1991–94.

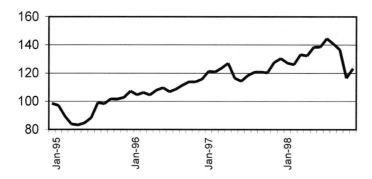

Figure 2.3   Yen/usd rate for the period 1995–98.

currency manager with a 4-year horizon, the strategic currency hedge at the start of 1995 would have been a fully hedged yen exposure.

A strategic currency benchmark uses a long-term currency model to project a particular currency into the horizon of the portfolio. In the above illustration, this period is 4 years. At this stage, we assume that a long-term currency model exists, enabling the projection of particular currencies to the end of a holding period of at least 1 year.

From Figures 2.2 and 2.3, although the exchange rate follows a well-defined direction, that is, yen appreciation in Figure 2.2 and yen depreciation in Figure 2.3, there are subperiods wherein the currency follows a short-term trending that opposes the long-term directional move. For example, from Figure 2.3, one can discern a period of short-term yen appreciation in mid-1997, as well as in late 1998. These subperiods provide opportunities for achieving excess returns via tactical currency bets. In this section, we concentrate on the long-term strategic benchmark problem. In Section 4, we discuss a generic currency model that addresses the issue of tactical management.

Pure currency portfolios take advantage of the dual role of money: as an asset class that is capable of earning a return during its holding period, and as a means of exchange that is capable of generating returns from currency movements. Currency portfolios have been prominent in the form of currency overlays in investment management as well as in the form of currency speculative attacks on weak currencies, particularly of emerging markets. We present a strategic pure currency portfolio that benefits from scenario optimization, both in the nonlinear programming and in the minimax framework. The strategic portfolio that we describe does not have the properties of currency overlay in terms of hedge ratios because we do not include noncurrency assets. Also, it does not have the properties of currency speculation in terms of short-term short positions aimed at potential devaluations. Instead, the strategic portfolio is a currency asset allocation that aims to identify the medium- to long-term optimal currency exposures. We choose a basket of currencies in view of multiple scenarios and show how an investor in currencies can benefit from optimization and, in particular, from minimax. In this section, we explore continuous minimax as a tool for foreign exchange optimization.

For a pure currency portfolio composed of m^{cur} currencies, we wish to find the portfolio mix that maximizes the expected return[3] and minimizes the risk from the point of view of a specific base currency:

$$\min_{x \in \Re^{m^{cur}}} \{-\alpha \langle x, r \rangle + (1 - \alpha) \langle x, Cx \rangle\} \qquad (2.1)$$

$$\langle 1, x \rangle = 1$$

$$x \geq 0$$

where

$$r_i = (1 + r_i^c)(1 + r_i^e) - 1, \quad \forall i \in [1, ..., m^{cur}]$$

1 is the unit vector, with unit elements, and $C \in \Re^{m^{cur} \times m^{cur}}$ is the covariance matrix of returns r_i. The formulation uses a trade-off variable $\alpha \in [0, 1]$ chosen to reflect the importance of return versus risk. By varying α between zero and one, we can find a set of portfolios that define the efficient frontier. Here, $r \in \Re^{m^{cur}}$ whose element r_i is the expected return from investing in currency i, composed of the expected cash return r_i^c and the expected currency appreciation r_i^e. Empirical evidence suggests that the second component, the return from currency appreciation, is of a higher order compared to cash returns. Although both components can be generally regarded as uncertain, cash return is relatively less risky. Regarding cash return, an investor can lock into a foreign time deposit and get the contracted return or invest in money market instruments or treasury bills. Being at the short end of the yield curve, these instruments have almost certain outcomes at the end of the investment period. Regarding currency return, the investor is facing a riskier investment: the volatility in the foreign exchange markets is the major source of concern for currency portfolio investors. It is of course the volatility of currencies that enables the investor to reap sizeable gains in the foreign exchange markets, provided she is in or out of a currency at the right time. It is the timing issue that only very few investors might manage to get right consistently. The highly uncertain nature of currencies and the highly attractive potential rewards make investors search for decision tools that might enable them to make decisions in the right direction.

In the following discussion we concentrate on the modeling of scenarios for currencies. We present a framework using long-term generic currency models that have been employed in various forms. First, we need to look at how

---

[3] The maximization is formulated as a minimization of negative expected return.

foreign exchange returns are calculated:

$$r_i^e = \frac{e_{i,T_H}}{e_{i,0}} - 1 \qquad (2.2)$$

where $e_{i,t}$ is the spot exchange rate at time t in terms of units of the base currency[4] per unit of foreign currency i, with $t = T_H$ representing the horizon date. When the allocation decision is to be made at time $t = 0$, the spot exchange rate $e_{i,0}$ is known but the spot exchange rate at the horizon, e_{i,T_H}, is unknown. Various models of exchange rate determination attempt to point-forecast e_{i,T_H} but with limited success. The lack of success of point-forecasting e_{i,T_H} does not necessarily render these models useless. Perhaps what is required is a completely different approach: moving away from point-forecasting towards range-forecasting may provide a wider perspective in estimating e_{i,T_H}. A long-term generic model might be

$$e_{i,T_H} = g_i(F_i) + y_i \qquad (2.3)$$

where the horizon's spot exchange rates are modeled by a function of a vector of factors F_i, plus an estimation error y_i. The function may be suitable for estimation by regression. It may be constructed using economic variables such as price indices, growth and trade figures, money supply, interest rates, etc., that apply to both the foreign country and the base country. We shall not discuss the details of F_i. The emphasis is in the relaxation of a generic point-forecast framework using range forecast within the minimax approach.

Let the estimation error y_i vary between some investor-defined upper and lower bounds as

$$y_i^{\text{lower}} \le y_i \le y_i^{\text{upper}}.$$

The bounds on y_i allow us to find the corresponding values of e_{i,T_H}. Consequently, the return on currency i is bounded, where the bounds are determined by the extreme values of y_i through the following relationship:

$$r_i^e = \frac{g_i(F_i) + y_i}{e_{i,0}} - 1.$$

The minimax framework yields the optimal allocation between currencies that cushions against the worst-case outcome for each currency return. We adopt m^{cur} currencies, including the base currency. We use $x \in \Re^{m^{cur}}$ as the minimizing variable and we define the maximizing variable as $y \in \Re^{m^{cur}}$.

$$\min_{x \in \Re^{m^{cur}}} \max_{y \in \Re^{m^{cur}}} \left\{ -\alpha \sum_{i=1}^{m^{cur}} x_i \left((1 + r_i^c) \left(\frac{g_i(F_i) + y_i}{e_{i,0}} \right) - 1 \right) + (1 - \alpha)\langle x, Cx \rangle \right\}$$

$$(2.4)$$

---

[4] The base currency is the home currency of the investor.

subject to

$$y_i^{\text{lower}} \leq y_i \leq y_i^{\text{upper}}, \quad i = 1,\dots,m^{cur}$$

$$\langle 1,x \rangle = 1$$

$$x \geq 0.$$

These ranges need to be refined further to ensure consistency, that is, preserving the triangulation property of currencies in order to avoid any trivial possibility of cross-currency inconsistency. This means that all exchange rates moving between lower bounds (and similarly, all upper bounds) should satisfy triangulation. To illustrate for three currencies,

$$\left(\frac{\text{usd}}{\text{euro}} \right)\left(\frac{\text{yen}}{\text{usd}} \right)\left(\frac{\text{euro}}{\text{yen}} \right) = 1. \tag{2.5}$$

This triangulation requirement is satisfied by

$$e_{i,t}\left(\frac{1}{e_{j,t}} \right)(e_{ij,t}) = 1$$

for any time t, and where $e_{ij,t}$ is the cross exchange rate between currencies i and j. From a usd-based perspective, $e_{i,t}$ is quoted as units of US dollar per unit of foreign currency i, while $e_{ij,t}$ is quoted as units of currency i per unit currency j. Consequently, when forecasting for the horizon's exchange rates, not only do we need a long-term relation as in (2.3) but also a long-term model for forecasting the cross rate. Let

$$e_{ij,T_H} = g_{ij}(F_{ij}) + y_{ij} \tag{2.6}$$

be a model of the cross exchange rate between currencies i and j. It is modeled by a function of a vector of factors F_{ij}, plus an estimation error y_{ij}. As in the modeling of straightforward exchange rates, F_{ij} may be constructed using economic variables such as price indices, growth and trade figures, money supply, interest rates, etc., that apply to both foreign countries, but this time without any reference to the base country.

Let the estimation error y_{ij} vary between some investor-defined upper and lower bounds as $y_{ij}^{\text{lower}} \leq y_{ij} \leq y_{ij}^{\text{upper}}$. The minimax formulation that incorporates the triangulation requirement is given by

$$\min_{x \in \Re^{m^{cur}}} \max_{y \in \Re^{m^{cur}}} \left\{ -\alpha \sum_{i=1}^{m^{cur}} x_i \left((1 + r_i^c)\left(\frac{g_i(F_i) + y_i}{e_{i,0}} \right) - 1 \right) + (1 - \alpha)\langle x, Cx \rangle \right\} \tag{2.7}$$

subject to

$$y_i^{\text{lower}} \le y_i \le y_i^{\text{upper}}, \quad i = 1, \dots, m^{cur}$$

$$y_{ij}^{\text{lower}} \le y_{ij} \le y_{ij}^{\text{upper}}, \quad i = 1, \dots, m^{cur} - 1, \ j = 1, \dots, m^{cur} - 1, \ i \ne j$$

$$\left(g_i(F_i) + y_i\right)\left(\frac{1}{g_j(F_j) + y_j}\right)\left(g_{ij}(F_{ij}) + y_{ij}\right) = 1, \quad i \ne j$$

$$\langle 1, x \rangle = 1$$

$$x \ge 0.$$

We note that for the second type of constraint, the number of unique cross-currency constraints is given by the number of combinations from $(m^{cur} - 1)$ taken two at a time (see CN 4).

3 STRATEGIC CURRENCY MANAGEMENT 2: CURRENCY OVERLAY

The term *currency overlay* is used in fund management to mean the application of a tactical currency trading system to supplement and/or complement the performance of a strategic currency benchmark. The overlay covers two distinct systems: the long-term management of currency exposures using strategic benchmarks, and the short- and medium-term management using tactical trading systems. This section deals with strategic benchmarking for overlay purposes. Sections 4 and 5 deal with tactical trading systems and Section 6 discusses the interplay between the strategic benchmark and the tactical trading system to provide a fuller picture of currency overlay.

While Section 2 concentrates on finding the strategic currency benchmark for a pure currency portfolio, this section is focused on finding the strategic benchmark for currency overlay purposes. We consider general asset portfolios, mainly the international component, and the problem we address is that of hedging the currency exposure by virtue of holding those international assets.

To describe currency overlay, we need to present it in the context of an international portfolio and a manager's base currency. The manager would have to consider separate hedges corresponding to the foreign currency exposures where the exposures come from the original allocations to either equity, fixed income or both. The important inputs to the analysis of optimal hedges are equity, fixed income and currency models. It is the interplay between the total return and the total risk of the equity and fixed income, from the point of view of the base currency, that determines the optimal hedges. We first show how the optimal hedges are analyzed in the expected value framework and then show how these are analyzed in the minimax framework.

For an asset portfolio with n assets, the portfolio mix $x \in \Re^n$ that maximizes the expected return and minimizes the risk from the point of view of a specific base currency is given by

$$\min_{x \in \Re^n} \{-\alpha \langle x, r \rangle + (1 - \alpha)\langle x, Cx \rangle\} \qquad (3.1)$$

subject to

$$\langle 1, x \rangle = 1$$

$$x \geq 0$$

where

$$r_i = (1 + r_i^a)(1 + r_i^e) - 1, \quad \forall i \in [1, ..., n].$$

Here, $r \in \Re^n$ whose element r_i is the expected return from investing in foreign asset i, composed of the expected local asset return r_i^a and the expected currency appreciation r_i^e. It is useful to recall at this stage that the covariance matrix C is based on r_i, and not on either r_i^a or r_i^e. By comparing (3.1) with (2.1), it can be observed that the cash return for a pure currency portfolio has been replaced by the local asset return for an asset portfolio. The investment in the foreign asset introduces an extra level of uncertainty, that is, the uncertainty of returns on the local asset. Not only do we need a long-term model for the currency, but we also need a long-term model for the foreign asset. In general, the modeling of equity requires a different skill from the modeling of fixed income. This means that the analysis of optimal hedge ratios require three different skills: on long-term equity modeling, on long-term fixed income modeling and on long-term currency modeling. With regard to the required equity model, one need only account for the equity market and not for individual stocks as it is the long-term path of the market that needs to be estimated and not the specific paths of individual stocks. Similarly for fixed income, we need a model that reflects significant yield curve shifts and are not concerned about relative values of individual fixed income assets.

Let r_i^a below represent the local return on asset a_i, be it fixed income or equity:

$$r_i^a = \frac{a_{i,T_H}}{a_{i,0}} - 1.$$

A long-term model for the local return of the foreign asset may be given by

$$a_{i,T_H} = g_i^a(F_i^a) + y_i^a \qquad (3.2)$$

where the value of the asset at the horizon is modeled by a function of a vector of factors F_i^a, with an estimation error y_i^a. In this model, as in (2.3), the function may be suitable for estimation by regression techniques. It may be constructed using economic variables such as interest rates, credit and liquidity indices, money supply, growth and trade figures, and capital flows in the case of fixed income. As for equity, the function may include submodels of dividend flow, capitalization growth and investment flow. We do not discuss the details of F_i^a. The emphasis is on the analysis of optimal hedge ratios.

The local asset return becomes

$$r_i^a = \frac{g_i^a(F_i^a) + y_i^a}{a_0^i} - 1$$

and (3.1) can be expressed as

$$\min_{x \in \Re^n} \left\{ -\alpha \left(\sum_{i=1}^n x_i \left(\left(\frac{g_i^a(F_i^a) + y_i^a}{a_{i,0}} \right) \left(\frac{g_i(F_i) + y_i}{e_{i,0}} \right) - 1 \right) \right) + (1 - \alpha)\langle x, Cx \rangle \right\}$$

$$(3.3)$$

subject to

$$\langle 1, x \rangle = 1$$

$$x \geq 0.$$

Again, a distinction needs to be made between fixed income and equity for the modeling of the local return of the foreign asset. By solving (3.3), we arrive at an optimal mix of foreign assets without any hedging strategy, that is, the assets are unhedged. However, in identifying a strategic currency exposure, the formulation should address the issue of optimal hedge ratios.

Neither (3.1) nor (3.3) are valid models for finding the optimal hedge ratio. While these formulations yield optimal allocations of foreign assets on an unhedged basis, they do not address the problem of finding the hedge ratio. The purpose of currency hedge ratio analysis is to study an existing international portfolio and find hedges against the different currency exposures so that the returns of the local assets are preserved or insulated against various currency depreciations. This means that for an existing international portfolio, the allocation of foreign assets is already known. It is just the allocation to the currency exposure, that is, the currency hedge ratio, that needs to be solved.

Recall that the total return on a foreign asset, from the point of view of the base currency, is

$$r_i = (1 + r_i^a)(1 + r_i^e) - 1. \tag{3.4}$$

This applies to an unhedged asset, that is, the currency return contributes to the asset's total return. If, for the same asset, its currency risk is hedged away, then its total return becomes

$$r_i = r_i^a(1 + r_i^e) + r_i^f \tag{3.5}$$

where r_i^f is the return on the forward exchange rate. This normally translates to a cost, that is, the premium paid for removing currency risk. For the analysis of currency hedge ratios one may simplify the problem by assuming that r_i^f is equal to the interest rate differential:

$$r_i^f = r_i^{c,\text{base}} - r_i^{c,\text{foreign}}. \tag{3.6}$$

Here, we adhere to the notation from (2.1) where we use r_i^c to represent the expected cash return, and use the superscripts foreign and base.

From the point of view of the base currency, there are two synthetic assets for each original foreign asset characterized by different risk and return properties. The first synthetic asset is the unhedged version whose return is given by (3.4) and its corresponding risk is the volatility of those unhedged returns. The second synthetic asset is the hedged version whose return is given by (3.5), and its corresponding risk is the volatility of those hedged returns.

Let $\hat{X}^u \in \mathfrak{R}^n$ be the unknown allocation to the n synthetic unhedged assets, and $\hat{X}^h \in \mathfrak{R}^n$ be the unknown allocation to the n synthetic hedged assets. Also, let $\hat{C} \in \mathfrak{R}^{2n \times 2n}$ be the covariance matrix for the $2n$ synthetic unhedged and hedged assets. $\hat{C}$ may be estimated by synthesizing the historical time series of both the synthetic unhedged and hedged assets and then calculating the covariance in the standard manner. We assume that the original allocation to the foreign assets is given by $W \in \mathfrak{R}^n$ which is a known allocation. The values of the elements of W are not changed by currency hedging. The maximization of the expected total return on the international portfolio and the minimization of its risk is given by

$$\min_{\hat{X}^u \in \mathfrak{R}^n} \max_{y \in \mathfrak{R}^n} \left\{ \begin{matrix} -\alpha \left(\begin{matrix} \sum_{i=1}^n \hat{x}_i^u \left(\left(\dfrac{g_i^a(F_i^a) + y_i^a}{a_{i,0}} \right) \left(\dfrac{g_i(F_i) + y_i}{e_{i,0}} \right) - 1 \right) \\ + \sum_{j=1}^n \hat{x}_j^h \left(\left(\dfrac{g_j^a(F_j^a) + y_j^a}{a_{j,0}} - 1 \right) \left(\dfrac{g_j(F_j) + y_j}{e_{j,0}} \right) + \left(r_j^{c,\text{base}} - r_j^{c,\text{foreign}} \right) \right) \end{matrix} \right) \\ + (1-\alpha) \left\langle \left[\hat{X}^u, \hat{X}^h \right], \hat{C} \left[\hat{X}^u, \hat{X}^h \right] \right\rangle \end{matrix} \right\}$$

$$\tag{3.7}$$

subject to

$$\langle 1, \hat{X}^u \rangle + \langle 1, \hat{X}^h \rangle = 1$$

$$\hat{x}_i^u + \hat{x}_i^h = w_i, \quad \text{for all foreign assets } i$$

$$\hat{X}^u \geq 0$$

$$\hat{X}^h \geq 0$$

$$\hat{X}^u = \begin{bmatrix} x_1^u \\ \vdots \\ x_n^u \end{bmatrix}, \quad \hat{X}^h = \begin{bmatrix} x_1^h \\ \vdots \\ x_n^h \end{bmatrix}, \quad W = \begin{bmatrix} w_1 \\ \vdots \\ w_n \end{bmatrix}.$$

Additionally, g_i and y_i are chosen to ensure the following condition is satisfied:

$$\left(g_i(F_i) + y_i\right)\left(\frac{1}{g_j(F_j) + y_j}\right)\left(g_{ij}(F_{ij}) + y_{ij}\right) = 1, \quad i \neq j$$

The last condition is again the triangulation requirement for currencies, similar to the constraint found in Section 2. The solutions to (3.7) are the optimal allocations to unhedged and hedged assets. The foreign currency exposure is given by the summation of all original allocations W that refer to the same foreign country. The optimal hedge ratio for a particular currency then becomes the ratio of the sum of all hedged allocations to the sum of all original allocations that refer to the same foreign country. The currency hedge ratio is given by

$$h = \frac{\sum \hat{x}_j^h}{\sum w_k} \tag{3.8}$$

where j is a counter that refers to synthetic hedged assets of one particular currency and k refers to the actual foreign asset of the same currency. For each currency, (3.8) can be calculated to find the optimal currency hedge ratios. Thus, for each point on the efficient frontier, a set of optimal hedge ratios can be determined. The choice of which point on the efficient frontier, and consequently which set of hedge ratios are relevant, depends on the portfolio manager. Formulation (3.7) is the mean-variance solution to the hedge ratio problem.

We present the robust formulation of the currency hedge ratio problem. Recall that the modeling of asset returns and currency returns involve the estimation of an equilibrium value based on a vector of factors with some specific estimation error. By providing ranges for the estimation error, one finds a more robust solution to the hedge ratio problem. In the minimax formulation below, we depart from expected value optimization to the minimization of the worst case. The worst case is a scenario (or a set of scenarios corresponding to multiple maximizers) that lies within some range of the estimation error. The minimax equivalent of (3.7) is given by

$$\min_{\substack{\hat{x}^u \in \Re^n \\ \hat{x}^h \in \Re^n}} \max_{\substack{y \in \Re^n \\ y^a \in \Re^n}} \left\{ \begin{array}{l} \left[-\alpha \left(\begin{array}{l} \left(\sum_{i=1}^{n} \hat{x}_i \left(\left(\frac{g_i^a(F_i^a) + y_i^a}{a_{i,0}} \right) \left(\frac{g_i(F_i) + y_i}{e_{i,0}} \right) - 1 \right) \right) \\ + \sum_{j=1}^{n} \hat{x}_j \left(\left(\frac{g_j^a(F_j^a) + y_j^a}{a_{j,0}} - 1 \right) \left(\frac{g_j(F_j) + y_j}{e_{j,0}} \right) + \left(r_j^{c,\text{base}} - r_j^{c,\text{foreign}} \right) \right) \right) \end{array} \right) \right. \\ \left. + (1 - \alpha) \left\langle \left[\hat{x}^u, \hat{x}^h \right], \hat{C} \left[\hat{x}^u, \hat{x}^h \right] \right\rangle \right] \end{array} \right\}$$

$$(3.9)$$

subject to

$$y_i^{\text{lower}} \leq y_i \leq y_i^{\text{upper}}, \quad i = 1, \ldots, n$$

$$y_j^{a,\text{lower}} \leq y_j^a \leq y_j^{a,\text{upper}}, \quad j = 1, \ldots, n$$

where n is the number of foreign assets,

$$\langle 1, \hat{X}^u \rangle + \langle 1, \hat{X}^h \rangle = 1$$

$$\hat{x}_i^u + \hat{x}_i^h = w_i, \quad \text{for all foreign assets } i$$

$$\hat{X}^u \geq 0$$

$$\hat{X}^h \geq 0$$

$$\hat{X}^u = \begin{bmatrix} x_1^u \\ \vdots \\ x_n^u \end{bmatrix}, \quad \hat{X}^h = \begin{bmatrix} x_1^h \\ \vdots \\ x_n^h \end{bmatrix}, \quad W = \begin{bmatrix} w_1 \\ \vdots \\ w_n \end{bmatrix}.$$

The currency forecast error y_i associated with each asset i clearly comes from the m^{cur} currency models only. This means that the first constraint of (3.9) on each y_i is the result of the constraints imposed by the m^{cur} currency models, that is,

$$y_k^{\text{lower}} \leq y_k \leq y_k^{\text{upper}}, \quad k = 1, \ldots, m^{cur}$$

Additionally, the following condition on the cross currencies of the original m^{cur} currency models must be satisfied:

$$y_{kl}^{\text{lower}} \leq y_{kl} \leq y_{kl}^{\text{upper}}, \quad k = 1, \ldots, m^{cur} - 1, \quad l = 1, \ldots, m^{cur} - 1, \quad k \neq l$$

$$k = 1, \ldots, m^{cur} - 1$$

$$l = 1, \ldots, m^{cur} - 1$$

(see CN 4), and

$$(g_k(F_k) + y_k)\left(\frac{1}{g_l(F_l) + y_l}\right)(g_{kl}(F_{kl}) + y_{kl}) = 1, \quad k \neq l.$$

The currency hedge ratio is again given by (3.8) which is used to compute, for each currency, the optimal hedge ratios. For each point on the efficient frontier, a set of optimal hedge ratios can be determined. Similar to the mean-variance version, the choice of which point on the efficient frontier, and consequently which set of hedge ratios are relevant, depends on the portfolio manager.

Formulations (3.7) and (3.9) are respectively the mean-variance and the minimax versions that yield the optimal hedge ratio per currency represented in the international portfolio. Either formulation gives the strategic currency benchmark, that is, the currency hedge ratios, that provide a guide to the long-term hedging requirements of the international portfolio. The hedge ratio varies across currencies. For a usd-based investor, a period of dollar strength may result in high hedge ratios, while a period of dollar weakness may result in low hedge ratios. Once the hedge ratios have been established, these represent the strategic hedges that need to be put in place to guard against long-term currency risk.

We present the tactical management of currency that deals with short-term currency risk. Sections 4 and 5 consider the tactical re-balancing of the currency exposure with a view to enhancing further the total returns from the international portfolio by seeking to gain short-term returns on the currency components.

4 A GENERIC CURRENCY MODEL FOR TACTICAL MANAGEMENT

The management of short-term fluctuations in currencies can add extra currency returns on top of the returns from a long term currency benchmark. In Sections 2 and 3, we construct strategic currency benchmarks that address the long-term issue of currency management and currency hedging. In the present section, we consider a tactical system that addresses short-term fluctuations in currencies. We discuss a generic currency model that produces a tactical signal indicating whether a portfolio should be long or short in a particular currency. Deviating from the usual definition, we adopt a currency model that generates a projection (i.e., a forecast) of an exchange rate in the future. Here, a model not only generates a forecast but also translates this into an implied return and then into an appropriate portfolio re-balancing recommendation. Hereafter, we refer to a currency model as a signal-generating model whose positive signal in a currency means holding a long position in that currency, with the size of the position equal to the size of the signal, and negative signal means holding a short position.

The horizon of the model may vary between 3 and 6 months, depending on the dominance of factors within the model. As can be seen in Figure 4.1, switches in the direction of the recommendation happen within this time frame. This suggests that the model attempts to capture the trending nature of the currency.

In reality, currency models do not possess a strong enough predictive power to claim consistent out-performance in the foreign exchange markets. In the figure, we see a more typical currency model, superimposed on the

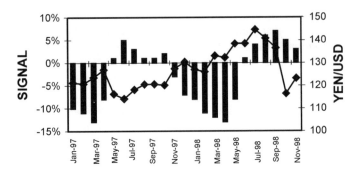

Figure 4.1 Yen/usd (line) against signals (bar) generated by a model.

nonsmoothed yen/usd. From the figure, we see that the model does not consistently lead the currency. The model's signal can vary in magnitude from 0% to over 100%, depending on the particular application, but if used in the context of benchmark-tracking, magnitudes in the order of 10% are more typical. Moreover, the signals of the model change over time, perhaps a reflection of the shift in dominance of factors within the model. Such a model may be used in the currency markets if its overall performance yields a sufficiently high return per unit of risk taken.

The generic currency model used in this section requires some clarification. Let M_i be a currency model for currency i, producing a signal $s_{i,t}$ at time t. The signal may be a complex average of signals from different factors that make up M_i. The particular use of this generic currency model is for capturing short-term movements in a currency. Let $\mathbf{M} \in \Re^{m^{cur}}$ be a multicurrency model comprising m^{cur} currencies, producing a vector of signals $S_t \in \Re^{m^{cur}}$, at time t. The interaction between the m^{cur} currencies would become apparent later when we show that any two pairs within this multicurrency framework may produce signals of opposite signs: this implies a simultaneous buying and selling of US dollars as a cross hedge (see CN 2).

A signal $s_{i,t}$ produced by a currency model would change over time. Let $\delta_{i,t} = s_{i,t} - s_{i,t-1}$ be the change in signal value from time $t - 1$ to t. $\delta_{i,t}$ represents a trade recommendation that should be implemented to shift the currency holding from $s_{i,t-1}$ to $s_{i,t}$. Let $\Delta_t \in \Re^{m^{cur}}$ be the vector of trades at time t. We adopt the US dollar as the base currency, and assume that a currency model will produce signals from the perspective of that currency. In other words, a trade recommendation $\delta_{i,t}$ of, say, 1% for a particular currency means buying 1% of that currency and selling 1% of US dollars.

The time series of signals $s_{i,t}$ produced by currency model M_i are accompanied by a time series of trades $\delta_{i,t}$. Depending on the way factors in the model shift their individual signals, $\delta_{i,t}$ can vary in direction several times within the modeling horizon. Recall that $\delta_{i,t}$ is the trade recommendation that needs to be implemented at time t in order to shift the holding from $s_{i,t-1}$ to $s_{i,t}$. We note that $\delta_{i,t}$ is generally smaller in magnitude than $s_{i,t-1}$. During times when $s_{i,t-1}$ implies holding a long position in a particular currency, a $\delta_{i,t}$ of similar sign to $s_{i,t-1}$ suggests an increase in the long position, whereas a $\delta_{i,t}$ of opposite sign to $s_{i,t-1}$ suggests a decrease in the long position. In the generic currency model, $s_{i,t-1}$ is the main source of profit (loss), while $\delta_{i,t}$ augments or mitigates the profit (loss), depending on its direction relative to $s_{i,t-1}$.

The generic currency model M_i, through its time series of signals $s_{i,t}$, generates currency returns for the currency manager. Positive returns accumulate during periods when the model is forecasting correctly. Similarly, negative returns accumulate when the model mis-forecasts. It is during periods when mis-forecasts dominate that we would wish to intervene to adapt the management of currency portfolios that attempts to minimize the accumulation of losses to these portfolios. We present the minimax framework for this purpose in Section 5.

5 THE MINIMAX FRAMEWORK

5.1 Single Currency Framework

We assume that the generic currency model presented in Section 4 results in a reasonable overall positive return for the currency manager. He/she continues to trade on the basis of the recommendations from the model, but recognizes the risk of incurring negative returns in following it. The return $r_{i,t}$ from holding a long position (see CN 3) in a particular currency i at time t, expressed as a function of $r_{i,t}^e$, from a usd-based perspective, is given by[5]

$$r_{i,t}(r_{i,t}^e) = s_{i,t-1} r_{i,t}^e$$

$$r_{i,t}^e = \left(\frac{e_{i,t}}{e_{i,t-1}} - 1 \right)$$

where $r_{i,t}^e$ is the raw currency return due to a shift in the spot exchange rate, and $e_{i,t}$ is the exchange rate for currency i against the US dollar, quoted as units of US dollar per unit of currency i, at time t. For a model that produces daily signals, $r_{i,t}$ represents the daily currency return for holding the particular position $s_{i,t-1}$ in currency i from the previous day.

On a forward-looking basis, the potential return, $r_{i,t+1}$, based on both the previous day's signal $s_{i,t-1}$ and the current day's trade recommendation $\delta_{i,t}$, is given by

---

[5] For ease of exposition, we concentrate on the currency return that excludes the forward bias. The use of the forward market in currency management introduces a forward bias in the currency return; this bias is generally small compared to the currency return based on spot values.

$$r_{i,t+1}(r^e_{i,t+1}) = (s_{i,t-1} + \delta_{i,t})r^e_{i,t+1}$$

$$r^e_{i,t+1} = \left(\frac{E(e_{i,t+1})}{e_{i,t}} - 1\right)$$

where the source of uncertainty is $e_{i,t+1}$ and $E(\cdot)$ denotes the expectations operator. In the multicurrency context, the return from all the currency exposures in the portfolio, $r^P_t \in \Re^1$, at time t is given by

$$r^P_t = \sum_i r_{i,t}(r^e_{i,t})$$

and the cumulative return r^P for a period of time is given by

$$r^P = \sum_t r^P_t.$$

This is the accumulated portfolio return over time which is important in assessing the overall health of a currency portfolio. Similarly, on a forward-looking basis, the potential portfolio return, $r_{i,t+1}$, at time t based on both the previous day's signal $s_{i,t-1}$ and the current day's trade recommendation $\delta_{i,t}$, for all currency i is given by

$$r^P_{t+1} = \sum_i r_{i,t+1}(r^e_{i,t+1}).$$

The potential cumulative return is therefore the sum of r^P and r^P_{t+1}.

A manager can attempt to improve overall performance by concentrating on the component currencies that contribute to the overall return. We concentrate initially on single currencies and later on multicurrencies. For a particular currency i, the risk of a negative return on any day is generally compensated by the overall positive return over a number of days that the currency manager can get by following the model's recommendations. In the context of the generic currency model described in Section 4, we assume that the horizon of the model is between 3 and 6 months. This means that for the currency manager, an overall positive return is expected in any 3–6 month period, but the risk of high negative returns within the period remains. The manager would attempt to minimize the occurrence of negative returns by considering the current performance of that currency component in the portfolio and the sign of the trade recommendation $\delta_{i,t}$ relative to the signal $s_{i,t-1}$. The time series of $s_{i,t}$ determines the performance over a period of the currency component within the portfolio. It is thus important to assess the accumulation of returns, whether positive or negative, in order to judge whether $\delta_{i,t}$, with a particular sign or direction relative to $s_{i,t-1}$, is a favorable trade, or not. Both $s_{i,t-1}$ and $\delta_{i,t}$ are component outputs of the generic currency model. Thus, it is assumed that the manager is not able to ignore $\delta_{i,t}$, mainly because it is a necessary variable for shifting the signal from $s_{i,t-1}$ to $s_{i,t}$. However, the manager can implement an overriding trade that would mitigate any potential increase in cumulative loss as a consequence of either $s_{i,t-1}$ or $\delta_{i,t}$, or both. Let $\zeta_{i,t}$ represent such an overriding trade, referred to as an *overlay trade recommendation*, that aims to minimize any potential cumulative loss. Consider the problem

$$\min_{\zeta_{i,t} \in \mathbb{R}^1} \left\{ \left(\left(\left(\sum_{j=t_0}^{t} r_{i,j} \right) + r_{i,t+1}(r_{i,t+1}^e) \right)^- + \left(\zeta_{i,t} r_{i,t+1}^e \right) \right)^2 \right\} \tag{5.1}$$

subject to

$$\zeta_{i,t}^{\text{lower}} \leq \zeta_{i,t} \leq \zeta_{i,t}^{\text{upper}}.$$

In (5.1), $(\cdot)^- = \min(\cdot, 0)$, t_0 is some predefined starting time for the calculation of cumulative P&L (profit and loss). The constraint represents size restrictions on $\zeta_{i,t}$. This formulation attempts to minimize the potential cumulative loss in view of both $s_{i,t-1}$ and $\delta_{i,t}$, as well as the expected move in the currency spot exchange rate $E(e_{i,t+1})$. In (5.1), the first term is the cumulative P&L with negative values only, that is, the running loss. If the cumulative P&L is positive, that is, the portfolio is accumulating profits, then the solution is $\zeta_{i,t} = 0$. Hence the overlay trade does not interfere with the current performance of the generic currency model. However, if the cumulative P&L is negative, that is, the portfolio is accumulating losses, then the solution is a $\zeta_{i,t} \neq 0$, which would be positive if the expected return on the currency is positive and would be negative is the expected return on the currency is negative.

The formulation does not require a budget constraint because this is not applicable in currency overlay. The upper and lower bound constraints on $\zeta_{i,t}$ are chosen such that the buying or selling of a currency is within any guidelines on hedging imposed by the investor or self-imposed by the fund manager. The allocation to different currencies does not necessarily add up to some predefined budget, as such a budget is not relevant.

The formulation given in (5.1) has a trivial solution for a single currency problem. If the first term is nonzero, then the optimizer will seek the upper limit or the lower limit on $\zeta_{i,t}$, depending on the expected currency return. This basic formulation is presented to illustrate the purpose of the overlay trade. In later formulations, we extend (5.1) to cover the multicurrency problem.

It is important to note that the model trade recommendation $\delta_{i,t}$ implies a currency spot exchange rate $\bar{E}(e_{i,t+1})$, where $\bar{E}(\cdot)$ denotes the implied expected value. A discrepancy between $E(e_{i,t+1})$ and $\bar{E}(e_{i,t+1})$ may not necessarily result in a nonzero value for the overlay trade $\zeta_{i,t}$. If there is such a discrepancy, and if such discrepancy means opposite currency expectations, then the overlay trade would only be activated if the contribution of $\bar{E}(e_{i,t+1})$ within $r_{i,t+1}(r_{i,t+1}^e)$ is such that the running cumulative P&L results in a negative value. So long as the running cumulative P&L is positive, then the formulation will not intrude in the default currency management as indicated by $s_{i,t-1}$ and $\delta_{i,t}$.

A key criterion for the success of the above formulation would be the estimation of $E(e_{i,t+1})$. In order for this formulation to recognize an increasing accumulation of loss due to the series $s_{i,t}$, the historic cumulative return, $\sum_t r_{i,t}$, has to be part of the minimization process, and the overall sign of $s_{i,t}$ during the period has to be analyzed in terms of its implied currency return. It may be prudent for the manager to set $E(e_{i,t+1})$ equal to the implied currency return from $s_{i,t}$ when $s_{i,t}$ is producing positive returns, and similarly, to set it opposite the implied currency return from $s_{i,t}$ when $s_{i,t}$ is producing negative returns. The limitation of the above formulation is on the dependency of the solution on the estimation of $E(e_{i,t+1})$ which may be driven by technical movements in the currency rather than being a reflection of a basic mis-forecasting by the model. This limitation can be addressed by the minimax formulation below, where we move away from an estimation of $E(e_{i,t+1})$ to an estimation of a feasible range for $e_{i,t+1}$.

In the minimax framework, we wish to minimize the maximum potential cumulative loss due to the model recommendations. Consider the following problem related to currency i:

$$\min_{\zeta_{i,t}\in\mathfrak{R}^1} \max_{e_{i,t+1}\in\mathfrak{R}^1} \left\{ \left(\left(\left(\sum_{j=t_0}^{t} r_{i,j}\right) + (s_{i,t-1} + \delta_{i,t})\left(\frac{e_{i,t+1} - e_{i,t}}{e_{i,t}}\right)\right)^{-} \right.\right.$$
$$\left.\left. + \left(\zeta_{i,t}\left(\frac{e_{i,t+1} - e_{i,t}}{e_{i,t}}\right)\right)^2\right)\right\}$$

(5.2)

subject to

$$e_{i,t+1}^{\text{lower}} \le e_{i,t+1} \le e_{i,t+1}^{\text{upper}}$$

$$\zeta_{i,t}^{\text{lower}} \le \zeta_{i,t} \le \zeta_{i,t}^{\text{upper}}$$

that is, upper and lower bounds on $e_{i,t+1}$ and $\zeta_{i,t}$. The estimation of the upper and lower bounds on $e_{i,t+1}$, as in the estimation of $E(e_{i,t+1})$, depends on the implied currency returns from series $s_{i,t}$. However, whereas $E(e_{i,t+1})$ is a point estimate, the upper and lower bounds produce a range estimate. The possibility of refining the estimation of this range provides a flexible formulation.

5.2 Single Currency Framework with Transaction Costs

We present the equivalent formulations of (5.1) and (5.2) in view of transaction costs. These formulations, although simple in the single currency framework, lead to complications in the multicurrency framework.

Problem (5.1) with transaction costs is given by (see CN 5)

$$\min_{\zeta_{i,t}\in\Re^1}\left\{\left(\left(\left(\sum_{j=t_0}^{t}r_{i,j}\right)+r_{i,t+1}(r_{i,t+1}^e)\right)^{-}+(\zeta_{i,t}r_{i,t+1}^e)-\sum_{j=t_0}^{t}k_i\left|\delta_{i,j}+\zeta_{i,j}\right|\right)^2\right\}$$

(5.3)

subject to

$$\zeta_{i,t}^{\text{lower}}\leq\zeta_{i,t}\leq\zeta_{i,t}^{\text{upper}}$$

where $k_i > 0$, that is, the cost[6] of transacting in currency i. The last term refers to the cumulative transaction cost from t_0 to t. This term has a negative sign to indicate that it is a cost. While the costs due to $\delta_{i,\cdot}$ do not affect the objective function, its inclusion provides a more complete picture of the accumulation of costs that may impact the cumulative P&L. Thus, the term $\left|\delta_{i,j}+\zeta_{i,j}\right|$ is constant in (5.3) up to $j=t-1$ and is only optimized for $j=t$.

Problem (5.2) with transaction costs is given by (see CN 5)

$$\min_{\zeta_{i,t}\in\Re^1}\max_{e_{i,t+1}\in\Re^1}\left\{\left(\left(\left(\sum_{j=t_0}^{t}r_{i,j}\right)+(s_{i,t-1}+\delta_{i,t})\left(\frac{e_{i,t+1}-e_{i,t}}{e_{i,t}}\right)\right)^{-}\right.\right.$$

$$\left.\left.+\left(\zeta_{i,t}\left(\frac{e_{i,t+1}-e_{i,t}}{e_{i,t}}\right)\right)\right)-\sum_{j=t_0}^{t}k_i\left|\delta_{i,j}+\zeta_{i,j}\right|\right)^2\right\}$$

(5.4)

subject to

$$e_{i,t+1}^{\text{lower}}\leq e_{i,t+1}\leq e_{i,t+1}^{\text{upper}}$$

$$\zeta_{i,t}^{\text{lower}}\leq\zeta_{i,t}\leq\zeta_{i,t}^{\text{upper}}.$$

5.3 Multicurrency Framework

We present the multicurrency equivalent of (5.1), and show that certain complications arise when considering multiple currencies. Let $Z_t \in \Re^{m^{cur}}$ be a vector whose elements $\zeta_{i,t}$ represent a manager's overlay trade recommendations in m^{cur} currencies at time t.

The multicurrency equivalent of (5.1) is given by

$$\min_{Z_t\in\Re^{m^{cur}}}\left\{\left(\left(\sum_{i}\left(\left(\sum_{j=t_0}^{t}r_{i,j}\right)+r_{i,t+1}(r_{i,t+1}^e)\right)\right)^{-}+\sum_{i}(\zeta_{i,t}r_{i,t+1}^e)\right)^2\right\}$$

(5.5)

[6] Transaction costs depend on the currency traded and the volume of trade, and they could vary from 10 basis points to 100 basis points, where a basis point is a 100th of a percentage point.

subject to

$$\zeta_{i,t}^{\text{lower}} \leq \zeta_{i,t} \leq \zeta_{i,t}^{\text{upper}}, \quad i = 1, \ldots, m^{cur}$$

and where

$$Z_t = \begin{bmatrix} \zeta_{1,t} \\ \vdots \\ \zeta_{m^{cur},t} \end{bmatrix}.$$

By expanding the components of the objective function, we have

$$\min_{Z_t \in \mathfrak{R}^{m^{cur}}} \left\{ \left(\left(\sum_i \left(\left(\sum_{j=t_0}^t r_{i,j} \right) + (s_{i,t-1} + \delta_{i,t}) \left(\frac{E(e_{i,t+1}) - e_{i,t}}{e_{i,t}} \right) \right) \right)^{-} \right. \right.$$

$$\left. \left. + \sum_i \left(\zeta_{i,t} \left(\frac{E(e_{i,t+1}) - e_{i,t}}{e_{i,t}} \right) \right) \right)^2 \right\} \tag{5.6}$$

subject to

$$\zeta_{i,t}^{\text{lower}} \leq \zeta_{i,t} \leq \zeta_{i,t}^{\text{upper}}, \quad i = 1, \ldots, m^{cur}$$

and where

$$Z_t = \begin{bmatrix} \zeta_{1,t} \\ \vdots \\ \zeta_{m^{cur},t} \end{bmatrix}.$$

Similarly, the multicurrency formulation of (5.2) is given by

$$\min_{Z_t \in \mathfrak{R}^{m^{cur}}} \max_{\hat{e}_{t+1} \in \mathfrak{R}^{m^{cur}}} \left\{ \left(\left(\sum_i \left(\left(\sum_{j=t_0}^t r_{i,j} \right) + \left(s_{i,t-1} + \delta_{i,t} \right) \left(\frac{e_{i,t+1} - e_{i,t}}{e_{i,t}} \right) \right) \right)^{-} \right. \right.$$

$$\left. \left. + \sum_i \left(\zeta_{i,t} \left(\frac{e_{i,t+1} - e_{i,t}}{e_{i,t}} \right) \right) \right)^2 \right\} \tag{5.7}$$

subject to

$$\hat{e}_{t+1}^{\text{lower}} \leq \hat{e}_{t+1} \leq \hat{e}_{t+1}^{\text{upper}}$$

$$e_{ij,t+1}^{\text{lower}} \leq e_{ij,t+1} \leq e_{ij,t+1}^{\text{upper}}, \quad i = 1, \ldots, m^{cur} - 1$$

$$j = 1, \ldots, m^{cur} - 1, \quad i \neq j \text{ (see CN 4)}$$

$$e_{i,t+1}\left(\frac{1}{e_{j,t+1}}\right)\left(e_{ij,t+1}\right) = 1, \quad i \neq j$$

$$\hat{e}_{t+1} = \begin{bmatrix} e_{1,t+1} \\ \vdots \\ e_{m^{cur},t+1} \end{bmatrix}, \quad \hat{e}_{t+1}^{lower} = \begin{bmatrix} e_{1,t+1}^{lower} \\ \vdots \\ e_{m^{cur},t+1}^{lower} \end{bmatrix}, \quad \hat{e}_{t+1}^{upper} = \begin{bmatrix} e_{1,t+1}^{upper} \\ \vdots \\ e_{m^{cur},t+1}^{upper} \end{bmatrix}$$

$$\zeta_{i,t}^{lower} \leq \zeta_{i,t} \leq \zeta_{i,t}^{upper}, \quad i = 1, \ldots, m^{cur}$$

where

$$Z_t = \begin{bmatrix} \zeta_{1,t} \\ \vdots \\ \zeta_{m^{cur},t} \end{bmatrix}.$$

The second and third constraints refer to the triangulation requirement within a multicurrency framework. It is important to note that for the tactical currency minimax formulations, we do not use any currency model to define the upper and lower bounds on the currencies. Neither do we use a long-term currency model to define a cross exchange rate to satisfy triangulation requirements of currencies. In the tactical management of currencies using minimax, all we need are predefined upper and lower bounds on the currency, and an explicit triangulation requirement in terms of future exchange rates. The second and third constraints limit the choice of the maximizing variable to within reasonable values.

Formulations (5.6) and (5.7) are the simple multicurrency expansions of (5.1) and (5.2). However, these may result in massive transaction costs because they do not account for cross hedging between currencies that do not involve transactions via the base currency. It is therefore important to include transaction costs in the formulation.

5.4 Multicurrency Framework with Transaction Costs

Consider the equivalent of (5.6) when transaction costs are included (see CN 5):

$$\min_{Z_t \in \mathfrak{R}^{m^{cur}}} \left\{ \left[\left(\left(\sum_i \left(\left(\sum_{j=t_0}^t r_{i,j} \right) + (s_{i,t-1} + \delta_{i,t}) \left(\frac{E(e_{i,t+1}) - e_{i,t}}{e_{i,t}} \right) \right) \right) \right. \right. \right.$$
$$\left. \left. \left. + \sum_i \left(\zeta_{i,t} \left(\frac{E(e_{i,t+1}) - e_{i,t}}{e_{i,t}} \right) \right) \right)^{-2} \right] - \sum_i \left(\sum_{j=t_0}^t k_i \left| \delta_{i,j} + \zeta_{i,j} \right| \right) \right\}$$

(5.8)

subject to the same constraints as (5.6) and with $K \in \mathfrak{R}^{m^{cur}}$ denoting the vector of transaction costs, that is,

$$K = \begin{bmatrix} k_1 \\ \vdots \\ k_{m^{cur}} \end{bmatrix}.$$

Similarly, consider the equivalent of (5.7) when transaction costs are included (see CN 5):

$$\min_{Z_t \in \mathfrak{R}^{m^{cur}}} \max_{\hat{e}_{t+1} \in \mathfrak{R}^{m^{cur}}} \left\{ \left[\left(\left(\sum_i \left(\left(\sum_{j=t_0}^t r_{i,j} \right) + (s_{i,t-1} + \delta_{i,t}) \left(\frac{e_{i,t+1} - e_{i,t}}{e_{i,t}} \right) \right) \right) \right. \right. \right.$$
$$\left. \left. \left. + \sum_i \left(\zeta_{i,t} \left(\frac{e_{i,t+1} - e_{i,t}}{e_{i,t}} \right) \right) \right)^{-2} \right] - \sum_i \left(\sum_{j=t_0}^t k_i \left| \delta_{i,j} + \zeta_{i,j} \right| \right) \right\}$$

(5.9)

subject to the same constraints as (5.7) and with the transaction cost vector $K \in \mathfrak{R}^{m^{cur}}$ defined as in (5.8).

As transaction costs are summed for all currencies without any consideration for cross hedging capabilities, the currency manager could potentially buy and sell US dollars, and incur transaction costs, when there is no need for such transactions. He/she can reduce the number of transactions by taking advantage of any existing cross hedging opportunities.

To illustrate the benefit from cross hedging, consider two foreign currencies A and B, for which the currency model recommends a buy trade of currency A accompanied by a sell trade of the US dollar, as well as a sell trade of currency B accompanied by a buy trade of the US dollar. By following the model's trade recommendations, the currency manager incurs transaction costs on two US dollar trades, when one transaction involving a buy trade of currency A accompanied by a sell trade of currency B would yield the same result but with less cost.

5.5 Worst-case Scenario

The minimax formulation given in (5.9) is subject to constraints on the values that the future exchange rates may take. These constraints define the worst-case scenario within which the currencies in the portfolio take on future values that would result in the worst cumulative loss for the portfolio.

In the context of the generic currency model described in Section 4, the signals and trade recommendations for each currency imply a distribution of future currency returns accompanied by an implied distribution of future values of the currency. The expected currency return and the dispersion about this expected value drives the model to generate a signal of a particular direction and magnitude. By extracting the distribution of future values of the currencies from the currency models, one can define the upper and lower bounds for the constraints in (5.9). To illustrate these constraints, let us assume that the currency model for yen generates a signal to hold a positive yen position, and a trade recommendation to buy more yen. Because the signal and the trade recommendation have the same direction, the currency model is effectively proposing an increase in the yen holding. This further implies that the yen is expected to appreciate within the horizon of the model. Depending on the success rate of the model in forecasting the movement of the yen, the distribution about the expected yen appreciation may vary. If the model results in low forecast errors, then the dispersion about the expected value may be tight; similarly, if the model results in high forecast errors, then the dispersion may be wide. The estimated mean $\mu_{yen,t+1}$ and standard deviation $\sigma_{yen,t+1}$ of the future values of the yen can be used to define the upper bound and lower bounds on the yen, both defined by two standard deviations from the mean:

$$e_{yen,t+1}^{lower} \leq e_{yen,t+1} \leq e_{yen,t+1}^{upper}$$

where

$$e_{yen,t+1}^{lower} = \mu_{yen,t+1} - 2\sigma_{yen,t+1} \qquad (5.10)$$

$$e_{yen,t+1}^{upper} = \mu_{yen,t+1} + 2\sigma_{yen,t+1}.$$

By repeating this process for all currencies in the model, one is able to define the ranges of future values for each currency. As in Section 2, the triangulation property of currencies has to be preserved; all lower and upper bounds should satisfy triangulation. This is ensured by the second and third constraints in (5.7).

The definition of the worst case can be refined further by considering the implied distributions of returns, and implied distributions of future values of exchange rates, for each factor in the model. In the generic currency model, the ultimate signal generated may be a result of an aggregation of signals from different factors that make up the model. To the extent that each of the factors within the model have varying forecasting capabilities throughout the model's history, one can utilize the information contained within each factor to define the upper and lower bounds on the future values of the currencies.

Assume that the currency model for our yen illustration generates a signal to hold a positive yen position which is dominated by two conflicting factors. A positive factor generates a signal to hold a positive yen position. A negative factor generates a signal in the opposite direction. The aggregation of the signals produced by each factor results in the overall model signal of holding a positive yen position. The constraints on the future values of the currencies, to be used in the minimax formulation, can be defined by considering any two dominant conflicting factors in the currency model. The estimated mean of the positive factor $^{+}\mu_{\text{yen},t+1}$ of the future values of the yen can be used to define the upper bound on the yen, while the estimated mean of negative factor $^{-}\mu_{\text{yen},t+1}$ can be used to define the lower bound.

Consider the following bounding:

$$e^{\text{lower}}_{\text{yen},t+1} \leq e_{\text{yen},t+1} \leq e^{\text{upper}}_{\text{yen},t+1}$$

where

$$e^{\text{lower}}_{\text{yen},t+1} = {}^{-}\mu_{\text{yen},t+1} \tag{5.11}$$

$$e^{\text{upper}}_{\text{yen},t+1} = {}^{+}\mu_{\text{yen},t+1}.$$

Alternatively, one can extend the definition of the bounds by using the standard deviation of the positive factor, $^{+}\sigma_{\text{yen},t+1}$, and the standard deviation of the negative factor, $^{-}\sigma_{\text{yen},t+1}$, as

$$e^{\text{lower}}_{\text{yen},t+1} \leq e_{\text{yen},t+1} \leq e^{\text{upper}}_{\text{yen},t+1}$$

where

$$e^{\text{lower}}_{\text{yen},t+1} = ({}^{-}\mu_{\text{yen},t+1}) - ({}^{-}\sigma_{\text{yen},t+1}) \tag{5.12}$$

$$e^{\text{upper}}_{\text{yen},t+1} = ({}^{+}\mu_{\text{yen},t+1}) + ({}^{+}\sigma_{\text{yen},t+1}).$$

By defining the bounds as in (5.11) or (5.12), one recognizes that the conflicting factors within the model provide a clue as to the worst case. The positive factor that dominates the model signal would have a higher implied expected future value for the yen compared to that of the total signal and the negative factor that dominates the opposite direction. But as the negative factor is overwhelmed by the positive factor, it would have a lower implied future value compared to that of the total signal. It is the existence of a dominant opposing factor, in this illustration given by the negative factor, that may be backed by low forecasting errors that should be seriously considered when defining the worst case. Consider a situation where the negative factor in fact yields lower forecasting errors in the short term compared to the positive factor. If the overall model signal recommends a positive yen holding, while the negative factor within that model recommends a short holding of the yen, we would expect an accumulation of losses to the currency portfolio. By defining the lower bound based on the expected yen depreciation, as signaled by the negative factor, one is accounting for the conflicting nature of factors within the model. Repeating this process for all currencies in the model leads to the definition of the ranges of future values for each currency. As before, these ranges have to be refined further to ensure consistency in the sense that the triangulation property of currencies has to be preserved.

5.6 A Momentum-based Minimax Strategy

The minimax formulation (5.9) can be further enhanced by considering the range of values of $\zeta_{i,t}$ that would promote a more acceptable potential loss from the currency portfolio. In that formulation, the constraints on the future values of the currencies determine the worst case combination of exchange rates. By incorporating new constraints on the minimax overlay trade recommendations, one is promoting the search for the best case $\zeta_{i,t}$ that would cushion the portfolio against the worst-case scenario.

The growth or the accumulation of loss to the portfolio provides information about the suitability of the model's trade recommendation $\delta_{i,t}$. A negative growth implies an increasing loss and an increasingly unfavorable forecast error, while a positive growth implies an increasingly healthy balance sheet and an increasingly favorable forecast error. The momentum of cumulative loss thus provides some information as to whether the model's trade recommendation $\delta_{i,t}$ should be implemented, or increasingly overlayed with an opposing $\zeta_{i,t}$.

Consider the momentum-based minimax strategy given by (see CN 4):

$$
\min_{Z_t \in \mathfrak{R}^{m^{cur}}} \max_{\hat{e}_{t+1} \in \mathfrak{R}^{m^{cur}}} \left\{ \left[\left(\left(\sum_i \left(\left(\sum_{j=t_0}^{t} r_{i,j} \right) + (s_{i,t-1} + \delta_{i,t}) \left(\frac{e_{i,t+1} - e_{i,t}}{e_{i,t}} \right) \right) \right)^{-2} \right) \right. \right.
$$
$$
\left. \left. + \sum_i \left(\zeta_{i,t} \left(\frac{e_{i,t+1} - e_{i,t}}{e_{i,t}} \right) \right) - \sum_i \left(\sum_{j=t_0}^{t} k_i \left| \delta_{i,j} + \zeta_{i,j} \right| \right) \right] \right\}
$$

(5.13)

subject to

$$
\hat{e}_{t+1}^{lower} \leq \hat{e}_{t+1} \leq \hat{e}_{t+1}^{upper}
$$

$$
e_{ij,t+1}^{lower} \leq e_{ij,t+1} \leq e_{ij,t+1}^{upper}, \quad i = 1,...,m^{cur} - 1
$$

$$j = 1, \ldots, m^{cur} - 1, \quad i \neq j \text{ (see CN 4)}$$

$$e_{i,t+1}\left(\frac{1}{e_{j,t+1}}\right)(e_{ij,t+1}) = 1, \quad i \neq j$$

$$\hat{e}_{t+1} = \begin{bmatrix} e_{1,t+1} \\ \vdots \\ e_{m^{cur},t+1} \end{bmatrix}, \quad \hat{e}_{t+1}^{lower} = \begin{bmatrix} e_{1,t+1}^{lower} \\ \vdots \\ e_{m^{cur},t+1}^{lower} \end{bmatrix}, \quad \hat{e}_{t+1}^{upper} = \begin{bmatrix} e_{1,t+1}^{upper} \\ \vdots \\ e_{m^{cur},t+1}^{upper} \end{bmatrix}$$

$$\zeta_{i,t}^{lower} \leq \zeta_{i,t} \leq \zeta_{i,t}^{upper}, \quad i = 1, \ldots, m^{cur}$$

where

$$Z_t = \begin{bmatrix} \zeta_{1,t} \\ \vdots \\ \zeta_{m^{cur},t} \end{bmatrix}$$

and transaction cost vector $K \in \Re^{m^{cur}}$ is as defined in (5.8).

Additionally, we have

$$\zeta_{i,t}^{upper} = f\left(\tau_i^1, \tau_i^2, \{r_{i,t}\}\right)$$

$$\zeta_{i,t}^{lower} = -\zeta_{i,t}^{upper}.$$

The variables τ_i^1 and τ_i^2 are predefined time periods for estimating the momentum of cumulative loss and $\{r_{i,t}\}$ is the time series of currency returns. These τ variables can be modeled separately from the currency model. The above formulation is also subject to triangulation constraints on the currencies.

We now describe the function that defines the upper and lower bounds on $\zeta_{i,t}$. Between $s_{i,t-1}$ and $\delta_{i,t}$, it is $s_{i,t-1}$ that substantially contributes to the profit or loss of a portfolio. The cumulative profit or loss, likewise, is driven by $\{s_{i,t}\}$, that is, the time series of signals. A positive growth, or an accumulation of profit, suggests that, in the short term, $\{s_{i,t}\}$ reflects a return distribution with a positive expected currency return, and perhaps accompanied by a small dispersion.

Assuming that the short-term history of performance is an accumulation of profit, then an increasing accumulation, or the momentum of positive growth could itself be treated as a synthetic factor, separate from the factors that make up the generic currency model. In the case of an accumulation of loss, the momentum of negative growth could similarly be treated. Let τ_i^1 be a long time period, say 65 days, and let τ_i^2 be a short time period, say 5 days. One can approximate an increasing accumulation by taking the average return over these time periods and comparing the short-period average against the long-period average. Let A_1 and A_2 be defined as

$$A_1 = \frac{\sum\limits_{t=-\tau_i^1}^{0} r_{i,t}}{\tau_i^1} \quad \text{and} \quad A_2 = \frac{\sum\limits_{t=-\tau_i^2}^{0} r_{i,t}}{\tau_i^2}.$$

Here, A_1 represents the average return for the long time period, while A_2 is for the short time period. Then

$$\zeta_{i,t}^{\text{upper}} = \begin{cases} w(A_2 - A_1) & \text{if } A_2 \geq A_1 \\ w(A_1 - A_2) & \text{if } A_1 > A_2 \end{cases} \tag{5.14}$$

$$\zeta_{i,t}^{\text{lower}} = -\zeta_{i,t}^{\text{upper}}$$

where w is a weight that determines the size or magnitude of the signal based on some predefined rule. The modeling of the τ variables to find the most appropriate time periods, as well as the modeling of the weight w, can be structured to be consistent with the generation of signals by the synthetic factor. The long-term performance of the synthetic factor should be comparable to the long-term performance of the other factors that make up the generic currency model.

The importance of bounding $\zeta_{i,t}$ as in (5.13) is that the aggressiveness[7] of such a decision variable is in line with the aggressiveness of the generic currency model. This makes for a tractable and systematic management of currency bets.

5.7 A Risk-controlled Minimax Strategy

The minimax formulation in Section 5.6 attempts to control the risk of extreme deviations from the portfolio's chosen benchmark by bounding the size of the overall trade recommendations. Formulation (5.9) can alternatively be risk-controlled by actively minimizing the tracking error that may result in following any trade recommendation. In (5.9), the constraints on the future values of the currencies determine the worst case combination of exchange rates. By incorporating a risk component into the formulation, one is promoting the search for the best case $\zeta_{i,t}$ that would cushion the portfolio against the worst-case scenario, subject to an acceptable level of tracking error (*TE*).

Consider the risk-controlled minimax strategy given by (see CN 4):

---

[7] Aggressiveness is the term used to refer to the parameterization of a currency model signal that determines the size or magnitude of a currency bet that needs to be implemented in order to achieve a desired performance profile.

$$\min_{Z_t \in \mathfrak{R}^{m^{cur}}} \max_{\hat{e}_{t+1} \in \mathfrak{R}^{m^{cur}}} \left\{ \left[\left(\left(\sum_i \left(\left(\sum_{j=t_0}^{t} r_{i,j} \right) + \left(s_{i,t-1} + \delta_t^i \right) \left(\frac{e_{i,t+1} - e_{i,t}}{e_{i,t}} \right) \right) \right) \right)^{-2} \right. \right. $$
$$\left. \left. + \sum_i \left(\zeta_{i,t} \left(\frac{e_{i,t+1} - e_{i,t}}{e_{i,t}} \right) \right) \right) + \sum_i \left(\sum_{j=t_0}^{t} k_i \left| \delta_{i,j} + \zeta_{i,j} \right| \right) \right] \right\}$$

(5.15)

subject to

$$\hat{e}_{t+1}^{lower} \leq \hat{e}_{t+1} \leq \hat{e}_{t+1}^{upper}$$

$$e_{ij,t+1}^{lower} \leq e_{ij,t+1} \leq e_{ij,t+1}^{upper}, \quad i = 1,...,m^{cur} - 1, \quad j = 1,...,m^{cur} - 1,$$

$$i \neq j \text{ (see CN 4)}$$

$$e_{i,t+1} \left(\frac{1}{e_{j,t+1}} \right) (e_{ij,t+1}) = 1, \quad i \neq j$$

$$\hat{e}_{t+1} = \begin{bmatrix} e_{1,t+1} \\ \vdots \\ e_{m^{cur},t+1} \end{bmatrix}, \quad \hat{e}_{t+1}^{lower} = \begin{bmatrix} e_{1,t+1}^{lower} \\ \vdots \\ e_{m^{cur},t+1}^{lower} \end{bmatrix}, \quad \hat{e}_{t+1}^{upper} = \begin{bmatrix} e_{1,t+1}^{upper} \\ \vdots \\ e_{m^{cur},t+1}^{upper} \end{bmatrix}$$

$$\zeta_{i,t}^{lower} \leq \zeta_{i,t} \leq \zeta_{i,t}^{upper}, \quad i = 1,...,m^{cur}$$

where

$$Z_t = \begin{bmatrix} \zeta_{1,t} \\ \vdots \\ \zeta_{m^{cur},t} \end{bmatrix}$$

and transaction cost vector $K \in \mathfrak{R}^{m^{cur}}$ is as defined in (5.8).

Additionally, a constraint on risk exposure is included:

$$\langle (S_t + \Delta_t + Z_t), \bar{C}(S_t + \Delta_t + Z_t) \rangle \leq (TE)^2$$

where $TE \in \mathfrak{R}^1$ is an acceptable level of tracking error defined by the manager, $\bar{C} \in \mathfrak{R}^{m^{cur} \times m^{cur}}$ is the covariance matrix of currency returns, and $S_t + \Delta_t + Z_t$ is defined as

$$(S_t + \Delta_t + Z_t) = \begin{bmatrix} s_{1,t} + \delta_{1,t} + \zeta_{1,t} \\ \vdots \\ s_{m^{cur},t} + \delta_{m^{cur},t} + \zeta_{m^{cur},t} \end{bmatrix}.$$

With the above formulation, the possibility of having a reasonable benchmark tracking performance is increased. If the realized future values of the exchange rates move outside their predefined limits used during optimization, then the constraint on tracking error would serve the purpose of controlling the volatility of the portfolio.

6 THE INTERPLAY BETWEEN THE STRATEGIC BENCHMARK AND TACTICAL MANAGEMENT

Earlier sections of this chapter indicate the need for a tactical system in order for the portfolio to benefit from short- to medium-term fluctuations in currencies. In this section, we discuss the interplay between the strategic currency benchmark and the tactical currency trades. We consider the dominance of the strategic benchmark and the constraints it imposes on the tactical management of currencies.

As in earlier sections, the term currency overlay is used in fund management to mean the application of the tactical currency trades to supplement and/ or complement the performance of the strategic currency benchmark. Because the strategic currency benchmark is designed to cover the risk of long-term depreciation in a currency, the corresponding currency hedge ratios drive the long-term performance of the benchmark. The strategic benchmark performance over the long term dominates the overall absolute performance of the portfolio currency exposure. In contrast, the excess performance provided by the tactical system is relatively small in proportion to the absolute performance of the strategic benchmark. Not only does the strategic benchmark impact the overall absolute performance, it also constrains the potential performance of the tactical system. The constraints imposed by the strategic benchmark limits the implementation of tactical trades.

We illustrate the above constraint using a hypothetical yen exposure. Suppose that after the analysis of the strategic currency benchmark, the resulting hedge ratio for the yen exposure in an international portfolio is a 100% hedging recommendation. After the implementation of this hedge, the international portfolio essentially is devoid of any yen exposure. This total elimination of the yen exposure puts a constraint on the tactical system. This means that the tactical recommendation, either coming from $\delta_{i,t}$, or $\zeta_{i,t}$, or both, is restricted in terms of moving the yen exposure. If the tactical recommendation is a sell trade for the yen, then the tactical trade cannot be implemented because there is no longer any yen exposure to sell. Only buy trades can be accommodated by the tactical system for this fully hedged yen example.

As another illustration, suppose that after the analysis of the strategic currency benchmark, the resulting hedge ratio for the yen exposure is a 50% hedging recommendation. After the implementation of this hedge, the international portfolio has essentially halved its yen exposure. Again, if the tactical recommendation is a sell trade for the yen, then the tactical trade can be implemented up to the remaining yen exposure, and not more. The constraint imposed by the strategic benchmark limits the potential performance of the tactical system, but does not fully eliminate it.

In the implementation of currency overlay, currency managers have sought to regain some of the potential performance from their tactical systems by advocating some loosening of these constraints imposed by the strategic currency benchmark. In the illustration of the 100% hedging of the yen exposure, a loosening of the constraint may take the form of allowing net short positions in yen. In the illustration for the yen exposure, if the tactical recommendation is a sell trade for the yen, then the tactical trade can be implemented up to the allowable short yen exposure. The portfolio could then potentially benefit from the performance of this trade which would otherwise have been foregone if the constraint has not been loosened.

7 CURRENCY MANAGEMENT USING MINIMAX AND OPTIONS

As with the use of options for managing asset portfolios from Chapter 9, the use of currency options for currency management is also attributed partly to their insurance capability. Currency managers in the import-export field are active users of currency options in view of the suitability of these for hedging expected cashflows from trade transactions. The periods within which these cashflows are expected to happen are short to medium term, and the use of options are deemed appropriate for providing the needed insurance policy for the term. However, the use of currency options by managers managing an international portfolio's currency exposure takes a different form to that used by managers in the import-export field. In this case, options are held in the short and medium term up to the maturity of the option. Options are used as insurance providers, as well as being held for very short-term periods when they are used as return enhancers complementing tactical currency systems.

Currency managers who provide overlay services for managing the currency exposure of international portfolios complement their overlay products by taking long or short positions in various currency options. The market provides a wide range of options: plain vanilla, asian, barrier, capped call, floored put, collar, lookback and quanto, to name a few. Additionally, combinations of options provide pay-out profiles that can be used for return enhancement. The reader is referred to Hull (1997) for a comprehensive discussion on these options.

The generic currency model discussed in Section 4 and the tactical currency systems discussed in Section 5 do not preclude the use of options. However, a tactical formulation that includes options may not provide a practical solution for subscribers to currency overlay. This is due to the following three reasons. Firstly, a large majority of overlay subscribers would not allow the use of options. Those who would allow options tend to restrict their use to very specific conditions on the currency pairs, or on the type of options, or that positions should be long only, or that a very small currency exposure can be managed using options. Secondly, currency overlay managers attempt to diversify their product range by offering option-based currency management distinct from tactical currency systems. This prevents an active promotion of options within existing tactical systems. Lastly, data availability restrict the simulations that currency managers can do in searching for option-based strategies that may complement their existing tactical systems. This is an important restriction in trading systems development as well-defined excess return and tracking error profiles are essential in the world of benchmark-based currency management.

In Chapter 9, a minimax formulation that incorporates the use of options is presented as an enhanced portfolio management tool where insurance is provided by an optimal choice of out-of-the-money options. Such a framework cannot be adapted for tactical currency systems in Section 5 due to the very short-term nature of these systems. However, option-based currency overlay systems would have the ability to tailor options of varying horizons and they may be more amenable to minimax formulations.

8 CONCLUDING REMARKS

In this chapter we discussed the need for currency management, mainly from the point of view of international portfolios where currency hedging is a critical issue in preserving the returns from the foreign assets that comprise the portfolio. We then subdivided the work of managing the currency exposure of an international portfolio in two ways: through a strategic currency management system that deals with the long-term direction of currencies, and through a tactical currency management system that deals with short-term fluctuations in particular currencies.

The strategic currency management system identifies a long-term currency benchmark that provides the overall direction or bias of the currency hedge that needs to be implemented. The tactical currency management system identifies a short-term currency bet that improves on the already-implemented long-term currency benchmark. This short-term currency bet, as provided by a currency model signal, ensures that short-term currency fluctuations are utilized to the benefit of the portfolio. The excess returns that can be generated via a tactical currency management system supplement the returns that can be achieved from the strategic currency hedge.

We presented minimax formulations for both the strategic and the tactical systems, and ways of identifying and evaluating worst-case scenarios. As currencies are constrained to move in relation to other currencies, the definition of worst-case scenarios are similarly constrained by the triangulation properties of exchange rates.

References

Eaker, M.R. and D.M. Grant (1990). "Currency Hedging Strategies for Internationally Diversified Equity portfolios", *Journal of Portfolio Management*, Fall, 30–32.

Eun, C.S. and B.G. Resnick (1985). "Currency Factor in International Portfolio Diversification", *Columbia Journal of World Business*, Summer, 45–53.

Eun, C.S. and B.G. Resnick (1988). "Exchange Rate Uncertainty, Forward Contracts and International Portfolio Selection", *Journal of Finance*, 43, 197–215.

Hauser, S. and A. Levy (1991). "Optimal Forward Coverage of International Fixed-income Portfolios", *Journal of Portfolio Management*, Summer, 54–59.

Hull, J. C. (1997). *Options, Futures and Other Derivatives*, Prentice Hall, London.

Jorion, P. (1989). "Asset Allocation with Hedged and Unhedged Foreign Stocks and Bonds", *Journal of Portfolio Management*, 49–54.

Levy, H. (1981). "Optimal Portfolio of Foreign Currencies with Borrowing and Lending", *Journal of Money, Credit and Banking*, 13, 325–341.

Perold, A.F. and E.C. Schulman (1988). "The Free Lunch in Currency Hedging: Implications for Investment Policy and Performance Standards", *Financial Analysts Journal*, 45–50.

Rosenberg, M.R. (1996). *Currency Forecasting: A Guide to Fundamental and Technical Models of Exchange Rate Determination*, Irwin, London.

Rustem, B. (1995). "Computing Optimal Multicurrency Mean-variance Portfolios", *Journal of Economic Dynamics and Control*, 19, 901–908.

APPENDIX: CURRENCY FORECASTING

Currency forecasting can be categorized into two major classes: fundamental-based modeling and technical analysis. This appendix gives a brief overview of these models, following the comprehensive discussion in Rosenberg (1996). For further details on these models, the reader is encouraged to refer to Rosenberg (1996) and references therein.

Forecasting models fall into two general categories: fundamental models and technical models. Associated with these are forecasting horizons that generally fall into three general categories: long-term forecasting where the emphasis is on structural and macro-economic forces that determine the equilibrium level of exchange rates, medium-term forecasting where an analysis of economic or business cycles may provide an insight into the cyclical position of exchange rates relative to the long-term equilibrium level, and short-term forecasting where the emphasis is on the analysis of speculative forces. While fundamental-based models appear to have a relative advantage in the medium- and long-term forecasting domains, technical models appear to have their relative advantage in the short-term domain. We give below a brief description of common fundamental models as well as technical models.

Most fundamental models attempt to estimate the long-run equilibrium exchange rate level or path that the exchange rate will gravitate towards in the long run, and perhaps oscillate about in the medium run. In models based on purchasing power parity, it is assumed that nominal exchange rates would converge to a fair value that reflects differences in national inflation rates. In external balance-based models, it is assumed that nominal exchange rates would converge to a fair value that is consistent with the attainment of a balanced current account.

Fundamental models that concentrate on the medium term fall in the general categories of asset-market models, monetary models, currency-substitution models and portfolio-balance models. In asset-market models of exchange rate determination, the supply of and demand for financial assets determine the medium-term trend that exchange rates take. In a monetary model, the supply of and demand for money determine the equilibrium exchange rate. In currency-substitution models, the anxiety of a nation in the local currency value erosion amplifies the volatility of the exchange rate and contributes to a perceived potential devaluation or depreciation in the currency. In the portfolio-balance models, the supply of and demand for money, as well as for bonds or government debt, determine exchange rate movements over medium-term periods. Fundamental models also consider the effect of economic variables such as interest rate differentials, fiscal policy changes and central bank intervention.

Technical analysis has gained popularity due to its relative success in forecasting in the short term. However, it has been criticized as a long-term model. Despite this apparent shortcoming of technical analysis, market participants, particularly traders, use various models of technical analysis. These fall into two general categories: trend-following, where the model ascertains whether a trend is developing, and contrarian, where the model ascertains whether a trend is due for correction. Whether trend-following or contrarian, technical analysis can be subcategorized in terms of the technique used: charting, use of neural networks, signal processing and statistical or mathematical processes. Furthermore, within the domain of charting, a deeper categorization is possible in terms of the indicators produced by the charting analysis. These indicators generally fall under any of the following: moving average indicators, pattern recognition, oscillator indicators, divergence indicators, or trend indicators.

The increasing trend in the use of technical analysis has been reinforced by the relative failure of fundamental models in generating short-term returns. However, market participants, particularly investors, realize that total reliance on a technical approach to currency forecasting can be very risky when false technical signals resulting from weak trending markets give rise to huge losses. Investors tend to avoid a strong reliance on technical signals especially when fundamental signals do not support or reinforce those signals. Additionally, investors tend to look not only at the short term, where technical models are relatively more useful, but at the medium and long term as well, where fundamental models are relatively more useful. There is a need to address the balance between the use of technical and fundamental models in order for market participants to minimize the risk of incurring currency losses due to mis-forecasting. Indeed, there has been a tendency to base a long-term currency view on fundamental models and a tendency to base a short-term currency view on technical models, and a tendency to weight any aggregation of signals are on the basis of the relative importance of making a long-term view as opposed to a short-term view.

COMMENTS AND NOTES

CN 1: Hedging of Currency Risk

Hedging is the technical term in finance to refer to the implementation of a strategy to mitigate any potential unfavorable outcome from holding a position. In the context of holding a currency portfolio or an international portfolio with currency exposures, hedging refers to the strategy of eliminating all or part of the potential negative return if a currency moves against the investor. The concept of base currency is very important in ascertaining the appropriate hedge. For a usd-based investor who invests in a foreign country's equity market, the currency risk comes from having to translate the gains (or losses) from the equity market into equivalent gains (or losses) in US dollar terms. If the foreign currency depreciates relative to the base currency, then the equivalent gain (or losses) in US dollar terms gets eroded. Generally, the hedging of currency risk involves the use of a forward currency contract that stipulates the exchange rate to apply to a particular nominal amount of the foreign currency for exchange back to the base currency at a future time. In complex hedging strategies, forward, swap, option, and spot transactions may be employed.

CN 2: Cross Hedging

A cross hedge refers to the implementation of a currency hedge when the currencies involved do not include the base currency. Any potential depreciation of the first currency relative to the second currency is mitigated by selling the first currency and buying the second currency. A cross hedge does not necessarily mean an improvement in the overall risk exposure of a portfolio from the point of view of the base currency. The reason for this lack of certainty is that a cross hedge has to depend on the movement of the bought currency, in this case the second currency, relative to the base currency.

CN 3: Long Position versus Short Position in a Currency

The terms "long" and "short" a currency refer to the holding of a foreign currency. A long position means that the investor owns the currency; this currency may physically reside in a deposit account or it may be invested in an asset denominated in that currency. A short position means that the investor does not own the currency but has sold the currency. This is possible in a situation where the investor enters into a forward contract to sell the currency even if she does not physically have notes and coins, or assets to back the currency.

CN 4: Cross-currency Constraints

The number of cross-currency constraints is given by the combination of $(m^{cur} - 1)$ currencies taken two at a time, that is,

$$^{(m^{cur}-1)}C^2 = \frac{(m^{cur} - 1)!}{((m^{cur} - 1) - 2)!2!}.$$

CN 5: Implementing the Transaction Cost Term

The overall transaction cost depends on the magnitude of $|\delta_{i,t} + \zeta_{i,t}|$. This term can be incorporated within the setting of the quadratic programming formulation using a simple reformulation. We note that in the formulations where a transaction cost term appear, only the variable part, $\zeta_{i,t}$, is considered.

Let

$$\delta_{i,t} + \zeta_{i,t} = x_{i,t}^+ - x_{i,t}^-, \quad \text{with } x_{i,t}^+, x_{i,t}^- \geq 0$$

$$|\delta_{i,t} + \zeta_{i,t}| = x_{i,t}^+ + x_{i,t}^-, \quad \text{and } x_{i,t}^+ = 0 \text{ if } x_{i,t}^- > 0$$

$$\text{and } x_{i,t}^- = 0 \text{ if } x_{i,t}^+ > 0.$$

Thus, the transaction cost component

$$\sum_t - k_i |\delta_{i,t} + \zeta_{i,t}|$$

is replaced in the objective by

$$\sum_t - k_i(x_{i,t}^+ + x_{i,t}^-) + cx_{i,t}^+ x_{i,t}^-$$

with added constraints

$$\delta_{i,t} + \zeta_{i,t} = x_{i,t}^+ - x_{i,t}^- \quad \text{and} \quad x_{i,t}^+, x_{i,t}^- \geq 0.$$

We assume that $c > 0$ is chosen to be sufficiently large to ensure $x_{i,t}^+ \times x_{i,t}^- = 0$.

Index